T0245446

CAMBRIDGE LIBRARY COLLECTION

Books of enduring scholarly value

Life Sciences

Until the nineteenth century, the various subjects now known as the life sciences were regarded either as arcane studies which had little impact on ordinary daily life, or as a genteel hobby for the leisured classes. The increasing academic rigour and systematisation brought to the study of botany, zoology and other disciplines, and their adoption in university curricula, are reflected in the books reissued in this series.

Himalayan Journals

Sir Joseph Hooker (1817–1911) was one of the greatest British botanists and explorers of the nineteenth century. He succeeded his father, Sir William Jackson Hooker, as Director of the Royal Botanic Gardens, Kew, and was a close friend and supporter of Charles Darwin. His journey to the Himalayas and India was undertaken between 1847 and 1851 to collect plants for Kew, and his account, published in 1854, was dedicated to Darwin. Hooker collected some 7,000 species in India and Nepal, and carried out surveys and made maps which proved of economic and military importance to the British. He was arrested by the Rajah of Sikkim, but the British authorities secured his release by threatening to invade, and annexing part of the small kingdom. Volume 2 continues Hooker's description of Nepal and Sikkim (including his brief imprisonment) and his return to Calcutta to begin his journey back to Great Britain.

Cambridge University Press has long been a pioneer in the reissuing of out-of-print titles from its own backlist, producing digital reprints of books that are still sought after by scholars and students but could not be reprinted economically using traditional technology. The Cambridge Library Collection extends this activity to a wider range of books which are still of importance to researchers and professionals, either for the source material they contain, or as landmarks in the history of their academic discipline.

Drawing from the world-renowned collections in the Cambridge University Library, and guided by the advice of experts in each subject area, Cambridge University Press is using state-of-the-art scanning machines in its own Printing House to capture the content of each book selected for inclusion. The files are processed to give a consistently clear, crisp image, and the books finished to the high quality standard for which the Press is recognised around the world. The latest print-on-demand technology ensures that the books will remain available indefinitely, and that orders for single or multiple copies can quickly be supplied.

The Cambridge Library Collection will bring back to life books of enduring scholarly value (including out-of-copyright works originally issued by other publishers) across a wide range of disciplines in the humanities and social sciences and in science and technology.

Himalayan Journals

Or, Notes of a Naturalist in Bengal,
the Sikkim and Nepal Himalayas,
the Khasia Mountains, &c

VOLUME 2

JOSEPH DALTON HOOKER

CAMBRIDGE
UNIVERSITY PRESS

CAMBRIDGE UNIVERSITY PRESS

Cambridge, New York, Melbourne, Madrid, Cape Town,
Singapore, São Paolo, Delhi, Tokyo, Mexico City

Published in the United States of America by Cambridge University Press, New York

www.cambridge.org
Information on this title: www.cambridge.org/9781108029360

This edition first published 1854
This digitally printed version 2011

ISBN 978-1-108-02936-0 Paperback

Pl. VII.

JDH delt

L Gruner dirt

Snow beds at 13,000 feet, in the Th'lonok Valley;
with Rhododendrons in blossom. Kinchin-junga
in a distance

London John Murray Decbr. 1853.

HIMALAYAN JOURNALS;

OR,

NOTES OF A NATURALIST

IN BENGAL, THE SIKKIM AND NEPAL HIMALAYAS,
THE KHASIA MOUNTAINS, &c.

BY JOSEPH DALTON HOOKER, M.D., R.N., F.R.S.

WITH MAPS AND ILLUSTRATIONS.

IN TWO VOLUMES.—VOL. II.

LONDON:
JOHN MURRAY, ALBEMARLE STREET.
1854.

The author reserves to himself the right of authorising a translation of this work.

CONTENTS.

CHAPTER XVIII.

CHAPTER XIX.

* In the body of the work, the latter part of this chapter has been accidentally numbered
XXVIII., and the three succeeding XXIX., &c.

LIST OF ILLUSTRATIONS.

LITHOGRAPHIC VIEWS.

WOOD ENGRAVINGS.

LIST OF ILLUSTRATIONS.

HIMALAYAN JOURNALS.

CHAPTER XVIII.

AFTER my return from the Terai, 1 was occupied during the month of April in preparations for an expedition to the loftier parts of Sikkim. The arrangements were the same as for my former journey, except with regard to food, which it was necessary should be sent out to me at intervals ; for we had had ample proof that the resources of the country were not equal to provisioning a party of from forty to fifty men, even had the Dewan been favourable to my travelling, which was clearly not the case.

Dr. Campbell communicated to the Rajah my intention of starting early in May for the upper Teesta valley, and,

in the Governor-General's name, requested that he would
facilitate my visiting the frontier of Sikkim, north-east of
Kinchinjunga. The desired permission was, after a little
delay, received; which appeared to rouse the Dewan to
institute a series of obstructions to my progress, which
caused so many delays that my exploration of the country
was not concluded till October, and I was prevented
returning to Dorjiling before the following Christmas.

Since our visit to the Rajah in December, no Vakeel
(agent) had been sent by the Durbar to Dorjiling, and
consequently we could only communicate indirectly with
his Highness, while we found it impossible to ascertain
the truth of various reports promulgated by the Dewan,
and meant to deter me from entering the country. In
April, the Lassoo Kajee was sent as Vakeel, but, having on
a previous occasion been dismissed for insolence and
incapacity, and again rejected when proposed by the
Dewan at Bhomsong, he was refused an audience; and he
encamped at the bottom of the Great Rungeet valley,
where he lost some of his party through fever. He retired
into Sikkim, exasperated, pretending that he had orders to
delay my starting, in consequence of the death of the heir
apparent; and that he was prepared to use strong measures should I cross the frontier.

No notice was taken of these threats: the Rajah was
again informed of my intended departure, unless his
own orders to the contrary were received through a proper
accredited agent, and I left Dorjiling on the 3rd of May,
accompanied by Dr. Campbell, who insisted on seeing me
fairly over the frontier at the Great Rungeet river.

Arrangements were made for supplies of rice following
me by instalments; our daily consumption being 80 lbs.,
a man's load. After crossing into Sikkim, I mustered my

party at the Great Rungeet river. I had forty-two in all, of whom the majority were young Lepchas, or Sikkim-born people of Tibetan races: all were active and cheerful looking fellows; only one was goitred, and he had been a salt-trader. I was accompanied by a guard of five Sepoys, and had a Lepcha and Tibetan interpreter. I took but one personal servant, a Portuguese half-caste (John Hoffman by name), who cooked for me: he was a native of Calcutta, and though hardy, patient, and long-suffering, and far better-tempered, was, in other respects, very inferior to Clamanze, who had been my servant the previous year, and who, having been bred to the sea, was as handy as he was clever; but who, like all other natives of the plains, grew intolerably weary of the hills, and left me.

The first part of my route lay over Tendong, a very fine mountain, which rises 8,613 feet, and is a conspicuous feature from Dorjiling, where it is known as Mount Ararat. The Lepchas have a curious legend of a man and woman having saved themselves on its summit, during a flood that once deluged Sikkim. The coincidence of this story with the English name of Ararat suggests the probability of the legend being fabulous; but I am positively assured that it is not so, but that it was current amongst the Lepchas before its English name was heard of, and that the latter was suggested from the peculiar form of its summit resembling that given in children's books as the resting-place of the ark.

The ascent from the Great Rungeet (alt. 818 feet) is through dry woods of Sal and Pines (*P. longifolia*). I camped the first night at the village of Mikk (alt. 3,900 feet) and on the following day ascended to Namtc (alt. 5,600 feet).

On the route I was met by the Lama of Silokfoke Goompa. Though a resident on the Lassoo Kajee's estates, he politely brought me a present, at the same time apologising for not waiting till I had encamped, owing to his excessive fat, which prevented his climbing: I accepted his excuses, though well aware that his real reason was that he wished to pay his respects, and show his good feeling, in private. Besides his ordinary canonicals, he carried a tall crozier-headed staff; and had a curious horn slung round his neck, full of amulets; it was short, of a transparent red colour, and beautifully carved, and was that of the small cow of Lhassa, which resembles the English species, and is not a yak (it is called "Tundro").

Namtchi was once a place of considerable importance; and still possesses a mendong, with six rows of inscribed slabs; a temple, and a Lama attached thereto: the latter· waited on me soon after I had encamped, but he brought no present, and I was not long kept in suspense as to his motives. These people are poor dissemblers; if they intend to obstruct, they do it clumsily and hesitatingly : in this instance the Lama first made up to my people, and, being coolly received, kept gradually edging up to my tent-door, where, after an awkward salute, he delivered himself with a very bad grace of his mission, which was from the Lassoo Kajee to stop my progress. I told him I knew nothing of the Lassoo Kajee or his orders, and should proceed on the following morning: he then urged the bad state of the roads, and advised me to wait two days till he should receive orders from the Rajah; upon which I dismissed him.

Soon afterwards, as I sat at my tent-door, looking along the narrow bushy ridge that winds up the mountain, I saw twenty or thirty men rapidly descending the rocky path: they were Lepchas, with blue and white striped garments, bows and quivers, and with their long knives gleaming in the sun: they seemed to be following a figure in red Lama costume, with a scarlet silk handkerchief wound round his head, its ends streaming behind him. Though expecting this apparition to prove the renowned Kajee and his myrmidons, coming to put a sudden termination to my progress, I could not help admiring the exceeding picturesqueness of the scenery and party. My fears were soon dissipated by my men joyfully shouting, "The Tchebu Lama! the Tchebu Lama!" and I soon recognised the rosy face and twinkling eyes of my friend of Bhomsong, the only man of intelligence about the Rajah's court, and the one whose services as Vakeel were particularly wanted at Dorjiling.

He told me that the Lassoo Kajee had orders (from whom, he would not say) to stop my progress, but that I should proceed nevertheless, and that there was no objection to my doing so; and he despatched a messenger to the Rajah, announcing my progress, and requesting him to send me a guide, and to grant me every facility, asserting that he had all along fully intended doing so.

On the following morning the Lama proceeded to Dorjiling, and I continued the ascent of Tendong, sending my men round the shoulder to Temi in the Teesta valley, where I proposed to pass the night. The road rapidly ascends by a narrow winding path, covered with a loose forest of oaks, rhododendrons, and various shrubs, not found at equal elevations on the wetter Dorjiling ranges: amongst them the beautiful laburnum-like *Piptanthus Nepalensis*, with golden blossoms, was conspicuous. Enormous blocks of white and red stratified quartz and slate, some 20 and even 40 yards long, rest on the narrow ridge at 7000 feet elevation. The last ascent is up a steep rounded cone with a broad flat top, covered with dwarf bamboo, a few oaks, laurels, magnolias, and white-flowered rhododendron trees (*R. argenteum*), which obstructed the view. I hung the barometers near one of the many chaits on the summit, where there is also a rude temple, in which worship is performed once a year. The elevation is 8,671 feet by my observations.* The geological formation of Tendong in some measure accounts for its peculiar form. On the conical summit are hard quartzoze porphyries, which have apparently forced up the gneiss and slates, which dip in all directions from the top, and are full of injected veins of quartz. Below 7000 feet, mica-schist prevails, always inclined at a very high angle; and I found jasper near Namtchi, with other indications of Plutonic action.

The descent on the north side was steep, through a rank vegetation, very different from that of the south face. The oaks are very grand, and I measured one (whose trunk was decayed, and split into three, however), which I found to be 49 feet in girth at 5 feet from the ground. Near Temi (alt. 4,770 feet) I gathered the fruit of *Kadsura*, a climbing plant allied to Magnolia, bearing round heads of large fleshy red drupes, which are pleasantly acid and much eaten; the seeds are very aromatic.

From Temi the road descends to the Teesta, the course of which it afterwards follows. The valley was fearfully hot, and infested with mosquitos and peepsas. Many fine plants grew in it:† I especially noticed *Aristolochia saccata*,

---

* 8,663 by Col. Waugh's trigonometrical observations.

† Especially upon the broad terraces of gravel, some of which are upwards of a mile long, and 200 feet above the stream: they are covered with boulders of rock, and are generally opposite feeders of the river.

which climbs the loftiest trees, bearing its curious pitcher-shaped flowers near the ground only ; its leaves are said to be good food for cattle. *Houttuynia*, a curious herb allied to pepper, grew on the banks, which, from the profusion of its white flowers, resembled strawberry-beds ; the leaves are eaten by the Lepchas. But the most magnificent plant of these jungles is *Hodgsonia*, (a genus I have dedicated to my friend, Mr. Hodgson), a gigantic climber allied to the gourd, bearing immense yellowish-white pendulous blossoms, whose petals have a fringe of buff-coloured curling threads, several inches long. The fruit is of a rich brown, like a small melon in form, and contains six large nuts, whose kernels (called "Katior-pot" by the Lepchas) are eaten. The stem, when cut, discharges water profusely from whichever end is held downwards. The "Took" (*Hydnocarpus*) is a beautiful evergreen tree, with tufts of yellow blossoms on the trunk : its fruit is as large as an orange, and is used to poison fish, while from the seeds an oil is expressed. Tropical oaks and Terminalias are the giants of these low forests, the latter especially, having buttressed trunks, appear truly gigantic ; one, of a kind called "Sung-lok," measured 47 feet in girth, at 5 feet, and 21 at 15 feet from the ground, and was fully 200 feet high. I could only procure the leaves by firing a ball into the crown. Some of their trunks lay smouldering on the ground, emitting a curious smell from the mineral matter in their ashes, of whose constituents an account will be found in the Appendix.

Birds are very rare, as is all animal life but insects, and a small fresh-water crab, *Thelphusa*, ("Ti-hi" of the Lepchas). Shells, from the absence of lime, are extremely scarce, and I scarcely picked up a single specimen : the most common are species of *Cyclostoma*.

The rains commenced on the 10th of May, greatly
increasing the discomforts of travelling, but moderating
the heat by drenching thunder-storms, which so soaked
the men's loads, that I was obliged to halt a day in the
Teesta valley to have waterproof covers made of platted
bamboo-work, enclosing Phrynium leaves. I was delighted
to find that my little tent was impervious to water, though
its thickness was but of one layer of blanket : it was a
single ridge with two poles, 7 feet high, 8 feet long, and
8 feet broad at the base, forming nearly an equilateral
triangle in front.

Bhomsong was looking more beautiful than ever in its
rich summer clothing of tropical foliage. I halted during
an hour of heavy rain on the spot where I had spent the
previous Christmas, and could not help feeling doubly
lonely in a place where every rock and tree reminded me
of that pleasant time. The isolation of my position, the
hostility of the Dewan, and consequent uncertainty of the
success of a journey that absorbed all my thoughts,
the prevalence of· fevers in the valleys I was traversing,
and the many difficulties that beset my path, all crowded
on the imagination when fevered by exertion and depressed
by gloomy weather, and my spirits involuntarily sank as I
counted the many miles and months intervening between
me and my home.

The little flat on which I had formerly encamped was now
covered with a bright green crop of young rice. The house
then occupied by the Dewan was now empty and unroofed ;
but the suspension bridge had been repaired, and its light
framework of canes, spanning the boiling flood of the
Teesta, formed a graceful object in this most beautiful
landscape. The temperature of the river was 58°, only 7°
above that of mid-winter, owing to the now melting snows.

I had rather expected to meet either with a guide, or with some further obstruction here, but as none appeared, I proceeded onwards as soon as the weather moderated.

Higher up, the scenery resembles that of Tchintam on the Tambur : the banks are so steep as to allow of no road,

PANDANUS.   SIKKIM SCREW-PINE.

and the path ascends from the river, at 1000 feet, to Lathiang village, at 4,800 feet, up a wild, rocky torrent that descends from Mainom to the Teesta.   The cliffs here are covered with wild plantains and screw-pines (*Pandanus*),

50 feet high, that clasp the rocks with cable-like roots, and bear one or two crowns of drooping leaves, 15 feet long: two palms, Rattan (*Calamus*) and *Areca gracilis*, penetrate thus far up the Teesta valley, but are scarcely found further.

From the village the view was superb, embracing the tropical gulley below, with the flat of Bhomsong deep down in the gorge, its bright rice-fields gleaming like emeralds amid the dark vegetation that surrounded it; the Teesta winding to the southward, the pine-clad rocky top of Mainom, 10,613 feet high, to the south-west, the cone of Mount Ararat far to the south, to the north black mountains tipped with snow, and to the east the magnificent snowy range of Chola, girdling the valley of the Ryott with a diadem of frosted silver. The coolies, each carrying upwards of 80 lb. load, had walked twelve hours that day, and besides descending 2000 feet, they had ascended nearly 4000 feet, and gone over innumerable ups and downs besides.

Beyond Lathiang, a steep and dangerous path runs along the east flank of Mainom, sometimes on narrow ledges of dry rock, covered with long grass, sometimes dipping into wooded gullies, full of *Edgeworthia Gardneri* and small trees of Andromeda and rhododendron, covered with orchids * of great beauty.

Descending to Gorh (4,100 feet), I was met by the Lama of that district, a tall, disagreeable-looking fellow, who informed me that the road ahead was impassable. The day being spent, I was obliged to camp at any rate; after which he visited me in full canonicals, bringing me a handsome present, but assuring me that he had no autho-

* Especially some species of *Sunipia* and *Cirrhopetalum*, which have not yet been introduced into England.

rity to let me advance. I treated him with civility, and regretted my objects being so imperative, and my orders so clear, that I was obliged to proceed on the following morning : on which he abruptly decamped, as I suspected, in order to damage the paths and bridges. He came again at daylight, and expostulated further ; but finding it of no use, he volunteered to accompany me, officiously offering me the choice of two roads. I asked for the coolest, knowing full well that it was useless to try and out-wit him in such matters. At the first stream the bridge was destroyed, but seeing the planks peeping through the bushes in which they had been concealed, I desired the Lama to repair it, which he did without hesitation. So it was at every point : the path was cumbered with limbs of trees, crossing-stones were removed from the streams, and all natural difficulties were increased. I kept constantly telling the Lama that as he had volunteered to show me the road, I felt sure he intended to remove all obstacles, and accordingly I put him to all the trouble I possibly could, which he took with a very indifferent grace. When I arrived at the swinging bridge across the Teesta, I found that the canes were loosened, and that slips of bamboo, so small as nearly to escape observation, were ingeniously placed low down over the single bamboo that formed the footing, intended to trip up the unwary passenger, and overturn him into the river, which was deep, and with a violent current. Whilst the Lama was cutting these, one of my party found a charcoal writing on a tree, announcing the speedy arrival from the Rajah of my old guide, Meepo ; and he shortly afterwards appeared, with instructions to proceed with me, though not to the Tibetan frontier. The lateness of the season, the violence of the rains, and the fears, on the Rajah's part, that I might

suffer from fever or accident, were all urged to induce
me to return, or at least only to follow the west branch of
the Teesta to Kinchinjunga.   These reasons failing, I was
threatened with Chinese interference on the frontier.   All
these objections I overruled, by refusing to recognise any
instructions that were not officially communicated to the
Superintendent of Dorjiling.

The Gorh Lama here took leave of me : he was a friend
of the Dewan, and was rather surprised to find that the
Rajah had sent me a guide, and now attempted to pass
himself off as my friend, pompously charging Meepo with
the care of me, and bidding me a very polite farewell.
I could not help telling him civilly, but plainly, what I
thought of him ; and so we parted.

Meepo was very glad to join my party again : he is a
thorough Lepcha in heart, a great friend of his Rajah and
of Tchebu Lama, and one who both fears and hates the
Dewan.   He assured me of the Rajah's good wishes and
intentions, but spoke with great doubt as to the probability
of a successful issue to my journey : he was himself ignorant of the road, but had brought a guide, whose appearance,
however, was against him, and who turned out to be sent
as a spy on us both.

Instead of crossing the Teesta here, we kept on for two
days up its west bank, to a cane bridge at Lingo, where the
bed of the river is still only 2000 feet above the sea, though
45 miles distant from the plains, and flowing in a valley
bounded by mountains 12,000 to 16,000 feet high.   The
heat was oppressive, from the closeness of the atmosphere,
the great power of the sun, now high at noon-day, and
the reflection from the rocks.   Leeches began to swarm as
the damp increased, and stinging flies of various kinds.
My clothes were drenched with perspiration during five

hours of every day, and the crystallising salt irritated the
skin.   On sitting down to rest, I was overcome with lan-
guor and sleep, and, but for the copious supply of fresh
water everywhere, travelling would have been intolerable.
The Coolies were all but naked, and were constantly
plunging into the pools of the rivers; for, though filthy in
their persons, they revel in cold water in summer.   They
are powerful swimmers, and will stem a very strong current, striking out with each arm alternately.   It is an animated sight when twenty or thirty of these swarthy
children of nature are disporting their muscular figures in
the water, diving after large fish, and sometimes catching
them by tickling them under the stones.

Of plants I found few not common at similar elevations
below Dorjiling, except another kind of Tree-fern,* whose
pith is eaten in times of scarcity. The India-rubber fig pene-
trates thus far amongst the mountains, but is of small size.
A Gentian, *Arenaria*, and some sub-alpine plants are met
with, though the elevation is only 2000 feet, and the whole
climate thoroughly tropical: they were annuals usually
found at 7000 to 10,000 feet elevation, and were growing
here on mossy rocks, cooled by the spray of the river, whose
temperature was only 56° 3.   My servant having severely
sprained his wrist by a fall, the Lepchas wanted to apply a
moxa, which they do by lighting a piece of puff-ball, or
Nepal paper that burns like tinder, laying it on the skin,
and blowing it till a large open sore is produced: they
shook their heads at .my treatment, which consisted in
transferring some of the leeches from our persons to the
inflamed part.

* *Alsophila spinulosa,* the "Pugjik" of the Lepchas, who eat the soft watery
pith: it is abundant in East Bengal and the Peninsula of India.. The other
Sikkim Tree-fern, *A. gigantea,* is far more common from the level of the plains to
6,500 elevation, and is found as far south as Java.

After crossing the Teesta by the cane bridge of Lingo, our route lay over a steep and lofty spur, round which the river makes a great sweep. On the ascent of this ridge we passed large villages on flats cultivated with buckwheat. The saddle is 5,500 feet high, and thence a rapid descent leads to the village of Singtam, which faces the north, and is 300 feet lower, and 3000 feet above the river, which is here no longer called the Teesta, but is known as the Lachen-Lachoong, from its double origin in the rivers of these names, which unite at Choongtam, twenty miles higher up. Of these, the source of the Lachen is in the Cholamoo lakes in Tibet; while the Lachoong rises on the south flank of Donkia mountain, both many marches north of my present position. At Singtam the Lachen-Lachoong runs westward, till joined by the Rihi from the north, and the Rinoong from the west, after receiving which it assumes the name of Teesta: of these affluents, the Rinoong is the largest, and drains the south-east face of Kinchinjunga and Pundim, and the north of Nursing: all which mountains are seen to the north-north-west of Singtam. The Rinoong valley is cultivated for several miles up, and has amongst others the village and Lamasery of Bah. Beyond this the view of black, rugged precipices with snowy mountains towering above them, is one of the finest in Sikkim. There is a pass in that direction, from Bah over the Tekonglah to the Thlonok valley, and thence to the province of Jigatzi in Tibet, but it is almost impracticable.

A race of wild men, called "Harrum-mo," are said to inhabit the head of the valley, living in the woods of a district called Mund-po, beyond Bah; they shun habitations, speak an unintelligible tongue, have more hair on the face than Lepchas, and do not plait that of their heads, but

Pl. VI.

J.D.H. delt.   John Murray, Albemarle Street, 1854.   W.L. Walton, lith.

Kinchinjunga from Singtam (Elevn. 5000 ft.) looking West.

wear it in a knot; they use the bow and arrow, and eat snakes and vermin, which the Lepchas will not touch. Such is the account I have heard, and which is certainly believed in Sikkim: similar stories are very current in half civilized countries; and if this has any truth, it possibly refers to the Chepangs,* a very remarkable race, of doubtful affinity and origin, inhabiting the Nepal forests.

At Singtam I was waited on by the Soubah of the district, a tall portly Bhoteea, who was destined to prove a most active enemy to my pursuits. He governs the country between Gorh and the Tibet frontier, for the Maha-Ranee (wife of the Rajah), whose dowry it is; and she being the Dewan's relative, I had little assistance to expect from her agent. His conduct was very polite, and he brought me a handsome offering for myself; but after delaying me a day on the pretext of collecting food for my people, of which I was in want, I was obliged to move on with no addition to my store, and trust to obtaining some at the next village, or from Dorjiling. Owing, however, to the increasing distance, and the destruction of the roads by the rains, my supplies from that place were becoming irregular: I therefore thought it prudent to reduce my party, by sending back my guard of Sepoys, who could be of no further use.

From this point the upper portion of the course of the Teesta (Lachen-Lachoong) is materially different from what it is lower down; becoming a boisterous torrent, as suddenly as the Tambur does above Mywa Guola. Its bed is narrower, large masses of rock impede its course, nor is there any place where it is practicable for rafts at any season; the only means of passing it being by cane bridges that are thrown across, high above the stream.

* Hodgson, in "Bengal Asiatic Society's Journal" for 1848.

The slope on either side of the valley is very steep ; that on the north, in particular, appearing too precipitous for any road, and being only frequented by honey seekers, who scale the rocks by cane ladders, and thus reach the pendulous bees'-nests, which are so large as in some instances to be conspicuous features at the distance of a mile. This pursuit appeared extremely perilous, the long thread-like canes in many places affording the only footing, over many yards of cliff : the procuring of this honey, however, is the only means by which many of the idle poor raise the rent which they must pay to the Rajah.

The most prominent effect of the steepness of the valleys is the prevalence of land-slips, which sometimes descend for 3000 feet, carrying devastation along their course : they are caused either by the melting of the snow-beds on the mountains, or by the action of the rains on the stratified rocks, and are much increased in effect and violence by the heavy timber-trees which, swaying forwards, loosen the earth at their roots, and give impetus to the mass. This phenomenon is as frequent and destructive as in Switzerland, where, however, more lives are lost, from the country being more populous, and from the people recklessly building in places particularly exposed to such accidents. A most destructive one had, however, occurred here the previous year, by which a village was destroyed, together with twelve of its inhabitants, and all the cattle. The fragments of rock precipitated are sometimes of enormous size, but being a soft mica-schist, are soon removed by weathering. It is in the rainy season that landslips are most frequent, and shortly after rain they are pretty sure to be heard far or near. I crossed the debris of the great one alluded to, on the first march beyond Singtam : the whole face of the mountain appeared more or less torn up for fully a mile, presenting a confused mass of white micaceous clay, full of angular masses of rock.   The path was very difficult and dangerous, being carried along the steep slope, at an angle, in some places, of 35°; and it was constantly shifting, from the continued downward sliding, and from the action of streams, some of which are large, and cut deep channels. In one I had the misfortune to lose my only sheep, which was carried away by the torrent. These streams were crossed by means of sticks and ricketty bamboos, and the steep sides (sometimes twenty or thirty feet high), were ascended by notched poles.

The weather continued very hot for the elevation (4000 to 5000 feet), the rain brought no coolness, and for the greater part of the three marches between Singtam and Chakoong, we were either wading through deep mud, or climbing over rocks.   Leeches swarmed in incredible pro-fusion in the streams and damp grass, and among the bushes: they got into my hair, hung on my eyelids, and crawled up my legs and down my back.   I repeatedly took upwards of a hundred from my legs, where the small ones used to collect in clusters on the instep: the sores which they produced were not healed for five months afterwards, and I retain the scars to the present day.   Snuff and tobacco leaves are the best antidote, but when marching in the rain, it is impossible to apply this simple remedy to any advantage.   The best plan I found to be rolling the leaves over the feet, inside the stockings, and powdering the legs with snuff.

Another pest is a small midge, or sand-fly, which causes intolerable itching, and subsequent irritation, and is in this respect the most insufferable torment in Sikkim; the minutest rent in one's clothes is detected by the acute senses of this insatiable bloodsucker, which is itself so small as to be barely visible without a microscope. We daily arrived at our camping-ground, streaming with blood, and mottled with the bites of peepsas, gnats, midges, and mosquitos, besides being infested with ticks.

As the rains advanced, insects seemed to be called into existence in countless swarms; large and small moths, cock-chafers, glow-worms, and cockroaches, made my tent a Noah's ark by night, when the candle was burning; together with winged ants, May-flies, flying earwigs, and many beetles, while a very large species of *Tipula* (daddy-long-legs) swept its long legs across my face as I wrote my journal, or plotted off my map. After retiring to rest and putting out the light, they gradually departed, except a few which could not find the way out, and remained to disturb my slumbers.

Chakoong is a remarkable spot in the bottom of the valley, at an angle of the Lachen-Lachoong, which here receives an affluent from Gnarem, a mountain 17,557 feet high, on the Chola range to the east.* There is no village, but some grass huts used by travellers, which are built close to the river on a very broad flat, fringed with alder, hornbeam, and birch : the elevation is 4,400 feet, and many European genera not found about Dorjiling, and belonging to the temperate Himalaya, grow intermixed with tropical plants that are found no further north. The birch, willow, alder, and walnut grow side by side with wild plantain, *Erythrina, Wallichia* palm, and gigantic bamboos : the *Cedrela Toona,* figs, *Melastoma, Scitamineæ,* balsams, *Pothos,* peppers, and gigantic climbing vines, grow mixed with brambles, speedwell, *Paris,* forget-me-not, and nettles

---

* This is called Black Rock in Col. Waugh's map. I doubt Gnarem being a generally known name : the people hardly recognise the mountain as sufficiently conspicuous to bear a name.

that sting like poisoned arrows. The wild English strawberry is common, but bears a tasteless fruit : its inferiority
is however counterbalanced by the abundance of a grateful
yellow raspberry. Parasitical Orchids (*Dendrobium nobile*,
and *densiflorum*, &c.), cover the trunks of oaks, while *Thalictrum* and *Geranium* grow under their shade *Monotropa*
and *Balanophora*, both parasites on the roots of trees (the
one a native of north Europe and the other of a tropical
climate), push their leafless stems and heads of flowers
through the soil together : and lastly, tree-ferns grow associated with the *Pteris aquilina* (brake) and *Lycopodium
clavatum* of our British moors ; and amongst mosses, the
superb Himalayan *Lyellia crispa*,* with the English *Funaria
hygrometrica*.

The dense jungles of Chakoong completely cover the
beautiful flat terraces of stratified sand and gravel, which
rise in three shelves to 150 feet above the river, and whose
edges appear as sharply cut as if the latter had but lately
retired from them. They are continuous with a line of
quartzy cliffs, covered with scarlet rhododendrons, and in
the holes of which a conglomerate of pebbles is found, 150
feet above the river. Everywhere immense boulders are
scattered about, some of which are sixty yards long : their
surfaces are water-worn into hollows, proving the river to
have cut through nearly 300 feet of deposit, which once
floored its valley. Lower down the valley, and fully
2000 feet above the river, I had passed numerous
angular blocks resting on gentle slopes where no landslips could possibly have deposited them ; and which I
therefore refer to ancient glacial action : one of these,

---

* This is one of the most remarkable mosses in the Himalaya mountains, and
derives additional interest from having been named after the late Charles Lyell,
Esq., of Kinnordy, the father of the most eminent geologist of the present day.

near the village of Niong, was nearly square, eighty feet
long, and ten high.

It is a remarkable fact, that this hot, damp gorge is
never malarious ; this is attributable to the coolness of the
river, and to the water on the flats not stagnating; for at
Choongtam, a march further north, and 1500 feet higher,
fevers and ague prevail in summer on similar flats, but
which have been cleared of jungle, and are therefore exposed
to the sun.

I had had constant headache for several mornings on
waking, which I did not fail to attribute to coming fever,
or to the unhealthiness of the climate ; till I accidentally
found it to arise from the wormwood, upon a thick couch
of the cut branches of which I was accustomed to sleep,
and which in dry weather produced no such effects.*

From Chakoong to Choongtam the route lay northwards,
following the course of the river, or crossing steep spurs of
vertical strata of mica-schist, that dip into the valley, and
leave no space between their perpendicular sides and the
furious torrent.   Immense landslips seamed the steep
mountain flanks; and we crossed with precipitation one
that extended fully 4000 feet (and perhaps much more) up
a mountain 12,000 feet high, on the east bank : it moves
every year, and the mud and rocks shot down by it were
strewn with the green leaves and twigs of shrubs, some of
the flowers on which were yet fresh and bright, while others
were crushed : these were mixed with gigantic trunks of
pines, with ragged bark and scored timbers.   The talus
which had lately been poured into the valley formed a gently
sloping bank, twenty feet high, over which the Lachen-

---

* This wormwood (*Artemisia Indica*) is one of the most common Sikkim plants
at 2000 to 6000 feet elevation, and grows twelve feet high : it is a favourite
food of goats.

Lachoong rolled, from a pool above, caused by the damming up of its waters.   On either side of the pool were cultivated terraces of stratified sand and pebbles, fifty feet high, whose alder-fringed banks, joined by an elegant cane bridge, were reflected in the placid water ; forming a little spot of singular quiet and beauty, that contrasted with the savage grandeur of the surrounding mountains, and the headstrong course of the foaming torrent below, amid whose deafening roar it was impossible to speak and be heard.

CANE-BRIDGE AND TUKCHAM MOUNTAIN.

The mountain of Choongtam is about 10,000 feet high ; it divides the Lachen from the Lachoong river, and terminates a lofty range that runs for twenty-two miles south from the lofty mountain of Kinchinjhow.   Its south exposed face is bare of trees, except clumps of pines towards the top, and is

very steep, grassy, and rocky, without water   It is hence
quite unlike the forest-clad mountains further south, and indicates a drier and more sunny climate.   The scenery much
resembles that of Switzerland, and of the north-west Himalaya, especially in the great contrast between the southern
and northern exposures, the latter being always clothed
with a dense vegetation.   At the foot of this very steep
mountain is a broad triangular flat, 5,270 feet above the
sea, and 300 feet above the river, to which it descends by
three level cultivated shelves.   The village, consisting of a
temple and twenty houses, is placed on the slope of the
hill.   I camped on the flat in May, before it became very
swampy, close to some great blocks of gneiss, of which
many lie on its surface : it was covered with tufts of sedge
(like *Carex stellulata*), and fringed with scarlet rhododendron, walnut, *Andromeda, Elæagnus* (now bearing pleasant
acid fruit), and small trees of a *Photinia*, a plant allied
to hawthorn, of the leaves of which the natives make
tea (as they do of *Gualtheria, Andromeda, Vaccinium*, and
other allied plants).   Rice, cultivated * in pools surrounded
by low banks, was just peeping above ground ; and scanty
crops of millet, maize, and buckwheat flourished on the
slopes around

The inhabitants of Choongtam are of Tibetan origin; few
of them had seen an Englishman before, and they flocked
out, displaying the most eager curiosity : the Lama and
Phipun (or superior officer) of the Lachoong valley
came to pay their respects with a troop of followers, and
there was lolling out of tongues, and scratching of ears, at
every sentence spoken, and every object of admiration.

---

* Choongtam is in position and products analogous to Lelyp, on the Tambur
(vol. i. p. 204).   Rice cultivation advances thus high up each valley, and at either
place Bhoteeas replace the natives of the lower valleys.

This extraordinary Tibetan salute at first puzzled me excessively, nor was it until reading MM. Huc and Gabet's travels on my return to England, that I knew of its being the *ton* at Lhassa, and in all civilised parts of Tibet.

As the valley was under the Singtam Soubah's authority, I experienced a good deal of opposition ; and the Lama urged the wrath of the gods against my proceeding. This argument, I said, had been disposed of the previous year, and I was fortunate in recognising one of my Changachelling friends, who set forth my kindly offices to the Lamas of that convent, and the friendship borne me by its monks, and by those of Pemiongchi. Many other modes of dissuading me were attempted, but with Meepo's assistance I succeeded in gaining my point. The difficulty and delays in remittance of food, caused by the landslips having destroyed the road, had reduced our provisions to a very low ebb ; and it became not only impossible to proceed, but necessary to replenish my stores on the spot. At first provisions enough were brought to myself, for the Rajah had issued orders for my being cared for, and having some practice among the villagers in treating rheumatism and goîtres, I had the power of supplying my own larder ; but I found it impossible to buy food for my people. At last, the real state of the case came out ; that the Rajah having gone to Choombi, his usual summer-quarters in Tibet, the Dewan had issued orders that no food should be sold or given to my people, and that no roads were to be repaired during my stay in the country ; thus cutting off my supplies from Dorjiling, and, in short, attempting to starve me out. At this juncture, Meepo received a letter from the Durbar, purporting to be from the Rajah, commanding my immediate return, on the grounds that I had been long enough in the country for my objects : it was not addressed to me, and I refused to receive it as an official communication; following up my refusal by telling Meepo that if he thought his orders required it, he had better leave me and return to the Rajah, as I should not stir without directions from Dr. Campbell, except forwards. He remained, however, and said he had written to the Rajah, urging him to issue stringent orders for my party being provisioned.

We were reduced to a very short allowance before the long-expected supplies came, by which time our necessities had almost conquered my resolution not to take by force of the abundance I might see around, however well I might afterwards pay. It is but fair to state that the improvident villagers throughout Sikkim are extremely poor in vegetable food at this season, when the winter store is consumed, and the crops are still green. They are consequently obliged to purchase rice from the lower valleys, which, owing to the difficulties of transport, is very dear;. and to obtain it they barter wool, blankets, musk, and Tibetan produce of all kinds. Still they had cattle, which they would willingly have sold to me, but for the Dewan's orders.

There is a great difference between the vegetation of Dorjiling and that of similar elevations near Choongtam situated far within the Himalaya: this is owing to the steepness and dryness of the latter locality, where there is an absence of dense forest, which is replaced by a number of social grasses clothing the mountain sides, many new and beautiful kinds of rhododendrons, and a variety of European genera,* which (as I have elsewhere noticed) are either

* *Deutzia, Saxifraga ciliata, Thalictrum, Euphorbia,* yellow violet, *Labiatæ, Androsace, Leguminosæ, Coriaria, Delphinium,* currant, *Umbelliferæ,* primrose, *Anemone, Convallaria, Roscoea, Mitella, Herminium, Drosera.*

wholly absent from the damper ranges of Dorjiling, or found there several thousand feet higher up. On the hill above Choongtam village, I gathered, at 5000 to 6000 feet, *Rhododendron arboreum* and *Dalhousiæ*, which do not generally grow at Dorjiling below 7,500 feet.* The yew appears at 7000 feet, whilst, on the outer ranges (as on Tonglo), it is only found at 9,500 to 10,000 feet; and whereas on Tonglo it forms an immense tall tree, with long sparse branches and slender drooping twigs, growing amongst gigantic magnolias and oaks, at Choongtam it is small and rigid, and much resembling in appearance our churchyard yew.† At 8000 feet the *Abies Brunoniana* is found; a tree quite unknown further south; but neither the larch nor the *Abies Smithiana* (Khutrow) accompanied it, they being confined to still more northern regions.

I have seldom had occasion to allude to snakes, which are rare and shy in most parts of the Himalaya; I, however, found an extremely venomous one at Choongtam; a small black viper, a variety of the cobra di capello,‡

* I collected here ten kinds of rhododendron, which, however, are not the social plants that they become at greater elevations. Still, in the delicacy and beauty of their flowers, four of them, perhaps, excel any others; they are, *R. Aucklandii*, whose flowers are five inches and a half in diameter; *R. Maddeni, R. Dalhousiæ*, and *R. Edgeworthii*, all white-flowered bushes, of which the two first rise to the height of small trees.

† The yew spreads east from Kashmir to the Assam Himalaya and the Khasia mountains; and the Japan, Philippine Island, Mexican, and other North American yews, belong to the same widely-diffused species. In the Khasia (its most southern limit) it is found as low as 5000 feet above the sea-level.

‡ Dr. Gray, to whom I am indebted for the following information, assures me that this reptile is not specifically distinct from the common Cobra of India; though all the mountain specimens of it which he has examined retain the same small size and dark colour. Of the other Sikkim reptiles which I procured, seven are *Colubridæ* and innocuous; five *Crotalidæ* are venomous, three of which are new species belonging to the genera *Parias* and *Trimesurus*. Lizards are not abundant, but I found at Choongtam a highly curious one, *Plestiodon Sikkimensis*, Gray; a kind of Skink, whose only allies are two North American congeners; and a species of *Agama* (a chameleon-like lizard) which in many

which it replaces in the drier grassy parts of the interior of Sikkim, the large cobra not inhabiting in the mountain regions. Altogether I only collected about twelve species in Sikkim, seven of which are venomous, and all are dreaded by the Lepchas. An enormous hornet (*Vespa magnifica*, Sm.), nearly two inches long, was here brought to me alive in a cleft-stick, lolling out its great thorn-like sting, from which drops of a milky poison distilled: its sting is said to produce fatal fevers in men and cattle, which may very well be the case, judging from that of a smaller kind, which left great pain in my hand for two days, while a feeling of numbness remained in the arm for several weeks. It is called Vok by the Lepchas, a common name for any bee: its larvæ are said to be greedily eaten, as are those of various allied insects.

Choongtam boasts a profusion of beautiful insects, amongst which the British swallow-tail butterfly (*Papilio Machaon*) disports itself in company with magnificent black, gold, and scarlet-winged butterflies, of the Trojan group, so typical of the Indian tropics. At night my tent was filled with small water-beetles (*Berosi*) that quickly put out the candle; and with lovely moths came huge cockchafers (*Encerris Griffithii*), and enormous and fœtid flying-bugs (of the genus *Derecterix*), which bear great horns on the thorax. The irritation of mosquito and midge bites, and the disgusting insects that clung with spiny legs to the blankets of my tent and bed, were often as effectual in banishing sleep, as were my anxious thoughts regarding the future.

important points more resembled an allied American genus than an Asiatic one. The common immense earth-worm of Sikkim, *Ichthyophis glutinosus*, is a native of the Khasia mountains, Singapore, Ceylon and Java. It is a most remarkable fact, that whereas seven out of the twelve Sikkim snakes are poisonous, the sixteen species I procured in the Khasia mountains are innocuous.

The temple at Choongtam is a poor wooden building, but contains some interesting drawings of Lhassa, with its extensive Lamaseries and temples ; they convey the idea of a town, gleaming, like Moscow, with gilded and copper roofs ; but on a nearer aspect it is found to consist of a mass of stone houses, and large religious edifices many stories high, the walls of which are regularly pierced with small square ornamented windows.*

There is nothing remarkable in the geology of Choongtam : the base of the hill consists of the clay and mica slates overlain by gneiss, generally dipping to the eastward ; in the latter are granite veins, containing fine tourmalines. Actinolites are found in some highly metamorphic gneisses, brought by landslips from the neighbouring heights. The weather in May was cloudy and showery, but the rain which fell was far less in amount than that at Dorjiling : during the day the sun's power was great ; but though it rose between five and six A.M., it never appeared above the lofty peaked mountains that girdle the valley till eight A.M.

---

* MM. Huc and Gabet's account of Lhassa is, I do not doubt, excellent as to particulars ; but the trees which they describe as magnificent, and girdling the city, have uniformly been represented to me as poor stunted willows, apricots, poplars, and walnuts, confined to the gardens of the rich. No doubt the impression left by these objects on the minds of travellers from tree-less Tartary, and of Sikkimites reared amidst stupendous forests, must be widely different. The information concerning Lhassa collected by Timkowski, "Travels of the Russian Mission to China" (in 1821) is greatly exaggerated, though containing much that is true and curious. The dyke to protect the city from inundations I never heard of ; but there is a current story in Sikkim that Lhassa is built in a lake-bed, which was dried up by a miracle of the Lamas, and that in heavy rain the earth trembles, and the waters bubble through the soil : a Dorjiling rain-fall, I have been assured, would wash away the whole city. Ermann (Travels in Siberia, i., p. 186), mentions a town (Klinchi, near Perm), thus built over subterraneous springs, and in constant danger of being washed away. MM. Huc and Gabet allude to the same tradition under another form. They say that the natives of the banks of the Koko-nor affirm that the waters of that lake once occupied a subterranean position beneath Lhassa, and that the waters sapped the foundations of the temples as soon as they were built, till withdrawn by supernatural agency.

Dark pines crest the heights around, and landslips score
their flanks with white seams below; while streaks of snow
remain throughout the month at 9000 feet above; and
everywhere silvery torrents leap down to the Lachen and
Lachoong.

J. D. H. delt.

JUNIPERUS RECURVA.

Height 30 feet. (See p 45.)

CHAPTER XIX.

FROM this place there were two routes to Tibet, each of
about six days' journey. One lay to the north-west up the
Lachen valley to the Kongra Lama pass, the other to the
east up the Lachoong to the Donkia pass. The latter river
has its source in small lakes in Sikkim, south of the Donkia
mountain, a shoulder of which the pass crosses, commanding
a magnificent view into Tibet. The Lachen, on the other
hand (the principal source of the Teesta), rises beyond
Sikkim in the Cholamoo lakes. The frontier at Kongra
Lama was described to me as being a political, and not a
natural boundary, marked out by cairns, standing on a
plain, and crossing the Lachen river. To both Donkia
and Kongra Lama I had every right to go, and was determined, if possible, to reach them, in spite of Meepo's
ignorance, our guide's endeavours to frighten my party

and mislead myself, and the country people's dread of incurring the Dewan's displeasure.

The Lachen valley being pronounced impracticable in the height of the rains, a month later, it behoved me to attempt it first, and it possessed the attraction of leading to a frontier described as far to the northward of the snowy Himalaya, on a lofty plateau, whose plants and animals were different from anything I had previously seen.

After a week the coolies arrived with supplies : they had been delayed by the state of the paths, and had consequently consumed a great part of my stock, reducing it to eight days' allowance. I therefore divided my party, leaving the greater number at Choongtam, with a small tent, and instructions to forward all food to me as it arrived. I started with about fifteen attendants, on the 25th of May, for Lamteng, three marches up the Lachen.

Descending the step-formed terraces, I crossed the Lachen by a good cane bridge. The river is a headstrong torrent, and turbid from the vast amount of earthy matter which it bears along ; and this character of extreme impetuosity, unbroken by any still bend, or even swirling pool, it maintains uninterruptedly at this season from 4000 to 10,000 feet. It is crossed three times, always by cane bridges, and I cannot conceive any valley of its nature to be more impracticable at such a season. On both sides the mountains rose, densely forest-clad, at an average angle of 35° to 40°, to 10,000 and 15,000 feet. Its extreme narrowness, and the grandeur of its scenery, were alike recalled to my mind, on visiting the Sachs valley in the Valais of Switzerland ; from which, however, it differs in its luxuriant forest, and in the slopes being more uniform and less broken up into those imposing precipices so frequent in Switzerland, but which are wanting in the temperate regions of the Sikkim Himalaya.

At times we scrambled over rocks 1000 feet above the river, or descended into gorges, through whose tributary torrents we waded, or crossed swampy terraced flats of unstratified shingle above the stream; whilst it was some-times necessary to round rocky promontories in the river, stemming the foaming torrent that pressed heavily against the chest as, one by one, we were dragged along by powerful Lepchas. Our halting-places were on flats close to the river, covered with large trees, and carpeted with a most luxuriant herbage, amongst which a wild buckwheat (*Polygonum**) was abundant, which formed an excellent spinach: it is called "Pullop-bi"; a name I shall here-after have occasion to mention with gratitude.

A few miles above Choongtam, we passed a few cottages on a very extensive terrace at Tumlong; but between this and Lamteng, the country is uninhabited, nor is it frequented during the rains. We consequently found that the roads had suffered, the little bridges and aids to climb precipices and cross landslips had been carried away, and at one place we were all but turned back. This was at the Taktoong river, a tributary on the east bank, which rushes down at an angle of 15°, in a sheet of silvery foam, eighteen yards broad. It does not, where I crossed it, flow in a deep gulley, having apparently raised its bed by an accumulation of enormous boulders; and a plank bridge was thrown across it, against whose slippery and narrow foot-boards the water dashed, loosening the supports on either bank, and rushing between their foundation stones.

My unwilling guide had gone ahead with some of the

* *Polygonum cymosum*, Wall. This is a common Himalayan plant, and is also found in the Khasia mountains.

coolies : I had suspected him all along (perhaps unjustly)
of avoiding the most practicable routes; but when I found
him waiting for me at this bridge, to which he sarcastically
pointed with his bow, I felt that had he known of it, to
have made difficulties before would have been a work of
supererogation. He seemed to think I should certainly turn
back, and assured me there was no other crossing (a statement I afterwards found to be untrue); so, comforting
myself with the hope that if the danger were imminent,
Meepo would forcibly stop me, I took off my shoes, and
walked steadily over : the tremor of the planks was like
that felt when standing on the paddle-box of a steamer,
and I was jerked up and down, as my weight pressed
them into the boiling flood, which shrouded me with
spray. I looked neither to the right nor to the left,
lest the motion of the swift waters should turn my
head, but kept my eye on the white jets d'eau springing
up between the woodwork, and felt thankful when fairly on
the opposite bank : my loaded coolies followed, crossing
one by one without fear or hesitation. The bridge was
swept into the Lachen very shortly afterwards.

Towards Lamteng, the path left the river, and passed
through a wood of *Abies Smithiana*.* Larch appears at
9000 feet, with *Abies Brunoniana*. An austere crab-apple,
walnut, and the willow of Babylon (the two latter perhaps
cultivated), yellow jessamine and ash, all scarce trees in
Sikkim, are more or less abundant in the valley, from
7000 to 8000 feet; as is an ivy, very like the English,
but with fewer and smaller yellow or reddish berries;

---

* Also called *A. Khutrow* and *Morinda*. I had not before seen this tree in the
Himalaya: it is a spruce fir, much resembling the Norway spruce in general
appearance, but with longer pendulous branches. The wood is white, and considered indifferent, though readily cleft into planks; it is called "Seh."

and many other plants,* not found at equal elevations on the outer ranges of the Himalaya.

Chateng, a spur from the lofty peak of Tukcham,† 19,472 feet high, rises 1000 feet above the west bank of the river; and where crossed, commands one of the finest alpine views in Sikkim. It was grassy, strewed with huge boulders of gneiss, and adorned with clumps of park-like pines: on the summit was a small pool, beautifully fringed with bushy trees of white rose, a white-blossomed apple, a *Pyrus* like *Aria*, another like mountain-ash, scarlet rhododendrons (*arboreum* and *barbatum*), holly, maples, and *Goughia*,‡ a curious evergreen laurel-like tree: there were also Daphnes, purple magnolia, and a pink sweet-blossomed *Sphærostema*. Many English water-plants§ grew in the water, but I found no shells; tadpoles, however, swarmed, which later in the season become large frogs. The "painted-lady" butterfly (*Cynthia Cardui*), and a pretty "blue" were flitting over the flowers, together with some great tropical kinds, that wander so far up these valleys, accompanying *Marlea*, the only sub-tropical tree that ascends to 8,500 feet in the interior of Sikkim.

The river runs close under the eastern side of the

---

* Wood-sorrel, a white-stemmed bramble, birch, some maples, nut, gigantic lily (*Lilium giganteum*), *Euphorbia, Pedicularis, Spiræa, Philadelphus, Deutzia, Indigofera,* and various other South Europe and North American genera.

† "Tuk" signifies head in Lepcha, and "cheam" or "chaum," I believe, has reference to the snow. The height of Tukcham has been re-calculated by Capt. R. Strachey, with angles taken by myself, at Dorjiling and Jillapahar, and is approximate only.

‡ This fine plant was named (Wight, "Ic. Plant.") in honour of Capt. Gough, son of the late commander-in-chief, and an officer to whom the botany of the peninsula of India is greatly indebted. It is a large and handsome evergreen, very similar in foliage to a fine rhododendron, and would prove an invaluable ornament on our lawns, if its hardier varieties were introduced into this country.

§ *Sparganium, Typha, Potamogeton, Callitriche, Utricularia,* sedges and rushes.

valley, which slopes so steeply as to appear for many miles almost a continuous landslip, 2000 feet high.

Lamteng village, where I arrived on the 27th of May, is quite concealed by a moraine to the south, which, with a parallel ridge on the north, forms a beautiful bay in the mountains, 8,900 feet above the sea, and 1000 above the Lachen. The village stands on a grassy and bushy flat, around which the pine-clad mountains rise steeply to the snowy peaks and black cliffs which tower above. It contains about forty houses, forming the winter-quarters of the inhabitants of the valley, who, in summer, move with their flocks and herds to the alpine pastures of the Tibet frontier. The dwellings are like those described at Wallanchoon, but the elevation being lower, and the situation more sheltered, they are more scattered; whilst on account of the dampness of the climate, they are raised higher from the ground, and the shingles with which they are tiled (made of *Abies Webbiana*) decay in two or three years. Many are painted lilac, with the gables in diamonds of red, black, and white: the roofs are either of wood or of the bark of *Abies Brunoniana*, held down by large stones : within they are airy and comfortable. They are surrounded by a little cultivation of buck-wheat, radishes, turnips, and mustard. The inhabitants, though paying rent to the Sikkim Rajah, consider themselves as Tibetans, and are so in language, dress, features, and origin: they seldom descend to Choongtam, but yearly travel to the Tibetan towns of Jigatzi, Kambajong, Giantchi, and even to Lhassa, having always commercial and pastoral transactions with the Tibetans, whose flocks are pastured on the Sikkim mountains during summer, and who trade with the plains of India through the medium of these villagers.

The snow having disappeared from elevations below

J. D. H. delt.

LAMTENG VILLAGE.

11,000 feet, the yaks, sheep, and ponies had just been driven 2000 feet up the valley, and the inhabitants were preparing to follow, with their tents and goats, to summer quarters at Tallum and Tungu. Many had goîtres and rheumatism, for the cure of which they flocked to my tent; dry-rubbing for the latter, and tincture of iodine for the former, gained me some credit as a doctor: I could, however, procure no food beyond trifling presents of eggs, meal, and more rarely, fowls.

On arriving, I saw a troop of large monkeys* gambolling in a wood of *Abies Brunoniana:* this surprised me, as I was not prepared to find so tropical an animal associated with a vegetation typical of a boreal climate. The only other quadrupeds seen here were some small earless rats, and musk-deer; the young female of which latter sometimes afforded me a dish of excellent venison; being, though dark-coloured and lean, tender, sweet, and short-fibred. Birds were scarce, with the exception of alpine pigeons (*Columba leuconota*), red-legged crows (*Corvus graculus*, L.), and the horned pheasant (*Meleagris Satyra*, L.) In this month insects are scarce, *Elater* and a black earwig being the most frequent: two species of *Serica* also flew into my tent, and at night moths, closely resembling European ones, came from the fir-woods. The vegetation in the neighbourhood of Lamteng is European and North American; that is to say, it unites the boreal and temperate floras of the east and west hemispheres; presenting also a few features peculiar to Asia. This is a subject of very great importance in physical geography; as a country combining the botanical characters of several others, affords materials for tracing the direction in which genera and

---

* *Macacus Pelops?* Hodgson. This is a very different species from the tropical kind seen in Nepal, and mentioned at vol. i. p. 278.

species have migrated, the causes that favour their migra-
tions, and the laws that determine the types or forms of
one region, which represent those of another. A glance at
the map will show that Sikkim is, geographically, peculiarly
well situated for investigations of this kind, being centri-
cally placed, whether as regards south-eastern Asia or the
Himalayan chain. Again, the Lachen valley at this spot is
nearly equi-distant from the tropical forests of the Terai
and the sterile mountains of Tibet, for which reason
representatives both of the dry central Asiatic and Siberian,
and of the humid Malayan floras meet there.

The mean temperature of Lamteng (about 50°) is that
of the isothermal which passes through Britain in lat. 52°,
and east Europe in lat. 48°, cutting the parallel of 45° in
Siberia (due north of Lamteng itself), descending to lat. 42°
on the east coast of Asia, ascending to lat. 48° on the west
of America, and descending to that of New York in the
United States. This mean temperature is considerably
increased by descending to the bed of the Lachen at 8000
feet, and diminished by ascending Tukcham to 14,000 feet,
which gives a range of 6000 feet of elevation, and 20° of
mean temperature. But as the climate and vegetation
become arctic at 12,000 feet, it will be as well to confine
my observations to the flora of 7000 to 10,000 feet; of
the mean temperature, namely, between 53° and 43°, the
isothermal lines corresponding to which embrace, on the
surface of the globe, at the level of the sea, a space varying
in different meridians from three to twelve degrees of
latitude.*   At first sight it appears incredible that such a
limited area, buried in the depths of the Himalaya, should
present nearly all the types of the flora of the north

---

* On the west coast of Europe, where the distance between these isothermal lines
is greatest, this belt extends almost from Stockholm and the Shetlands to Paris.

temperate zone; not only, however, is this the case, but space is also found at Lamteng for the intercalation of types of a Malayan flora, otherwise wholly foreign to the north temperate region.

A few examples will show this. Amongst trees the Conifers are conspicuous at Lamteng, and all are of genera typical both of Europe and North America : namely, silver fir, spruce, larch, and juniper, besides the yew : there are also species of birch, alder, ash, apple, oak, willow, cherry, bird-cherry, mountain-ash, thorn, walnut, hazel, maple, poplar, ivy, holly, Andromeda, *Rhamnus.* Of bushes ; rose, berberry, bramble, rhododendron, elder, cornel, willow, honeysuckle, currant, *Spiræa, Viburnum, Cotoneaster, Hippophae.* Herbaceous plants * are far too numerous to be enumerated, as a list would include most of the common genera of European and North American plants.

Of North American genera, not found in Europe, were *Buddleia, Podophyllum, Magnolia, Sassafras? Tetranthera, Hydrangea, Diclytra, Aralia, Panax, Symplocos, Trillium,* and *Clintonia.* The absence of heaths is also equally a feature in the flora of North America. Of European genera, not found in North America, the Lachen valley has *Coriaria, Hypecoum,* and various *Cruciferæ.* The Japanese and Chinese floras are represented in Sikkim by *Camellia, Deutzia, Stachyurus, Aucuba, Helwingia, Stauntonia, Hydrangea, Skimmia, Eurya, Anthogonium,* and *Enkianthus.* The Malayan by Magnolias, *Talauma,* many vacciniums and rhododendrons, *Kadsura, Goughia, Marlea,* both coriaceous and deciduous-leaved *Cælogyne, Oberonia, Cyrtosia, Calanthe,*

---

* As an example, the ground about my tent was covered with grasses and sedges, amongst which grew primroses, thistles, speedwell, wild leeks, *Arum, Convallaria, Callitriche, Oxalis, Ranunculus, Potentilla, Orchis, Chærophyllum, Galium, Paris,* and *Anagallis*; besides cultivated weeds of shepherd's-purse, dock, mustard, Mithridate cress, radish. turnip, *Thlaspi arvense,* and *Poa annua.*

and other orchids; *Ceropegia Parochetus, Balanophora,* and many *Scitamineæ;* and amongst trees, by *Engelhardtia, Goughia,* and various laurels.

Shortly after my arrival at Lamteng, the villagers sent to request that I would not shoot, as they said it brought on excessive rain,* and consequent damage to the crops. My necessities did not admit of my complying with their wish unless I could procure food by other means; and I at first paid no attention to their request. The people, however, became urgent, and the Choongtam Lama giving his high authority to the superstition, it appeared impolitic to resist their earnest supplication; though I was well aware that the story was trumped up by the Lama for the purpose of forcing me to return. I yielded on the promise of provisions being supplied from the village, which was done to a limited extent; and I was enabled to hold out till more arrived from Dorjiling, now, owing to the state of the roads, at the distance of twenty days' march. The people were always civil and kind: there was no concealing the fact that the orders were stringent, prohibiting my party being supplied with food, but many of the villagers sought opportunities by night of replenishing my stores. Superstitious and timorous, they regard a doctor with great veneration; and when to that is added his power of writing, drawing, and painting, their admiration knows no bounds: they flocked round my tent all day, scratching their ears, lolling out their tongues, making a clucking noise, smiling, and timidly peeping over my shoulder, but flying in alarm when my little dog resented their familiarity by snapping at their legs. The

---

* In Griffith's narrative of "Pemberton's Mission to Bhotan" ("Posthumous Papers, Journal," p. 283), it is mentioned that the Gylongs (Lamas) attributed a violent storm to the members of the mission shooting birds.

men spend the whole day in loitering about, smoking,
and spinning wool: the women in active duties; a few
were engaged in drying the leaves of a shrub (*Symplocos*)
for the Tibet market, which are used as a yellow dye;
whilst, occasionally, a man might be seen cutting a spoon
or a yak-saddle out of rhododendron wood.

During my stay at Lamteng, the weather was all but
uniformly· cloudy and misty, with drizzling rain, and a
southerly, or up-valley wind, during the day, which changed
to an easterly one at night : occasionally distant thunder
was heard.   My rain-gauges showed very little rain compared with what fell at Dorjiling during the same period ;
the clouds were thin, both sun and moon shining through
them, without, however, the former warming the soil:
hence my tent was constantly wet, nor did I once sleep
in a dry bed till the 1st of June, which ushered in the
month with a brilliant sunny day.   At night it generally
rained in torrents, and the roar of landslips and avalanches
was then all but uninterrupted for hour after hour : sometimes it was a rumble, at others a harsh grating sound,
and often accompanied with the crashing of immense timber-
trees, or the murmur of the distant snowy avalanches. The
amount of denudation by atmospheric causes is here quite
incalculable; and I feel satisfied that the violence of the
river at this particular part of its course (where it traverses
those parts of the valleys which are most snowy and
rainy), is proximately due to impediments thus accumulated in its bed.

It was sometimes clear at sunrise, and I made many
ascents of Tukcham, hoping for a view of the mountains
towards the passes ; but I was only successful on one
occasion, when I saw the table top of Kinchinjhow, the
most remarkable, and one of the most distant peaks of

dazzling snow which is seen from Dorjiling, and which, I
was told, is far beyond Sikkim, in Tibet.*   I kept up a
constant intercourse with Choongtam, sending my plants
thither to be dried, and gradually reducing my party as
our necessities urged my so doing; lastly, I sent back the
shooters, who had procured very little, and whose occupation was now gone.

On the 2nd of June, I received the bad news that a large
party of coolies had been sent from Dorjiling with rice,
but that being unable or afraid to pass the landslips,
they had returned: we had now no food except a kid, a
few handfuls of flour, and some potatos, which had been sent
up from Choongtam.   All my endeavours to gain information respecting the distance and position of the frontier
were unavailing; probably, indeed, the Lama and Phipun
(or chief man of the village), were the only persons who
knew; the villagers calling all the lofty pastures a few
marches beyond Lamteng "Bhote" or "Cheen" (Tibet).
Dr. Campbell had procured for me information by which
I might recognise the frontier were I once on it; but no
description could enable me to find my way in a country
so rugged and forest-clad, through tortuous and perpetually
forking valleys, along often obliterated paths, and under
cloud and rain.   To these difficulties must be added
the deception of the rulers, and the fact (of which I was
not then aware), that the Tibet frontier was formerly at
Choongtam; but from the Lepchas constantly harassing
the Tibetans, the latter, after the establishment of the
Chinese rule over their country, retreated first to Zemu
Samdong, a few hours walk above Lamteng, then to
Tallum Samdong, 2000 feet higher; and, lastly, to

Such, however, is not the case; Kinchinjhow is on the frontier of Sikkim,
though a considerable distance behind the most snowy of the Sikkim mountains.

Kongra Lama, 16,000 feet up the west flank of Kin-chinjhow.

On the third of June I took a small party, with my tent, and such provisions as I had, to explore up the river. On hearing of my intention, the Phipun volunteered to take me to the frontier, which he said was only two hours dis-tant, at Zemu Samdong, where the Lachen receives the Zemu river from the westward: this I knew must be false, but I accepted his services, and we started, accompanied by a large body of villagers, who eagerly gathered plants for me along the road.

The scenery is very pretty; the path crosses extensive and dangerous landslips, or runs through fine woods of spruce and *Abies Brunoniana*, and afterwards along the river-banks, which are fringed with willow (called "Lama"), and *Hippophae*. The great red rose (*Rosa macrophylla*), one of the most beautiful Himalayan plants, whose single flowers are as large as the palm of the hand, was blossoming, while golden *Potentillas* and purple primroses flowered by the stream, and *Pyrola* in the fir-woods.

Just above the fork of the valley, a wooden bridge (Samdong) crosses the Zemu, which was pointed out to me as the frontier, and I was entreated to respect two sticks and a piece of worsted stretched across it; this I thought too ridiculous, so as my followers halted on one side, I went on the bridge, threw the sticks into the stream, crossed, and asked the Phipun to follow; the people laughed, and came over: he then told me that he had authority to permit of my botanising there, but that I was in Cheen, and that he would show me the guard-house to prove the truth of his statement. He accordingly led me up a steep bank to an extensive broad flat, several hundred feet above the river, and forming a triangular base to the great spur which, rising steeply behind, divides the valley. This flat was marshy and covered with grass; and buried in the jungle were several ruined stone houses, with thick walls pierced with loopholes: these had no doubt been occupied by Tibetans at the time when this was the frontier.

The elevation which I had attained (that of the river being 8,970 feet) being excellent for botanising, I camped; and the villagers, contented with the supposed success of their strategy, returned to Lamteng.

My guide from the Durbar had staid behind at Lamteng, and though Meepo and all my men well knew that this was not the frontier, they were ignorant as to its true position, nor could we even ascertain which of the rivers was the Lachen.* The only routes I possessed indicated two paths northwards from Lamteng, neither crossing a river: and I therefore thought it best to remain at Zemu Samdong till provisions should arrive. I accordingly halted for three days, collecting many new and beautiful plants, and exploring the roads, of which five (paths or yak-tracks) diverged from this point, one on either bank of each river, and one leading up the fork.

On one occasion I ascended the steep hill at the fork; it was dry and rocky, and crowned with stunted pines. Stacks of different sorts of pine-wood were stored on the flat at its base, for export to Tibet, all thatched with the bark of *Abies Brunoniana*. Of these the larch (*Larix Griffithii*, "Sah"), splits well, and is the most durable of any; but the planks are small, soft, and white.† : The silver fir (*Abies Webbiana*, "Dunshing") also splits well; it is white, soft, and highly prized for durability. The wood

---

* The eastern afterwards proved to be the Lachen.

† I never saw this wood to be red, close-grained, and hard, like that of the old Swiss larch; nor does it ever reach so great a size.

of *Abies Brunoniana* ("Semadoong") is like the others in appearance, but is not durable; its bark is however very useful. The spruce (*Abies Smithiana*, "Seh") has also white wood, which is employed for posts and beams.* These are the only pines whose woods are considered very useful; and it is a curious circumstance that none produce any quantity of resin, turpentine, or pitch; which may perhaps be accounted for by the humidity of the climate.

Pinus longifolia (called by the Lepchas "Gniet-koong," and by the Bhoteeas "Teadong") only grows in low valleys, where better timber is abundant. The weeping blue juniper (*Juniperus recurva*, "Deschoo"), and the arboreous black one (called "Tchokpo" †) yield beautiful wood, like that of the pencil cedar,‡ but are comparatively scarce, as is the yew (*Taxus baccata*, "Tingschi"), whose timber is red. The "Tchenden," or funereal cypress, again, is valued only for the odour of its wood: *Pinus excelsa*, "Tongschi," though common in Bhotan, is, as I have elsewhere remarked, not found in east Nepal or Sikkim; the wood is admirable, being durable, close-grained, and so resinous as to be used for flambeaux and candles.

On the flat were flowering a beautiful magnolia with globular sweet-scented flowers like snow-balls, several balsams, with species of *Convallaria*, *Cotoneaster*, *Gentian*, *Spiræa*, *Euphorbia*, *Pedicularis*, and honeysuckle. On the hill-side were creeping brambles, lovely yellow, purple, pink, and

---

* These woods are all soft and loose in grain, compared with their European allies.

† This I have, vol. i. p. 256, referred to the *J. excelsa* of the north-west Himalaya, a plant which under various names is found in many parts of Europe and many parts of Europe and North America; but since then Dr. Thomson and I have had occasion to compare my Sikkim conifers with the north-west Himalayan ones, and we have found that this Sikkim species is probably new, and that *J. excelsa* is not found east of Nepal.

‡ Also a juniper, from Bermuda (*J. Bermudiana*).

white primroses, white-flowered *Thalictrum* and *Anemone*,
berberry, *Podophyllum*, white rose, fritillary, *Lloydia*, &c.
On the flanks of Tukcham, in the bed of a torrent, I gathered
many very alpine plants, at the comparatively low elevation of
10,000 feet, as dwarf willows, *Pinguicula*, (a genus not
previously found in the Himalaya), *Oxyria*, *Androsace*,
Tofieldia, *Arenaria*, saxifrages, and two dwarf heath-like
Andromedas.* The rocks were all of gneiss, with granite
veins, tourmaline, and occasionally pieces of pure plumbago.

Our guide had remained at Lamteng, on the plea of a
sore on his leg from leech-bites : his real object, however,
was to stop a party on their way to Tibet with madder
and canes, who, had they continued their journey, would
inevitably have pointed out the road to me. The villagers
themselves now wanted to proceed to the pasturing-grounds
on the frontier; so the Phipun sent me word that I might
proceed as far as I liked up the east bank of the Zemu. I
had explored the path, and finding it practicable, and
likely to intersect a less frequented route to the frontier
(that crossing the Tekonglah pass from Bah, see p. 13), I
determined to follow it. A supply of food arrived from Dorjiling on the 5th of June, reduced, however, to one bag of
rice, but with encouraging letters, and the assurance that
more would follow at once. My men. of whom I had
eight, behaved admirably, although our diet had for five
days chiefly consisted of *Polygonum* (" Pullop-bi "), wild

---

* Besides these, a month later, the following flowered in profusion: scarlet
Buddleia ! gigantic lily, yellow jasmine, *Aster*, *Potentilla*, several kinds of orchids,
willow-herb (*Epilobium*), purple *Roscoea*, *Neillia*, *Morina*, many grasses and *Umbelliferæ*. These formed a rank and dense herbaceous, mostly annual vegetation,
six feet high, bound together with *Cuscuta*, climbing *Leguminosæ*, and *Ceropegia*.
The great summer heat and moisture here favour the ascent of various tropical
genera, of which I found in August several *Orchideæ* (*Calanthe*, *Microstylis*, and
Cælogyne), also *Begonia*, *Bryonia*, *Cynanchum*, *Aristolochia*, *Eurya*, *Procris*, *Acanthaceæ*, and *Cyrtandraceæ*.

leeks ("Lagook"), nettles and *Procris* (an allied, and more succulent herb), eked out by eight pounds of Tibet meal ("Tsamba"), which I had bought for ten shillings by stealth from the villagers.   What concerned me most was the destruction of my plants by constant damp, and the want of sun to dry the papers; which reduced my collections to a tithe of what they would otherwise have been.

From Zemu Samdong the valley runs north-west, for two marches, to the junction of the Zemu with the Thlonok, which rises on the north-east flank of Kinchinjunga: at this place I halted for several days, while building a bridge over the Thlonok.   The path runs first through a small forest of birch, alder, and maple, on the latter of which I found *Balanophora* * growing abundantly: this species produces the great knots on the maple roots, from which the Tibetans form the cups mentioned by MM. Huc and Gabet.   I was so fortunate as to find a small store of these knots, cleaned, and cut ready for the turner, and hidden behind a stone by some poor Tibetan, who had never returned to the spot: they had evidently been there a very long time.

In the ravines there were enormous accumulations of ice, the result of avalanches; one of them crossed the river, forming a bridge thirty feet thick, at an elevation of only 9,800 feet above the sea.   This ice-bridge was 100 yards broad, and flanked by heaps of boulders, the effects of combined land and snowslips. These stony places were covered with a rich herbage of rhubarb, primroses, *Euphorbia, Sedum, Polygonum, Convallaria,* and a purple *Dentaria* ("Kenroop-bi") a cruciferous plant much eaten as a pot-herb.   In the pinewoods a large mushroom ("Onglau,"† Tibet.) was abundant,

---

* A curious leafless parasite, mentioned at vol. i. p. 133.

† *Cortinarius Emodensis* of the Rev. M. J. Berkeley, who has named and

which also forms a favourite article of food. Another pot-herb (to which I was afterwards more indebted than any) was a beautiful *Smilacina*, which grows from two to five feet high, and has plaited leaves and crowded panicles of white bell-shaped flowers, like those of its ally the lily of the valley, which it also resembles in its mucilaginous properties. It is called "Chokli-bi," * and its young flower-heads, sheathed in tender green leaves, form an excellent vegetable. Nor must I forget to include amongst the eatable plants of this hungry country, young shoots of the mountain-bamboo, which are good either raw or boiled, and may be obtained up to 12,000 feet in this valley. A species of *Asarum* (Asarabacca) grows in the pine-woods; a genus not previously known to be Himalayan. The root, like its English medicinal con-gener, has a strong and peculiar smell. At 10,000 feet *Abies Webbiana* commences, with a close undergrowth of a small twiggy holly. This, and the dense thicket of rhodo-dendron † on the banks of the river and edges of the wood, rendered the march very fatiguing, and swarms of midges kept up a tormenting irritation.

The Zemu continued an impetuous muddy torrent, whose hoarse voice, mingled with the deep grumbling noise ‡ of

described it from my specimens and drawings. It is also called "Yungla tchamo" by the Tibetans, the latter word signifying a toadstool. Mr. Berkeley informs me that the whole vast genus *Cortinarius* scarcely possesses a single other edible species; he adds that *C. violaceus* and *violaceo-cinereus* are eaten in Austria and Italy, but not always with safety.

* It is also found on the top of Sinchul, near Dorjiling.

† Of which I had already gathered thirteen kinds in this valley.

‡ The dull rumbling noise thus produced is one of the most singular pheno-mena in these mountains, and cannot fail to strike the observer. At night, especially, the sound seems increased, the reason of which is not apparent, for in these regions, so wanting in animal life, the night is no stiller than the day, and the melting of snow being less, the volume of waters must be somewhat, though not conspicuously, diminished. The interference of sound by heated currents of different density is the most obvious cause of the diminished reverberation during the day, to which Humboldt adds the increased tension of vapour, and possibly an echo from its particles.

the boulders rolling along its bed, was my lullaby for many nights. Its temperature at Zemu Samdong was 45° to 46° in June. At its junction with the Thlonok, it comes down a steep gulley from the north, foreshortened into a cataract 1000 feet high, and appearing the smaller stream of the two; whilst the Thlonok winds down from the snowy face of Kinchinjunga, which is seen up the valley, bearing W.S.W., about twenty miles distant. All around are lofty and rocky mountains, sparingly wooded with pines and larch, chiefly on their south flanks, which receive the warm, moist, up-valley winds; the faces exposed to the north being colder and more barren : exactly the reverse of what is the case at Choongtam, where the rocky and sunny south-exposed flanks are the driest.

My tent was pitched on a broad terrace, opposite the junction of the Zemu and Thlonok, and 10,850 feet above the sea. It was sheltered by some enormous transported blocks of gneiss, fifteen feet high, and surrounded by a luxuriant vegetation of most beautiful rhododendrons in full flower, willow, white rose, white-flowered cherry, thorn, maple and birch. Some great tuberous-rooted *Arums* * were very abundant; and the ground was covered with small pits, in which were large wooden pestles : these are used in the preparation of food from the arums, to which the miserable inhabitants of the valley have recourse in spring, when their yaks are calving. The roots are bruised with the pestles, and thrown into these holes with water. Acetous fermentation commences in seven or eight days, which is a sign that the acrid poisonous principle is dissipated : the pulpy, sour, and fibrous mass is then boiled and eaten; its nutriment

* Two species of *Arisæma,* called "Tong" by the Tibetans, and "Sinkree" by the Lepchas.

being the starch, which exists in small quantities, and which they have not the skill to separate by grating and washing. This preparation only keeps a few days, and produces bowel complaints, and loss of the skin and hair, especially when insufficiently fermented. Besides this, the " chokli-bi," and many other esculents, abounded here ; and we had great need of them before leaving this wild uninhabited region.

I repeatedly ascended the north flank of Tukcham along a watercourse, by the side of which were immense slips of rocks and snow-beds; the mountain-side being excessively steep.   Some of the masses of gneiss thus brought down were dangerously poised on slopes of soft shingle, and daily moved a little downwards.   All the rocks were gneiss and granite, with radiating crystals of tourmaline as thick as the thumb.   Below 12,000 to 13,000 feet the mountain-sides were covered with a dense scrub of rhododendron bushes, except where broken by rocks, landslips, and torrents : above this the winter's snow lay deep, and black rocks and small glaciers, over which avalanches were constantly falling with a sullen roar, forbade all attempts to proceed.   My object in ascending was chiefly to obtain views and compass-bearings, in which I was generally disappointed : once only I had a magnificent prospect of Kinchinjunga, sweeping down in one unbroken mass of glacier and ice, fully 14,000 feet high, to the head of the Thlonok river, whose upper valley appeared a broad bay of ice ; doubtless forming one of the largest glaciers in the Himalaya, and increased by lateral feeders that flow into it from either flank of the valley.   The south side of this (the Thlonok) valley is formed by a range from Kinchinjunga, running east to Tukcham, where it terminates : from it rises the beautiful mountain Liklo,* 22,582 feet

* D^2 of the peaks laid down in Colonel Waugh's " Trigonometrical Survey from

high, which, from Dorjiling, appears as a sharp peak, but is here seen to be a jagged crest running north and south. On the north flank of the valley the mountains are more sloping and black, with patches of snow above 15,000 feet, but little anywhere else, except on another beautiful peak (alt. 19,240 feet) marked D^3 on the map. This flank is also continuous from Kinchin; it divides Sikkim from Tibet, and runs north-east to the great mountain Chomiomo (which was not visible), the streams from its north flank flowing into the Arun river (in Tibet). A beautiful blue arch of sky spanned all this range, indicating the dry Tibetan climate beyond.

I made two futile attempts to ascend the Thlonok river to the great glaciers at the foot of Kinchinjunga, following the south bank, and hoping to find a crossing-place, and so to proceed north to Tibet. The fall of the river is not great at this part of its course, nor up to 12,000 feet, which was the greatest height I could attain, and about eight miles beyond my tents; above that point, at the base of Liklo, the bed of the valley widens, and the rhododendron shrubbery was quite impervious, while the sides of the mountain were inaccessible. We crossed extensive snow-beds, by cutting holes in their steep faces, and rounded rocks in the bed of the torrent, dragging one another through the violent current, whose temperature was below 40°.

On these occasions, the energy of Meepo, Nimbo (the chief of the coolies) and the Lepcha boys, was quite remarkable, and they were as keenly anxious to reach the holy country of Tibet as I could possibly be. It was

Dorjiling," I believe to be the "Liklo" of Dr. Campbell's itineraries from Dorjiling to Lhassa, compiled from the information of the traders (See "Bengal Asiatic Society's Journal" for 1848); the routes in which proved of the utmost value to me.

sometimes dark before we got back to our tents, tired, with torn clothes and cut feet and hands, returning to a miserable dinner of boiled herbs; but never did any of them complain, or express a wish to leave me. In the evenings and mornings they were always busy, changing my plants, and drying the papers over a sulky fire at my tent-door; and at night they slept, each wrapt in his own blanket, huddled together under a rock, with another blanket thrown over them all. Provisions reached us so seldom, and so reduced in quantity, that I could never allow more than one pound of rice to each man in a day, and frequently during this trying month they had not even that; and I eked out our meagre supply with a few ounces of preserved meats, occasionally "splicing the main brace" with weak rum and water.

At the highest point of the valley which I reached, water boiled at 191·3, indicating an elevation of 11,903 feet. The temperature at 1 p.m. was nearly 70°, and of the wet bulb 55°, indicating a dryness of 0·462, and dew point 47·0. Such phenomena of heat and dryness are rare and transient in the wet valleys of Sikkim, and show the influence here of the Tibetan climate.*

After boiling my thermometer on these occasions, I generally made a little tea for the party; a refreshment to which they looked forward with child-like eagerness. The fairness with which these good-hearted people used to divide the scanty allowance, and afterwards the leaves, which are greatly relished, was an engaging trait in their simple character: I have still vividly before me their sleek swarthy faces and twinkling Tartar eyes, as they lay

* I gathered here, amongst an abundance of alpine species, all of European and arctic type, a curious trefoil, the *Parochetus communis*, which ranges through 9000 feet of elevation on the Himalaya, and is also found in Java and Ceylon.

stretched on the ground in the sun, or crouched in the sleet and snow beneath some sheltering rock; each with his little polished wooden cup of tea, watching my notes and instruments with curious wonder, asking, "How high are we?" "How cold is it?" and comparing the results with those of other stations, with much interest and intelligence.

On the 11th June, my active people completed a most ingenious bridge of branches of trees, bound by withes of willow; by which I crossed to the north bank, where I camped on an immense flat terrace at the junction of the rivers, and about fifty feet above their bed. The first step or ascent from the river is about five feet high, and formed of water-worn boulders, pebbles, and sand, scarcely stratified: the second, fully 1000 yards broad, is ten feet high, and swampy. The uppermost is fifteen feet above the second, and is covered with gigantic boulders, and vast rotting trunks of fallen pines, buried in an impenetrable jungle of dwarf small-leaved holly and rhododendrons. The surface was composed of a rich vegetable mould, which, where clear of forest, supported a rank herbage, six to eight feet high.*

Our first discovery, after crossing, was of a good bridge across the Zemu, above its junction, and of a path leading

---

* This consisted of grasses, sedges, *Bupleurum*, rhubarb, *Ranunculus*, *Convallaria*, *Smilacina*, nettles, thistles, *Arum*, balsams, and the superb yellow *Meconopsis Nepalensis*, whose racemes of golden poppy-like flowers were as broad as the palm of the hand; it grows three and even six feet high, and resembles a small hollyhock; whilst a stately *Heracleum*, ten feet high, towered over all. Forests of silver fir, with junipers and larch, girdled these flats, and on their edges grew rhododendrons, scarlet *Spiræa*, several honeysuckles, white *Clematis*, and *Viburnum*. Ferns are much scarcer in the pine-woods than elsewhere in the forest regions of the Himalaya. In this valley (alt. 10,850 feet), I found only ten kinds; *Hymenophyllum*, *Lomaria*, *Cystopteris*, *Davallia*, two *Polypodia*, and several *Aspidia* and *Asplenia*. *Selaginella* ascends to Zemu Samdong (9000 feet). The *Pteris aquilina* (brake) does not ascend above 10,000 feet.

down to Zemu Samdong; this was, however, scarcely traceable up either stream. My men were better housed here in sheds: and I made several more ineffectual attempts to ascend the valley to the glaciers. The path, gradually vanishing, ran alternately through fir-woods, and over open grassy spots, covered with vegetation, amongst which the gigantic arum was plentiful, whose roots seemed to be the only attraction in this wet and miserable valley.

On my return one day, I found my people in great alarm, the Phipun having sent word that we were on the Tibet side of the rivers, and that Tibetan troops were coming to plunder my goods, and carry my men into slavery. I assured them he only wanted to frighten them; that the Cheen soldiers were civil orderly people; and that as long as Meepo was with us, there was no cause for fear. Fortunately a young musk-deer soon afterwards broke cover close to the tent, and its flesh wonderfully restored their courage: still I was constantly harassed by threats; some of my people were suffering from cold and bowel complaints, and I from rheumatism; while one fine lad, who came from Dorjiling, was delirious with a violent fever, contracted in the lower valleys, which sadly dispirited my party.

Having been successful in finding a path, I took my tent and a few active lads 1000 feet up the Zemu, camping on a high rock above the forest region, at 12,070 feet, hoping thence to penetrate northwards. I left my collections in the interim at the junction of the rivers, where the sheds and an abundance of firewood were great advantages for preserving the specimens. At this elevation we were quite free from midges and leeches (the latter had not appeared above 11,500 feet), but the weather continued so uniformly rainy and bad, that we could make no progress. I repeatedly followed the river for several miles, ascending to 13,300 feet; but though its valley widened, and its current was less rapid, the rhododendron thickets below, and the cliffs above, defeated all endeavours to reach the drier climate beyond, of which I had abundant evidence in the arch of brilliant blue that spanned the heavens to the north, beyond a black canopy of clouds that hid everything around, and poured down rain without one day's intermission, during the eight which I spent here.

BLACK JUNIPER (height sixty feet) AND YOUNG LARCH.
(See p. 45.)

CHAPTER XX.

My little tent was pitched in a commanding situation,
on a rock fifty feet above the Zemu, overlooking the course
of that river to its junction with the Thlonok.   The descent
of the Zemu in one thousand feet is more precipitous than
that of any other river of its size with which I am acquainted
in Sikkim, yet immediately above my camp it was more tranquil than at any part of its course onwards to the plains of
India, whether as the Zemu, Lachen or Teesta.   On the
west bank a fine mountain rose in steep ridges and shrubby
banks to 15,000 feet; on the east a rugged cliff towered
above the stream, and from this, huge masses of rock were
ever and anon precipitated into the torrent, with a
roar that repeatedly spread consternation amongst us.
During rains especially, and at night, when the chilled
atmospheric currents of air descend, and the sound is not

dissipated as in the day-time, the noise of these falls is sufficiently alarming. My tent was pitched near the base of the cliff, and so high above the river, that I had thought it beyond the reach of danger; but one morning I found that a large fragment of granite had been hurled during the night to my very door, my dog having had a very narrow escape. To what depth the accumulation at the base of this cliff may reach, I had no means of judging, but the rapid slope of the river-bed is mainly due to this, and to old moraines at the mouth of the valley below. I have seen few finer sights than the fall of these stupendous blocks into the furious torrent, along which they are carried amid feathery foam for many yards before settling to rest.

Across the Thlonok to the southwards, rose the magnificent mountain of Tukcham, but I only once caught a glimpse of its summit, which even then clouded over before I could get my instruments adjusted for ascertaining its height. Its top is a sharp cone, surrounded by rocky shoulders, that rise from a mass of snow. Its eastern slope of 8000 feet is very rapid (about 38°) from its base at the Zemu river to its summit.

Glaciers in the north-west Himalaya descend to 11,000 feet; but I could not discover any in these valleys even so low as 14,000 feet, though at this season extensive snow-beds remain unmelted at but little above 10,000 feet. The foot of the stupendous glacier filling the broad head of the Thlonok is certainly not below 14,000 feet; though being continuous with the perpetual snow (or névé) of the summit of Kinchinjunga, it must have 14,000 feet of ice, in perpendicular height, to urge it forwards.

All my attempts to advance up the Zemu were fruitless, and a snow bridge by which I had hoped to cross to the

opposite bank was carried away by the daily swelling
river, while the continued bad weather prevented any
excursions for days together. Botany was my only resource,
and as vegetation was advancing rapidly under the influence of the southerly winds, I had a rich harvest: for though
Compositæ, *Pedicularis*, and a few more of the finer Himalayan plants flower later, June is still the most glorious
month for show.

Rhododendrons occupy the most prominent place,
clothing the mountain slopes with a deep green mantle
glowing with bells of brilliant colours; of the eight or
ten species growing here, every bush was loaded with
as great a profusion of blossoms as are their northern
congeners in our English gardens. Primroses are next,
both in beauty and abundance ; and they are accompanied
by yellow cowslips, three feet high, purple polyanthus, and
pink large-flowered dwarf kinds nestling in the rocks, and
an exquisitely beautiful blue miniature species, whose
blossoms sparkle like sapphires on the turf. Gentians begin
to unfold their deep azure bells, aconites to rear their tall
blue spikes, and fritillaries and *Meconopsis* burst into
flower. On the black rocks the gigantic rhubarb forms
pale pyramidal towers a yard high, of inflated reflexed
bracts, that conceal the flowers, and over-lapping one
another like tiles, protect them from the wind and rain: a
whorl of broad green leaves edged with red spreads on the
ground at the base of the plant, contrasting in colour with
the transparent bracts, which are yellow, margined with
pink. This is the handsomest herbaceous plant in Sikkim:
it is called "Tchuka," and the acid stems are eaten
both raw and boiled; they are hollow and full of pure
water : the root resembles that of the medicinal rhubarb,
but it is spongy and inert; it attains a length of four

feet, and grows as thick as the arm. The dried leaves afford a substitute for tobacco; a smaller kind of rhubarb is however more commonly used in Tibet for this purpose; it is called "Chula."

The elevation being 12,080 feet, I was above the limit of trees, and the ground was covered with many kinds of small-flowered honeysuckles, berberry, and white rose.*

I saw no birds, and of animals only an occasional musk-deer. Insects were scarce, and quite different from what I had seen before; chiefly consisting of *Phryganea* (May-fly) and some *Carabidæ* (an order that is very scarce in the Himalaya); with various moths, chiefly *Geometræ*.

The last days of June (as is often the case) were marked by violent storms, and for two days my tent proved no protection; similar weather prevailed all over India, the barometer falling very low. I took horary observations of the barometer in the height of the storm on the 30th: the tide was very small indeed (·024 inch, between 9·50 A.M. and 4 P.M.), and the thermometer ranged between 47° and 57° 8, between 7 A.M. and midnight. Snow fell abundantly as low as 13,000 feet, and the rivers were much swollen, the size and number of the stones they rolled along producing a deafening turmoil. Only 3·7 inches of rain fell between the 23rd of June and the 2nd of July; whilst 21 inches fell at Dorjiling, and 6·7 inches at Calcutta. During the same period the mean temperature was 48°; extremes, $\frac{62°\ 0}{36°\ 5}$. The humidity was nearly at saturation-point, the wind southerly, very raw and cold, and drizzling rain constantly

---

* Besides these I found a prickly *Aralia*, maple, two currants, eight or nine rhododendrons, many *Sedums*, *Rhodiola*, white *Clematis*, red-flowered cherry, birch, willow, *Viburnum*, juniper, a few ferns, two *Andromedas*, *Menziesia*, and *Spiræa*. And in addition to the herbs mentioned above, may be enumerated *Parnassia*, many Saxifrages, *Soldanella*, *Draba*, and various other *Cruciferæ*, *Nardostachys*, (spikenard), *Epilobium*, *Thalictrum*, and very many other genera, almost all typical of the Siberian, North European, and Arctic floras.

fell. A comparison of thirty observations with Dorjiling gave a difference of 14° temperature, which is at the rate of 1° for every 347 feet of ascent.*

The temperature of these rivers varies extremely at different parts of their course, depending on that of their affluents. The Teesta is always cool in summer (where its bed is below 2000 feet), its temperature being 20° below that of the air; whereas in mid-winter, when there is less cloud, and the snows are not melting, it is only a few degrees colder than the air.† At this season, in descending from 12,000 feet to 1000 feet, its temperature does not rise 10°, though that of the air rises 30° or 40°. It is a curious fact, that the temperature of the northern feeders of the Teesta, in some parts of their course, rises with the increasing elevation! Of this the Zemu afforded a curious example: during my stay at its junction with the Thlonok it was 46°, or 6° warmer than that river; at 1100 feet higher it was 48°, and at 1100 feet higher still it was 49°! These observations were repeated in different weeks, and several times on the same day, both in ascending and descending, and always with the same result: they told, as certainly as if I had followed the river to its source, that it rose in a drier and comparatively sunny climate, and flowed amongst little snowed mountains.

---

* Forty-seven observations, comparative with Calcutta, gave 34°·8 difference, and if 5°5 of temperature be deducted for northing in latitude, the result is 1° for every 412 feet of ascent. My observations at the junction of the rivers alt. 10,850 feet), during the early part of the month, gave 1° to 304 feet, as the result of twenty-four observations with Dorjiling, and 1° to 394 feet, from seventy-four observations with Calcutta.

† During my sojourn at Bhomsong in mid-winter of 1848 (see v. i. p. 305), the mean temperature of the Teesta was 51°, and of the air 52°3; at that elevation the river water rarely exceeds 60° at midsummer. Between 4000 feet and 300 (the plains) its mean temperature varies about 10° between January and July; at 6000 feet it varies from 55° to 43° during the same period; and at 10,000 feet it freezes at the edges in winter and rises to 50° in July.

Meanwhile, the Lachen Phipun continued to threaten us, and I had to send back some of the more timorous of my party. On the 28th of June fifty men arrived at the Thlonok, and turned my people out of the shed at the junction of the rivers, together with the plants they were preserving, my boards, papers, and utensils. The boys came to me breathless, saying that there were Tibetan soldiers amongst them, who declared that I was in Cheen, and that they were coming on the following morning to make a clean sweep of my goods, and drive me back to Dorjiling. I had little fear for myself, but was anxious with respect to my collections: it was getting late in the day, and raining, and I had no mind to go down and expose myself to the first brunt of their insolence, which I felt sure a night of such weather would materially wash away. Meepo was too frightened, but Nimbo, my Bhotan coolie Sirdar, volunteered to go, with two stout fellows; and he accordingly brought away my plants and papers, having held a parley with the enemy, who, as I suspected, were not Tibetans. The best news he brought was, that they were half clad and without food; the worst, that they swaggered and bullied: he added, with some pride, that he gave them as good as he got, which I could readily believe, Nimbo being really a resolute fellow,* and accomplished in Tibet slang.

On the following morning it rained harder than ever, and the wind was piercingly cold. My timid Lepchas huddled behind my tent, which, from its position, was only to be stormed in front. I dismantled my little observatory, and packed up the instruments, tied my dog, Kinchin, to one of the tent-pegs, placed a line of stones opposite

* In East Nepal he drew his knife on a Ghorka sepoy; and in the following winter was bold enough to make his escape in chains from Tumloong.

the door, and seated myself on my bed on the ground, with my gun beside me.

The dog gave tongue as twenty or thirty people defiled up the glen, and gathered in front of my tent; they were ragged Bhoteeas, with bare heads and legs, in scanty woollen garments sodden with rain, which streamed off their shaggy hair, and furrowed their sooty faces: their whole appearance recalled to my mind Dugald Dalgetty's friends, the children of the mist.

They appeared nonplussed at seeing no one with me, and at my paying no attention to them, whilst the valiant Kinchin effectually scared them from the tent-door. When they requested a parley, I sent the interpreter to say that I would receive three men, and that only provided all the rest were sent down immediately; this, as I anticipated, was acceded to at once, and there remained only the Lachen Phipun and his brother. Without waiting to let him speak, I rated him soundly, saying, that I was ready to leave the spot when he could produce any proof of my being in Bhote (or Cheen), which he knew well I was not; that, since my arrival at Lachen, he had told me nothing but lies, and had contravened every order, both of the Rajah and of Tchebu Lama. I added, that I had given him and his people kindness and medicine, their return was bad, and he must go about his business at once, having, as I knew, no food, and I having none for him. He behaved very humbly throughout, and finally took himself off much discomfited, and two days afterwards sent men to offer to assist me in moving my things.

The first of July was such a day as I had long waited for to obtain a view, and I ascended the mountain west of my camp, to a point where water boiling at 185° 7 (air 42°), gave an elevation of 14,914 feet. On the top of the range,

about 1000 feet above this, there was no snow on the
eastern exposures, except in hollows, but on the west
slopes it lay in great fields twenty or thirty feet thick;
while to the north, the mountains all appeared destitute of
snow, with grassy flanks and rugged tops.

Drizzling mist, which had shrouded Tukcham all the
morning, soon gathered on this mountain, and prevented
any prospect from the highest point reached; but on the
ascent I had an excellent view up the Zemu, which opened
into a broad grassy valley, where I saw with the glass some
wooden sheds, but no cattle or people. To reach these,
however, involved crossing the river, which was now impossible; and I reluctantly made up my mind to return on the
morrow to Zemu Samdong, and thence try the other river.

On my descent to the Thlonok, I found that the herbaceous plants on the terraces had grown fully two feet during
the fortnight, and now presented almost a tropical luxuriance and beauty. Thence I reached Zemu Samdong in
one day, and found the vegetation there even more gay and
beautiful: the gigantic lily was in full flower, and scenting
the air, with the lovely red rose, called " Chirring " by the
Tibetans. *Neillia* was blossoming profusely at my old
camping-ground, to which I now returned after a month's
absence.

Soon after my arrival I received letters from Dr. Campbell,
who had strongly and repeatedly represented to the Rajah
his opinion of the treatment I was receiving; and this
finally brought an explicit answer, to the effect that his
orders had been full and peremptory that I should be
supplied with provisions, and safely conducted to the
frontier. With these came letters on the Rajah's part from
Tchebu Lama to the Lachen Phipun, ordering him to take
me to the pass, but not specifying its position; fortunately,

however, Dr. Campbell sent me a route, which stated the
pass to be at Kongra Lama, several marches beyond this,
and in the barren country of Tibet.

On the 5th of July the Singtam Soubah arrived from
Chola (the Rajah's summer residence) : he was charged to
take me to the frontier, and brought letters from his
highness, as well as a handsome present, consisting of
Tibet cloth, and a dress of China silk brocaded with gold :
the Ranee also sent me a basket of Lhassa sweetmeats,
consisting of Sultana raisins from Bokhara, sliced and
dried apricots from Lhassa, and *Diospyros* fruit from China
(called " Gubroon " by the Tibetans).   The Soubah
wanted to hurry me on to the frontier and back at once,
being no doubt instigated to do so by the Dewan's party,
and by his having no desire to spend much time in the
dreary lofty regions I wanted to explore.   I positively
refused, however, to start until more supplies arrived,
except he used his influence to provide me with food ; and
as he insisted that the frontier was at Tallum Samdong,
only one march up the Lachen, I foresaw that this move was
to be but one step forward, though in the right direction.
He went forward to Tallum at once, leaving me to follow.

The Lamteng people had all migrated beyond that point
to Tungu, where they were pasturing their cattle : I sent
thither for food, and procured a little meal at a very high
price, a few fowls and eggs ; the messenger brought back
word that Tungu was in Tibet, and that the villagers
ignored Kongra Lama.   A large piece of yak-flesh being
brought for sale, I purchased it ; but it proved the toughest
meat I ever ate, being no doubt that of an animal that had
succumbed to the arduous duties of a salt-carrier over the
passes : at this season, however, when the calves are not a
month old, it was in vain to expect better.

Large parties of women and children were daily passing my tent from Tungu, to collect arum-roots at the Thlonok, all with baskets at their backs, down to rosy urchins of six years old: they returned after several days, their baskets neatly lined with broad rhododendron leaves, and full of a nauseous-looking yellow acid pulp, which told forcibly of the extreme poverty of the people. The children were very fair; indeed the young Tibetan is as fair as an English brunette, before his perennial coat of smoke and dirt has permanently stained his face, and it has become bronzed and wrinkled by the scorching sun and rigorous climate of these inhospitable countries. Children and women were alike decked with roses, and all were good-humoured and pleasant, behaving with great kindness to one another, and unaffected politeness to me.

During my ten days' stay at Zemu Samdong, I formed a large collection of insects, which was in great part destroyed by damp : many were new, beautiful, and particularly interesting, from belonging to types whose geographical distribution is analogous to that of the vegetation. The caterpillar of the swallow-tail butterfly (*Papilio Machaon*), was common, feeding on umbelliferous plants, as in England ; and a *Sphynx* (like *S. Euphorbiæ*) was devouring the euphorbias ; the English *Cynthia Cardui* (painted-lady butterfly) was common, as were "sulphurs," "marbles," *Pontia* (whites), "blues," and *Thecla*, of British aspect but foreign species. Amongst these, tropical forms were rare, except one fine black swallow-tail. Of moths, *Noctuæ* and *Geometræ* abounded, with many flies and *Tipulæ*. *Hymenoptera* were scarce, except a yellow *Ophion*, which lays its eggs in the caterpillars above-mentioned. Beetles were most rare, and (what is remarkable) the wood-borers (*longicorns* and *Curculio*) particularly so. A large *Telephora* was very common, and had the usual propensity of its congeners for blood; *lamellicorns* were also abundant.

On the 11th of July five coolies arrived with rice: they had been twenty days on the road, and had been obliged to make great detours, the valley being in many places impassable. They brought me a parcel of English letters; and I started up the Lachen on the following day, with renewed spirits and high hopes. The road first crossed the Zemu and the spur beyond, and then ascended the west bank of the Lachen, a furious torrent for five or six miles, during which it descends 1000 feet, in a chasm from which rise lofty black pine-clad crags, topped by snowy mountains, 14,000 to 16,000 feet high. One remarkable mass of rock, on the east bank, is called "Sakya-zong" (or the abode of Sakya, often pronounced Thakya, one of the Boodhist Trinity); at its base a fine cascade falls into the river.

Above 11,000 feet the valley expands remarkably, the mountains recede, become less wooded, and more grassy, while the stream is suddenly less rapid, meandering in a broader bed, and bordered by marshes, covered with *Carex*, *Blysmus*, dwarf Tamarisk, and many kinds of yellow and red *Pedicularis*, both tall and beautiful. There are far fewer rhododendrons here than in the damper Zemu valley at equal elevations, and more Siberian, or dry country types of vegetation, as *Astragali* of several kinds, *Habenaria*, *Epipactis*, dandelion, and a caraway, whose stems (called in Tibet "Gzira") are much sought for as a condiment.* The Singtam Soubah and Lachen Phipun

---

* *Umbelliferæ* abound here; with sage, *Ranunculus*, *Anemone*, Aconites, *Halenia*, Gentians, *Panax*, *Euphrasia*, speedwell, *Prunella vulgaris*, thistles, bistort, *Parnassia*, purple orchis, *Prenanthes*, and *Lactuca*. The woody plants of this region are willows, birch, *Cotoneaster*, maple, three species of *Viburnum*, three of *Spiræa*, *Vaccinium*, *Aralia*, *Deutzia*, *Philadelphus*, rhododendrons, two junipers, silver fir, larch, three honeysuckles, *Neillia*, and a *Pieris*, whose white blossoms are so full of honey as to be sweet and palatable.

received me at the bridge (Samdong), at Tallum, and led me across the river (into Cheen they affirmed) to a pretty green sward, near some gigantic gneiss boulders, where I camped, close by the river, and 11,480 feet above the sea.

The village of Tallum consists of a few wretched stone huts, placed in a broad part of the valley, which is swampy, and crossed by several ancient moraines, which descend from the gulleys on the east flank.* The cottages are from four to six feet high, without windows, and consist of a single apartment, containing neither table, chair, stool, nor bed; the inmates huddle together amid smoke, filth, and darkness, and sleep on a plank; and their only utensils are a bamboo churn, copper, bamboo, and earthenware vessels, for milk, butter, &c.

Grassy or stony mountains slope upwards, at an angle of 20°,† from these flats to 15,000 feet, but no snow is visible, except on Kinchinjhow and Chomiomo, about fifteen miles up the valley. Both these are flat-topped, and dazzlingly white, rising into small peaks, and precipitous on all sides; they are grand, bold, isolated masses, quite unlike the ordinary snowy mountains in form, and far more imposing even than Kinchinjunga, though not above 22,000 feet in elevation.

Herbaceous plants are much more numerous here than in any other part of Sikkim; and sitting at my tent-door, I could, without rising from the ground, gather forty-three plants,‡ of which all but two belonged to English genera.

---

* I have elsewhere noticed that in Sikkim, the ancient moraines above 9000 feet are almost invariably deposited from valleys opening to the westward.

† At Lamteng and up the Zemu the slopes are 40° and 50°, giving a widely different aspect to the valleys.

‡ In England thirty is, on the average, the equivalent number of plants, which in favourable localities I have gathered in an equal space. In both cases many are seedlings of short-lived annuals, and in neither is the number a test of the luxuriance of the vegetation; it but shows the power which the different species exert in their struggle to obtain a place.

In the rich soil about the cottages were crops of dock, shepherd's-purse, *Thlaspi arvense*, *Cynoglossum* of two kinds (one used as a pot-herb), balsams, nettle, *Galeopsis*, mustard, radish, and turnip. On the neighbouring hills, which I explored up to 15,000 feet, I found many fine plants, partaking more or less of the Siberian type, of which *Corydalis*, *Leguminosæ*, *Artemisia*, and *Pedicularis*, are familiar instances. I gathered upwards of 200 species, nearly all belonging to north European genera. Twenty-five were woody shrubs above three feet high, and six were ferns; * sedges were in great profusion, amongst them three of British kinds: seven or eight were *Orchideæ*, including a fine *Cypripedium*.

The entomology of Tallum, like its botany, was Siberian, Arctic types occurring at lower elevations than in the wetter parts of Sikkim. Of beetles the honey-feeding ones prevailed, with European forms of others that inhabit yak-droppings.† Bees were common, both *Bombus* and *Andrœna*, but there were no wasps, and but few ants. Grasshoppers and other *Orthoptera* were rare, as were *Hemiptera*; *Tipula* was the common dipterous insect, with a small sand-fly: there were neither leeches, mosquitos, ticks, nor midges. Pigeons, red-legged crows, and hawks were the common birds; with a few waders in the marshes.

Being now fairly behind most of the great snow and rain-collecting mountains, I experienced a considerable

---

* *Cryptogramma crispa, Davallia,* two *Aspidia,* and two *Polypodia.* I gathered *ten* at the same elevation, in the damper Zemu valley (see p. 49, *note*). I gathered in this valley a new species of the remarkable European genus *Struthiopteris,* which has not been found elsewhere in the Himalaya.

† As *Aphodius* and *Geotrupes.* Predaceous genera were very rare, as *Carabus* and *Staphylinus,* so typical of boreal regions. *Coccinella* (lady-bird), which swarms at Dorjiling, does not ascend so high, and a *Clytus* was the only longicorn. *Bupretis, Elater,* and *Blaps* were found but rarely. Of butterflies, the *Machaon* seldom reaches this elevation, but the painted-lady, *Pontia, Colias, Hipparchia, Argynnis,* and *Polyommatus,* are all found.

change in the climate, which characterises all these rearward lofty valleys, where very little rain falls, and that
chiefly drizzle; but this is so constant that the weather
feels chilly, raw, and comfortless, and I never returned dry
from botanising. The early mornings were bright with
views northwards of blue sky and Kinchinjhow, while to the
south the lofty peak of Tukcham, though much nearer, was
seldom seen, and black cumuli and nimbi rolled up the steep
valley of the Lachen to be dissipated in mist over Tallum.
The sun's rays were, however, powerful at intervals during
the forenoon, whence the mean maximum temperature of
July occurred at about 10 A.M. The temperature of the
river was always high, varying with the heat of the day from
47° to 52°; the mean being 50°

These streams do not partake of the diurnal rise and fall,
so characteristic of the Swiss rivers and those of the western
Himalaya, where a powerful sun melts the glaciers by
day, and their head-streams are frozen by night. Here
the clouds alike prevent solar and nocturnal radiation, the
temperature is more uniform, and the corroding power of
the damp southerly wind that blows strongly throughout
the day is the great melting agent. One morning I saw a
vivid and very beautiful halo 20° degrees distant from the
sun's disc; it was no doubt caused by snow in the higher
regions of the atmosphere, as a sharp shower of rain fell
immediately afterwards: these are rare phenomena in
mountainous countries.

The Singtam Soubah visited me daily, and we enjoyed
long friendly conversations: he still insisted that the Yang-
choo (the name he gave to the Lachen at this place) was
the boundary, and that I must not go any further. His
first question was always "How long do you intend to
remain here? have you not got all the plants and stones

you want? you can see the sun much better with those brasses and glasses * lower down; it is very cold here, and there is no food : "—to all which I had but one reply, that I should not return till I had visited Kongra Lama. He was a portly man, and, I think, at heart good-natured : I had no difficulty in drawing him on to talk about Tibet, and the holy city of Teshoo Loombo, with its thousands of gilt temples, nunneries, and convents, its holiest of all the holy grand Lamas of Tibet, and all the wide Boodhist world besides. Had it even been politic, I felt it would be unfair to be angry with a man who was evidently in a false position between myself and his two rulers, the Rajah and Dewan; who had a wife and family on the smiling flanks of Singtam, and who longed to be soaking in the warm rain of Sikkim, drinking Murwa beer (a luxury unknown amongst these Tibetans) and gathering in his crops of rice, millet, and buckwheat. Though I may owe him a grudge for his subsequent violence, I still recal with pleasure the hours we spent together on the banks of the Lachen. In all matters respecting the frontier, his lies were circumstan-tial; and he further took the trouble of bringing country people to swear that this was Cheen, and that there was no such place as Kongra Lama. I had written to ask Dr. Campbell for a definite letter from Tchebu Lama on this point, but unfortunately my despatches were lost; the messenger who conveyed them missed his footing in crossing the Lachen, and escaped narrowly with life, while the turban in which the letters were placed was carried down the current.

Finally the Soubah tried to persuade my people that one so incorrigibly obstinate must be mad, and that they had better leave me. One day, after we had had a long discus-

* Alluding to the sextant, &c.

sion about the geography of the frontier, he inflamed my
curiosity by telling me that Kinchinjhow was a very holy
mountain; more so than its sister-peaks of Chumulari
and Kinchinjunga; and that both the Sikkim and Tibetan
Lamas, and Chinese soldiers, were ready to oppose my
approach to it. This led to my asking him for a sketch
of the mountains; he called for a large sheet of paper, and
some charcoal, and wanted to form his mountains of sand;
I however ordered rice to be brought, and though we had
but little, scattered it about wastefully. This had its effect:
he stared at my wealth, for he had all along calculated on
starving me out, and retired, looking perplexed and
crestfallen. Nothing puzzled him so much as my being
always occupied with such, to him, unintelligible pursuits;
a Tibetan "cui bono?" was always in his mouth: "What
good will it do *you?*" "Why should you spend weeks
on the coldest, hungriest, windiest, loftiest place on the
earth, without even inhabitants?" Drugs and idle
curiosity he believed were my motives, and possibly a
reverence for the religion of Boodh, Sakya, and Tsong-
kaba. Latterly he had made up his mind to starve me
out, and was dismayed when he found I could hold out
better than himself, and when I assured him that I should
not retrace my steps until his statements should be verified
by a letter from Tchebu; that I had written to him, and
that it would be at least thirty days before I could receive
an answer.

On the 19th of July he proposed to take me to Tungu,
at the foot of Kinchinjhow, and back, upon ponies, provided
I would leave my people and tent, which I refused to do.
After this I saw little of him for several days, and began
to fear he was offended, when one morning his attendant
came to me for medicine with a dismal countenance, and

in great alarm: he twisted his fingers together over his stomach to symbolise the nature of the malady which produced a commotion in his master's bowels, and which was simply the colic. I was aware that he had been reduced to feed upon "Tong" (the arum-root) and herbs, and had always given him half the pigeons I shot, which was almost the only animal food I had myself. Now I sent him a powerful dose of medicine; adding a few spoonfuls of China tea and sugar for friendship.

On the 22nd, being convalescent, he visited me, looking wofully yellow. After a long pause, during which he tried to ease himself of some weighty matter, he offered to take me to Tungu with my tent and people, and thence to Kongra Lama, if I would promise to stay but two nights. I asked whether Tungu was in Cheen or Sikkim; he replied that after great enquiry he had heard that it was really in Sikkim; "Then," said I, "we will both go to-morrow morning to Tungu, and I will stay there as long as I please:" he laughed, and gave in with apparent good grace.

After leaving Tallum, the valley contracts, passing over great ancient moraines, and again expanding wider than before into broad grassy flats. The vegetation rapidly diminishes in stature and abundance, and though the ascent to Tungu is trifling, the change in species is very great. The *Spiræa*, maple, *Pieris*, cherry, and larch disappear, leaving only willow, juniper, stunted birch, silver fir, white rose, *Aralia*, berberry, currant, and more rhododendrons than all these put together; *

* *Cyananthus*, a little blue flower allied to *Campanula*, and one of the most beautiful alpines I know, covered the turfy ground, with *Orchis, Pedicularis, Gentian, Potentilla, Geranium*, purple and yellow *Meconopsis*, and the *Artemisia* of Dorjiling, which ascends to 12,000 feet, and descends to the plains, having a range of 11,500 feet in elevation. Of ferns, *Hymenophyllum, Cistopteris*, and *Cryptogramma crispa* ascend thus high.

while mushrooms and other English fungi * grew amongst the grass.

Tungu occupies a very broad valley, at the junction of the Tungu-choo from the east, and the Lachen from the north. The hills slope gently upwards to 16,000 feet, at an average angle of 15°; they are flat and grassy at the

TUNGU VILLAGE.

base, and no snow is anywhere to be seen.† A stupendous rock, about fifty feet high, lay in the middle of the valley,

* One of great size, growing in large clumps, is the English *Agaricus comans*, Fr., and I found it here at 12,500 feet, as also the beautiful genus *Crucibulum*, which is familiar to us in England, growing on rotten sticks, and resembling a diminutive bird's nest with eggs in it.

† In the wood-cut the summit of Chomiomo is introduced, as it appears from a few hundred feet above the point of view.

broken in two : it may have been detached from a cliff, or
have been transported thither as part of an ancient moraine
which extends from the mouth of the Tungu-choo valley
across that of the Lachen. The appearance and position
of this great block, and of the smaller piece lying beside
it, rather suggest the idea of the whole mass having fallen
perpendicularly from a great height through a *crevasse* in
a glacier, than of its having been hurled from so considerable
a distance as from the cliffs on the flanks of the valley : it
is faithfully represented in the accompanying woodcut. A
few wooden houses were collected near this rock, and
several black tents were scattered about. I encamped at
an elevation of 12,750 feet, and was waited on by the
Lachen Phipun with presents of milk, butter, yak-flesh, and
curds ; and we were not long before we drowned old enmity
in buttered and salted tea.

On my arrival I found the villagers in a meadow, all
squatted cross-legged in a circle, smoking their brass and
iron pipes, drinking tea, and listening to a letter from the
Rajah, concerning their treatment of me. Whilst my men
were pitching my tent, I gathered forty plants new to me,
all of Tartarian types.* Wheat or barley I was assured
had been cultivated at Tungu when it was possessed by
Tibetans, and inhabited by a frontier guard, but I saw
no appearance of any cultivation. The fact is an important
one, as barley requires a mean summer temperature of 48°
to come to maturity. According to my observations, the

---

* More Siberian plants appeared, as *Astragali, Chenopodium, Artemisia,* some
grasses, new kinds of *Pedicularis, Delphinium,* and some small Orchids. Three
species of *Parnassia* and six primroses made the turf gay, mixed with saxifrages,
Androsace and *Campanula.* By the cottages was abundance of shepherd's-purse,
Lepidium, and balsams, with dock, *Galeopsis,* and *Cuscuta.* Several low dwarf
species of honeysuckle formed stunted bushes like heather; and *Anisodus,* a
curious plant allied to *Hyoscyamus,* whose leaves are greedily eaten by yaks, was
very common.

mean temperature of Tungu in July is upwards of 50°, and, by calculation, that of the three summer months, June, July, and August, should be about 46° 5. As, however, I do not know whether these cerealia were grown as productive crops, much stress cannot be laid upon the fact of their having been cultivated, for in a great many parts of Tibet the barley is annually cut green for fodder.

In the evening the sick came to me : their complaints, as usual, being rheumatism, ophthalmia, goîtres, cuts, bruises, and poisoning by Tong (*Arum*), fungi, and other deleterious vegetables. At Tallum I attended an old woman who dressed her ulcers with *Plantago* (plantain) leaves, a very common Scotch remedy ; the ribs being drawn out from the leaf, which is applied fresh : it is rather a strong application.

On the following morning I was awakened by the shrill cries of the Tibetan maidens, calling the yaks to be milked, " Toosh—toosh—toooosh," in a gradually higher key ; to which Toosh seemed supremely indifferent, till quickened in her movements by a stone or stick, levelled with unerring aim at her ribs ; these animals were changing their long winter's wool for sleek hair, and the former hung about them in ragged masses, like tow. Their calves gambolled by their sides, the drollest of animals, like ass-colts in their antics, kicking up their short hind-legs, whisking their bushy tails in the air, rushing up and down the grassy slopes, and climbing like cats to the top of the rocks.

The Soubah and Phipun came early to take me to Kongra Lama, bringing ponies, genuine Tartars in bone and breed. Remembering the Dewan's impracticable saddle at Bhomsong, I stipulated for a horse-cloth or pad, upon which I had no sooner jumped than the beast threw back his ears, seated himself on his haunches, and, to my

consternation, slid backwards down a turfy slope, pawing the
earth with his fore-feet as he went, and leaving me on the
ground, amid shrieks of laughter from my Lepchas. My
steed being caught, I again mounted, and was being led
forward, when he took to shaking himself like a dog till
the pad slipped under his belly, and I was again unhorsed.
Other ponies displayed equal prejudices against my mode
of riding, or having my weight anywhere but well on their
shoulders, being all-powerful in their fore-quarters ; and so
I was compelled to adopt the high demi-pique saddle with
short stirrups, which forced me to sit with my knees up
to my nose, and to grip with the calves of my legs and
heels.   All the gear was of yak or horse-hair, and the bit
was a curb and ring, or a powerful twisted snaffle.

The path ran N.N.W for two miles, and then crossed
the Lachen above its junction with the Nunee * from the
west : the stream was rapid, and twelve yards in breadth ;
its temperature was 48°. About six miles above Tungu,
the Lachen is joined by the Chomio-choo, a large affluent
from Chomiomo mountain.   Above this the Lachen meanders along a broad stony bed, and the path rises over a
great ancient moraine, whose level top is covered with
pools, but both that and its south face are bare, from
exposure to the south wind, which blows with fury through
this contracted part of the valley to the rarified atmosphere
of the lofty, open, and dry country beyond.   Its north slope,
on the contrary, is covered with small trees and brushwood, rhododendron, birch, honeysuckle, and mountain-
ash. These are the most northern shrubs in Sikkim, and
I regarded them with deep interest, as being possibly the

* I suspect there is a pass by the Nunee to the sheds I saw up the Zemu valley
on the 2nd of July, as I observed yaks grazing high up the mountains : the
distance cannot be great, and there is little or no snow to interfere.

last of their kind to be met with in this meridian, for many
degrees further north : perhaps even no similar shrubs occur
between this and the Siberian Altai, a distance of 1,500
miles.   The magnificent yellow cowslip (*Primula Sikkimensis*) gilded the marshes, and *Caltha,** *Trollius*, Anemone,
Arenaria, *Draba*, Saxifrages, Potentillas, Ranunculus, and
other very alpine plants abounded.

At the foot of the moraine was a Tibetan camp of broad,
black, yak-hair tents, stretched out with a complicated
system of ropes, and looking at a distance—(to borrow
M. Huc's graphic simile)—like fat-bodied, long-legged
spiders ! Their general shape is hexagonal, about twelve
feet either way, and they are stretched over six short posts,
and encircled with a low stone wall, except in front.   In
one of them I found a buxom girl, the image of good
humour, making butter and curd from yak-milk.   The
churns were of two kinds ; one being an oblong box of
birch-bark, or close bamboo wicker-work, full of branched
rhododendron twigs, in which the cream is shaken : she
good-naturedly showed me the inside, which was frosted
with snow-white butter, and alive with maggots.   The
other churn was a goat-skin, which was rolled about, and
shaken by the four legs.   The butter is made into great
squares, and packed in yak-hair cloths; the curd is eaten
either fresh, or dried and pulverised (when it is called
" Ts'cheuzip ").

Except bamboo and copper milk-vessels, wooden ladles,
tea-churn, and pots, these tents contained no furniture but
goat-skins and blankets, to spread on the ground as a bed.
The fire was made of sheep and goats'-droppings, lighted
with juniper-wood ; above it hung tufts of yaks'-hair, one

---

* This is the *C. scaposa*, n. sp. The common *Caltha palustris*, or "marsh
marigold" of England, which is not found in Sikkim, is very abundant in the
north-west Himalaya.

for every animal lost during the season,* by which means
a reckoning is kept. Although this girl had never before
seen a European, she seemed in no way discomposed at my
visit, and gave me a large slice of fresh curd.

Beyond this place (alt. 14,500 feet), the valley runs
up north-east, becoming very stony and desolate, with green
patches only by the watercourses : at this place, however,
thick fogs came on, and obscured all view. At 15,000 feet,
I passed a small glacier on the west side of the valley, the
first I had met with that descended nearly to the river,
during the whole course of the Teesta.

Five miles further on we arrived at the tents of the
Phipun, whose wife was prepared to entertain us with
Tartar hospitality : magnificent tawny Tibet mastiffs were
baying at the tent-door, and some yaks and ponies were
grazing close by. We mustered twelve in number, and
squatted cross-legged in a circle inside the tent, the Soubah
and myself being placed on a pretty Chinese rug. Salted and
buttered tea was immediately prepared in a tea-pot for us
on the mat, and in a great caldron for the rest of the party;
parched rice and wheat-flour, curd, and roasted maize†
were offered us, and we each produced our wooden cup,
which was kept constantly full of scalding tea-soup, which,
being made with fresh butter, was very good. The flour
was the favourite food, of which each person dexterously
formed little dough-balls in his cup, an operation I could
not well manage, and only succeeded in making a nauseous
paste, that stuck to my jaws and in my throat. Our

---

* The Siberians hang tufts of horse-hair inside their houses from superstitious
motives (Ermann's "Siberia," i., 281).

† Called "pop-corn" in America, and prepared by roasting the maize in an iron
vessel, when it splits and turns partly inside out, exposing a snow-white spongy
mass of farina. It looks very handsome, and would make a beautiful dish for
dessert.

hostess' hospitality was too *exigeant* for me, but the others seemed as if they could not drink enough of the scalding tea.

We were suddenly startled from our repast by a noise like loud thunder, crash following crash, and echoing through the valley. The Phipun got up, and coolly said, "The rocks are falling, it is time we were off, it will rain soon." The moist vapours had by this time so accumulated, as to be condensed in rain on the cliffs of Chomiomo and Kinchinjhow; which, being loosened, precipitated avalanches of rocks and snow. We proceeded amidst dense fog, soon followed by hard rain; the roar of falling rocks on either hand increasing as these invisible giants spoke to one another in voices of thunder through the clouds. The effect was indescribably grand: and as the weather cleared, and I obtained transient peeps of their precipices of blue ice and black rock towering 5000 feet above me on either hand, the feeling of awe produced was almost overpowering. Heavy banks of vapour still veiled the mountains, but the rising mist exposed a broad stony track, along which the Lachen wandered, split into innumerable channels, and enclosing little oases of green vegetation, lighted up by occasional gleams of sunshine. Though all around was enveloped in gloom, there was in front a high blue arc of cloudless sky, between the beetling cliffs that formed the stern portals of the Kongra Lama pass.

CHAPTER XXI.

WE reached the boundary between Sikkim and Tibet early
in the afternoon; it is drawn along Kongra Lama, which
is a low flat spur running east from Kinchinjhow towards
Chomiomo, at a point where these mountains are a few
miles apart, thus crossing the Lachen river: * it is marked
by cairns of stone, some rudely fashioned into chaits,
covered with votive rags on wands of bamboo. I made the
altitude by barometer 15,745 feet above the sea, and by
boiling water, 15,694 feet, the water boiling at 184.1°; the
temperature of the air between 2. 40. and 4 P.M. varied
from 41° 3 to 42° 5, the dew-point 39·8°; that of the

---

* The upper valley of the Lachen in Tibet, which I ascended in the following
October, is very open, flat, barren, and stony; it is bounded on the north by
rounded spurs from Chomiomo, which are continued east to Donkia, forming
a watershed to the Lachen on the south, and to the Arun on the north.

Lachen was 47°, which was remarkably high. We were bitterly cold; as the previous rain had wetted us through, and a keen wind was blowing up the valley. The continued mist and fog intercepted all view, except of the flanks of the great mountains on either hand, of the rugged snowy ones to the south, and of those bounding the Lachen to the north. The latter were unsnowed, and appeared lower than Kongra Lama, the ground apparently sloping away in that direction; but when I ascended them, three months afterwards, I found they were 3000 feet higher! a proof how utterly fallacious are estimates of height, when formed by the eye alone. My informants called them Peuka-t'hlo; "peu" signifies north in Tibetan, and "t'hlo" a hill in Lepcha.

Isolated patches of vegetation appeared on the top of the pass, where I gathered forty kinds of plants, most of them being of a tufted habit characteristic of an extreme climate; some (as species of *Caryophylleæ*) forming hemispherical balls on the naked soil; others* growing in matted tufts level with the ground. The greater portion had no woolly covering; nor did I find any of the cottony species of *Saussurea*, which are so common on the wetter mountains to the southward. Some most delicate-flowered plants even defy the biting winds of these exposed regions; such are a prickly *Meconopsis* with slender flower-stalks and four large blue poppy-like petals, a *Cyananthus* with a membranous bell-shaped corolla, and a fritillary. Other curious plants were a little yellow saxifrage with long runners (very like the arctic *S. flagellaris*, of Spitzbergen

---

* The other plants found on the pass were; of smooth hairless ones, *Ranunculus*, Fumitory, several species of *Stellaria, Arenaria, Cruciferæ, Parnassia, Morina*, saxifrages, *Sedum*, primrose, *Herminium, Polygonum, Campanula, Umbelliferæ*, grasses and *Carices:* of woolly or hairy ones, *Anemone, Artemisia, Myosotis*, *Draba, Potentilla*, and several *Compositæ*, &c.

and Melville Island), and the strong-scented spikenard (*Nardostachys*).

The rocks were chiefly of reddish quartz, and so was the base of Chomiomo. Kinchinjhow on the contrary was of gneiss, with granite veins : the strike of both was north-west, and the dip north-east 20° to 30°

We made a fire at the top with sheep's droppings, of which the Phipun had brought up a bagfull, and with it a pair of goat-skin bellows, which worked by a slit that was opened by the hand in the act of raising; when inflated, the hole was closed, and the skin pressed down, thus forcing the air through the bamboo nozzle : this is the common form of bellows throughout Tibet and the Himalaya.

After two hours I was very stiff and cold, and suffering from headache and giddiness, owing to the elevation ; and having walked about thirteen miles botanizing, I was glad to ride down. We reached the Phipun's tents about 6 P.M., and had more tea before proceeding to Tungu. The night was fortunately fine and calm, with a few stars and a bright young moon, which, with the glare from the snows, lighted up the valley, and revealed magnificent glimpses of the majestic mountains. As the moon sank, and we descended the narrowing valley, darkness came on, and with a boy to lead my sure-footed pony, I was at liberty uninterruptedly to reflect on the events of a day, on which I had attained the object of so many years' ambition. Now that all obstacles were surmounted, and I was returning laden with materials for extending the knowledge of a science which had formed the pursuit of my life, will it be wondered at that I felt proud, not less for my own sake, than for that of the many friends, both in India and at home, who were interested in my success ?

We arrived at Tungu at 9 P.M., my pony not having

stumbled once, though the path was rugged, and crossed by many rapid streams. The Soubah's little shaggy steed had carried his portly frame (fully fifteen stone weight) the whole way out and back, and when he dismounted, it shook itself, snorted, and seemed quite ready for supper.

On the following morning I was occupied in noting and arranging my collections, which consisted of upwards of 200 plants; all gathered above 14,000 feet elevation.[*] Letters arrived from Dorjiling with unusual speed, having been only seventeen days on the road : they were full of valuable suggestions and encouragement from my friends Hodgson, Campbell, and Tchebu Lama.

On the 26th of July the Phipun, who waited on me every morning with milk and butter, and whose civility and attentions were now unremitting, proposed that I should accompany him to an encampment of Tibetans, at the foot of Kinchinjhow. We mounted ponies, and ascended the Tunguchoo eastwards : it was a rapid river for the first thousand feet, flowing in a narrow gorge, between sloping, grassy, and rocky hills, on which large herds of yaks were feeding, tended by women and children, whose black tents were scattered about. The yak-calves left their mothers to run beside our ponies, which became unmanageable, being almost callous to the bit; and the whole party was sometimes careering over the slopes, chased by the grunting herds : in other places, the path was narrow and dangerous, when the sagacious animals proceeded with the utmost gravity and caution. Rounding one rocky spur, my pony stumbled, and pitched me forward : fortunately I lighted on the path.

---

* Amongst them the most numerous Natural orders and genera were, *Cruciferæ* 10; *Compositæ* 20; *Ranunculaceæ* 10; *Alsineæ* 9; *Astragali* 10; *Potentillæ* 8; grasses 12; *Carices* 15; *Pedicularis* 7; *Boragineæ* 7.

The rocks were gneiss, with granite veins (strike north-east, dip south-east): they were covered with *Ephedra*,* an *Onosma* which yields a purple dye, *Orchis,* and species of *Androsace ;* while the slopes were clothed with the spikenard and purple *Pedicularis,* and the moist grounds with yellow cowslip and long grass. A sudden bend in the valley opened a superb view to the north, of the full front of Kinchinjhow, extending for four or five miles east and west; its perpendicular sides studded with the immense icicles, which are said to have obtained for it the name of "jhow,"—the "bearded" Kinchin. Eastward a jagged spur stretches south, rising into another splendid mountain, called Chango-khang (the Eagle's crag), from whose flanks descend great glaciers, the sources of the Tunguchoo.

We followed the course of an affluent, -called the Cha-choo, along whose bed ancient moraines rose in successive ridges : on these I found several other species of European genera.† Over one of these moraines, 500 feet high, the path ascends to the plains of Palung, an elevated grassy expanse, two miles long and four broad, extending south-ward from the base of Kinchinjhow. Its surface, though very level for so mountainous a country, is yet varied with open valleys and sloping hills, 500 to 700 feet high : it is bounded on the west by low rounded spurs from Kinchinjhow, that form the flank of the Lachen valley ; while on the east it is separated from Chango-khang by the Chachoo, which cuts a deep east and west trench along the base of Kinchinjhow, and then turns south to the

---

* A curious genus of small shrubs allied to pines, that grows in the south of Europe. This species is the European *E. vulgaris ;* it inhabits the driest parts of north-west India. and ascends to 17,000 feet in Tibet, but is not found in the moist intervening countries.

† *Delphinium, Hypecoum, Sagina, Gymnandra, Artemisia, Caltha, Dracocephalum, Leontopodium.*

Tunguchoo. The course of the Chachoo, where it turns south, is most curious: it meanders in sickle-shaped curves along the marshy bottom of an old lake-bed, with steep shelving sides, 500 to 600 feet deep, and covered with juniper bushes.* It is fed by the glaciers of Kinchinjhow, and some little lakes to the east.

The mean height of Palung plains is 16,000 feet: they are covered with transported blocks, and I have no doubt their surface has been much modified by glacial action. I was forcibly reminded of them by the slopes of the Wengern Alp, but those of Palung are far more level. Kinchinjhow rises before the spectator, just as the Jungfrau, Mönch, and Eigher Alps do from that magnificent point of view.

On ascending a low hill, we came in sight of the Tibet camp at the distance of a mile, when the great mastiffs that guarded it immediately bayed; and our ponies starting off at full gallop, we soon reached an enclosure of stone dykes, within which the black tents were pitched. The dogs were of immense size, and ragged, like the yaks, from their winter coat hanging to their flanks in great masses; each was chained near a large stone, on and off which he leapt as he gave tongue; they are very savage, but great cowards, and not remarkable for intelligence.

The people were natives of Gearee and Kambajong, in the adjacent province of Dingcham, which is the loftiest, coldest, most windy and arid in Eastern Tibet, and in which are the sources of all the streams that flow to Nepal, Sikkim, and Bhotan on the one side, and into the Yaru-tsampu on the other. These families repair yearly to Palung, with their flocks, herds, and tents, paying tribute to the Sikkim

---

* These, which grow on an eastern exposure, exist at a higher elevation than any other bushes I have met with.

Rajah for the privilege: they arrive in June and leave in
September. Both men and women were indescribably
filthy; as they never wash, their faces were perfectly black
with smoke and exposure, and the women's with a pigment
of grease as a protection from the wind. The men were

LEPCHA GIRLS (THE OUTER FIGURES), AND TIBETAN WOMEN.

dressed as usual, in the blanket-cloak, with brass pipes, long
knives, flint, steel, and amulets; the women wore similar,
but shorter cloaks, with silver and copper girdles, trowsers,
and flannel boots. Their head-dresses were very remarkable.
A circular band of plaited yak's hair was attached to the
back hair, and encircled the head like a saint's glory,* at
some distance round it. A band crossed the forehead,

* I find in Ermann's "Siberia" (i., p. 210), that the married women of Yekater-
inberg wear a head-dress like an ancient glory covered with jewels, whilst the
unmarried ones plait their tresses. The same distinguished traveller mentions
having seen a lad of six years old suckled, amongst the Tungooze of East Siberia.

from which coins, corals, and turquoises, hung down to
the eyebrows, while lappets of these ornaments fell over
the ears. Their own hair was plaited in two tails, brought
over the shoulders, and fastened together in front; and a
little yellow felt cap, traversely elongated, so as not to
interfere with the shape of the glory, was perched on
the head. Their countenances were pleasing, and their
manners timid.

The children crawled half-naked about the tent, or
burrowed like moles in an immense heap of goats' and
sheep-droppings, piled up for fuel, upon which the family
lounged. An infant in arms was playing with a "coral,"
ornamented much like ours, and was covered with jewels
and coins. This custom of decorating children is very
common amongst half-civilised people; and the coral is,
perhaps, one of the last relics of a barbarous age that is
retained amongst ourselves. One mother was nursing her
baby, and churning at the same time, by rolling the goat-
skin of yak-milk about on the ground. Extreme poverty
induces the practice of nursing the children for years; and
in one tent I saw a lad upwards of four years of age
unconcernedly taking food from his aunt, and immediately
afterwards chewing hard dry grains of maize.

The tents were pitched in holes about two feet and a
half deep; and within them a wall of similar height was
built all round: in the middle was a long clay arched
fire-place, with holes above, over which the cauldrons were
placed, the fire being underneath. Saddles, horse-cloths,
and the usual accoutrements and implements of a nomade
people, all of the rudest description, hung about: there
was no bed or stool, but Chinese rugs for sleeping on.
I boiled water on the fire-place; its temperature (184° 5)
with that of the air (45° 5) gave an elevation of 15,867

feet. Barometric observations, taken in October, at a point considerably lower down the stream, made the elevation 15,620 feet, or a few feet lower than Kongra Lama pass.

A Lama accompanied this colony of Tibetans, a festival in honour of Kinchinjhow being annually held at a large chait hard by, which is painted red, ornamented with banners, and surmounted by an enormous yak's skull, that faces the mountain. The Lama invited me into his tent, where I found a wife and family. An extempore altar was at one end, covered with wafers and other pretty ornaments, made of butter, stamped or moulded with the fingers.* The tents being insupportably noisome, I preferred partaking of the buttered brick-tea in the open air; after which, I went to see the shawl-wool goats sheared in a pen close by. There are two varieties: one is a large animal, with great horns, called "Rappoo;"† the other smaller, and with slender horns, is called "Tsilloo." The latter yields the finest wool, but they are mixed for ordinary purposes. I was assured that the sheep (of which large flocks were grazing near) afford the finest wool of any. The animals were caught by the tail, their legs tied, the long winter's hair pulled out, and the remainder cut away with a broad flat knife, which was sharpened with a scythe-stone. The operation was clumsily performed, and the skin much cut.

Turnips are grown at Palung during the short stay of the people, and this is the most alpine cultivation in Sikkim: the seed is sown early in July, and the tubers are fit to be eaten in October, if the season is favourable.

---

* The extensive use of these ornaments throughout Tibet, on the occasion of religious festivals, is alluded to by MM. Huc and Gabet.

† This is the "Changra," and the smaller the "Chyapu" of Mr. Hodgson's catalogue. (See "British Museum Catalogue.")

They did not come to maturity this year, as I found on again visiting this spot in October; but their tops had afforded the poor Tibetans some good vegetables. The mean temperature of the three summer months at Palung is probably about 40°, an element of comparatively little importance in regulating the growth and ripening of vegetables at great elevations in Tibetan climates; where a warm exposure, the amount of sunshine, and of radiated heat, have a much greater influence.

During the winter, when these families repair to Kambajong, in Tibet, the flocks and herds are all stall-fed, with long grass, cut on the marshy banks of the Yaru. Snow is said to fall five feet deep at that place, chiefly after January; and it melts in April.

After tea, I ascended the hills overhanging the Lachen valley, which are very bare and stony; large flocks of sheep were feeding on them, chiefly upon small tufted sedges, allied to the English *Carex pilularis*, which here forms the greatest part of the pasture: the grass grows mixed with it in small tufts, and is the common Scotch mountain pasture-grass (*Festuca ovina*).

On the top of these hills, which, for barrenness, reminded me of the descriptions given of the Siberian steppes, I found, at 17,000 feet elevation, several minute arctic plants, with *Rhododendron nivale*, the most alpine of woody plants. On their sterile slopes grew a curious plant allied to the *Cherleria* of the Scotch Alps, forming great hemispherical balls on the ground, eight to ten inches across, altogether resembling in habit the curious Balsam-bog (*Bolax glebaria*) of the Falkland Islands, which grows in very similar scenes.*

---

* *Arenaria rupifraga*, Fenzl. This plant is mentioned by Dr. Thomson ("Travels in Tibet," p. 426) as common in Tibet, as far north as the Karakoram, at an

A few days afterwards, I again visited Palung, with the view of ascertaining the height of perpetual snow on the south face of Kinchinjhow; unfortunately, bad weather came on before I reached the Tibetans, from whom I obtained a guide in consequence. From this place a ride of about four miles brought me to the source of the Chachoo, in a deep ravine, containing the terminations of several short, abrupt glaciers,* and into which were precipitated avalanches of snow and ice. I found it impossible to distinguish the glacial ice from perpetual snow; the larger beds of snow where presenting a flat surface, being generally drifts collected in hollows, or accumulations that have fallen from above. when these accumulations rest on slopes they become converted into ice, and, obeying the laws of fluidity, flow downwards as glaciers. I boiled water at the most advantageous position I could select, and obtained an elevation of 16,522 feet.† It was snowing heavily at this time, and we crouched under a gigantic

elevation between 16,000 and 18,000 feet. In Sikkim it is found at the same level. Specimens of it are exhibited in the Kew Museum. As one instance illustrative of the chaotic state of Indian botany, I may here mention that this little plant, a denizen of such remote and inaccessible parts of the globe, and which has only been known to science a dozen years, bears the burthen of no less than six names in botanical works. This is the *Bryomorpha rupifraga* of Karelin and Kireloff (enumeration of Soongarian plants), who first described it from specimens gathered in 1841, on the Alatau mountains (east of Lake Aral). In Ledebour's "Flora Rossica" (i. p. 780) it appears as *Arenaria* (sub-genus *Dicranilla*) *rupifraga*, Fenzl, MS. In Decaisne and Cambessede's Plants of Jacquemont's "Voyage aux Indes Orientales," it is described as *Flourensia cœspitosa*, and in the plates of that work it appears as *Periandra cœspitosa;* and lastly, in Endlicher's "Genera Plantarum," Fenzl proposes the long new generic name of *Thylacospermum* for it. I have carefully compared the Himalayan and Alatau plants, and find no difference between them, except that the flower of the Himalayan one has 4 petals and sepals, 8 stamens, and 2 styles, and that of the Alatau 5 petals and sepals, 10 stamens, and 2—3 styles, characters which are very variable in allied plants. The flowers appear polygamous, as in the Scotch alpine *Cherleria*, which it much resembles in habit, and to which it is very nearly related in botanical characters.

* De Saussure's glaciers of the second order: see "Forbes' Travels in the Alps," p. 79.

† Temperature of boiling water, 183°, air 35°.

boulder, benumbed with cold. I had fortunately brought a small phial of brandy, which, with hot water from the boiling-apparatus kettle, refreshed us wonderfully.

The spur that divides these plains from the Lachen river, rises close to Kinchinjhow, as a lofty cliff of quartzy gneiss, dipping north-east 30° this I had noticed from the Kongra Lama side. On this side the dip was also to the northward, and the whole cliff was crossed by cleavage planes, dipping south, and apparently cutting those of the foliation at an angle of about 60°: it is the only decided instance of the kind I met with in Sikkim. I regretted not being able to examine it carefully, but I was prevented by the avalanches of stones and snow which were continually being detached from its surface.*

The plants found close to the snow were minute primroses, *Parnassia, Draba,* tufted wormwoods (*Artemisia*), saxifrages, gentian, small *Compositæ,* grasses, and sedges. Our ponies unconcernedly scraped away the snow with their hoofs, and nibbled the scanty herbage. When I mounted mine, he took the bit between his teeth, and

---

* I extremely regret not having been at this time acquainted with Mr. D. Sharpe's able essays on the foliation, cleavage, &c., of slaty rocks, gneiss, &c., in the Geological Society's Journal (ii. p. 74, and v. p. 111), and still more so with his subsequent papers in the Philosophical Transactions: as I cannot doubt that many of his observations, and in particular those which refer to the great arches in which the folia (commonly called strata) are disposed, would receive ample illustration from a study of the Himalaya. At vol. i. p. 309, I have distantly alluded to such an arrangement of the gneiss, &c., into arches, in Sikkim, to which my attention was naturally drawn by the writings of Professor Sedgwick (" Geolog. Soc. Trans.") and Mr. Darwin (" Geological Observations in South America ") on these obscure subjects. I may add that wherever I met with the gneiss, mica, schists, and slates, in Sikkim, very near one another, I invariably found that their cleavage and foliation were conformable. This, for example, may be seen in the bed of the great Rungeet, below Dorjiling, where the slates overlie mica schists, and where the latter contain beds of conglomerate. In these volumes I have often used the more familiar term of stratification, for foliation. This arises from my own ideas of the subject not having been clear when the notes were taken.

scampered back to Palung, over rocks and hills, through
bogs and streams; and though the snow was so
blinding that no object could be distinguished, he
brought me to the tents with unerring instinct, as straight
as an arrow.

Wild animals are few in kind and rare in individuals,
at Tungu and elsewhere on this frontier; though there is
no lack of cover and herbage. This must be owing to the
moist cold atmosphere; and it reminds me that a similar
want of animal life is characteristic of those climates at the
level of the sea, which I have adduced as bearing a great
analogy to the Himalaya, in lacking certain natural orders of
plants. Thus, New Zealand and Fuegia possess, the former
no land animal but a rat, and the latter very few indeed,
and none of any size. Such is also the case in Scotland
and Norway. Again, on the damp west coast of Tasmania,
quadrupeds are rare; whilst the dry eastern half of the
island once swarmed with opossums and kangaroos. A
few miles north of Tungu, the sterile and more lofty provinces of Tibet abound in wild horses, antelopes, hares,
foxes, marmots, and numerous other quadrupeds; although
their altitude, climate, and scanty vegetation are apparently
even more unsuited to support such numbers of animals
of so large a size than the karroos of South Africa, and
the steppes of Siberia and Arctic America, which similarly abound in animal life. The laws which govern the
distribution of large quadrupeds seem to be intimately
connected with those of climate; and we should have
regard to these considerations in our geological speculations, and not draw hasty conclusions from the absence
of the remains of large herbivora in formations disclosing a
redundant vegetation.

Besides the wild sheep found on these mountains, a species

of marmot * (" Kardiepieu " of the Tibetans) sometimes migrates in swarms (like the Lapland "Lemming") from Tibet as far as Tungu.   There are few birds but red-

TIBET MARMOT.

legged crows and common ravens.   Most of the insects belonged to arctic types, and they were numerous in individuals.†

The Choongtam Lama was at a small temple near Tungu during the whole of my stay, but he would not come to visit me, pretending to be absorbed in his devotions.   Passing one day by the temple, I found him catechising two young aspirants for holy orders.   He is one of the Dukpa sect, wore his mitre, and was seated cross-legged on the grass with his scriptures on his knees: he put questions to the boys, when he who answered best took the other some yards

* The *Lagopus Tibetanus* of Hodgson. I procured one that displayed an extraordinary tenacity of life : part of the skull was shot away, and the brain protruded; still it showed the utmost terror at my dog.

† As *Meloe,* and some flower-feeding lamellicorns. Of butterflies I saw blues (*Polyommatus*), marbled whites, *Pontia, Colias* and *Argynnis.* A small *Curculio* was frequent, and I found *Scolopendra,* ants and earthworms, on sunny exposures as high as 15,500 feet.

off, put him down on his hands and knees, threw a cloth over his back, and mounted; then kicking, spurring, and cuffing his steed, he was galloped back to the Lama and kicked off; when the catechising recommenced.

I spent a week at Tungu most pleasantly, ascending the neighbouring mountains, and mixing with the people, whom I found uniformly kind, frank, and extremely hospitable; sending their children after me to invite me to stop at their tents, smoke, and drink tea; often refusing any remuneration, and giving my attendants curds and yak-flesh. If on foot, I was entreated to take a pony; and when tired I never scrupled to catch one, twist a yak-hair rope over its jaw as a bridle, and throwing a goat-hair cloth upon its back (if no saddle were at hand), ride away whither I would. Next morning a boy would be sent for the steed, perhaps bringing an invitation to come and take it again. So I became fond of brick-tea boiled with butter, salt, and soda, and expert in the Tartar saddle; riding about perched on the shoulders of a rough pony, with my feet nearly on a level with my pockets, and my knees almost meeting in front.

On the 28th of July much snow fell on the hills around, as low as 14,000 feet, and half an inch of rain at Tungu;* the former soon melted, and I made an excursion to Chomiomo on the following day, hoping to reach the lower line of perpetual snow. Ascending the valley of the Chomiochoo, I struck north up a steep slope, that ended in a spur of vast tabular masses of quartz and felspar, piled like slabs in a stone quarry, dipping south-west 5° to 10°, and striking north-west. These resulted from the decomposition of gneiss, from which the layers of mica had been washed away, when the rain and frost splitting up the fragments,

---

* An inch and a half fell at Dorjiling during the same period.

the dislocation is continued to a great depth into the substance of the rock.

Large silky cushions of a forget-me-not grew amongst the rocks, spangled with beautiful blue flowers, and looking like turquoises set in silver: the *Delphinium glaciale** was also abundant, exhaling a rank smell of musk.   It indicates a very great elevation in Sikkim, and on my ascent far above it, therefore, I was not surprised to find water boil at 182° 6 (air 43°), which gives an altitude of 16,754 feet.

A dense fog, with sleet, shut out all view; and I did not know in what direction to proceed higher, beyond the top of the sharp, stony ridge I had attained.   Here there was no perpetual snow, which is to be accounted for by the nature of the surface facilitating its removal, the edges of the rocks which project through the snow, becoming heated, and draining off the water as it melts.

During my stay at Tungu, from the 23rd to the 30th of July, no day passed without much deposition of moisture, but generally in so light a form that throughout the whole time but one inch was registered in the rain-gauge; during the same time four inches and a half of rain fell at Dorjiling, and three inches and a half at Calcutta.   The mean temperature was 50° $\left(\frac{\text{Max. } 65°}{\text{Min. } 40° 7}\right)$; extremes, $\frac{65°}{38°}$.   The mean range (23° 3) was thus much greater than at Dorjiling, where it was only 8° 9.   A thermometer, sunk three feet, varied only a few tenths from 57° 6.   By twenty-five comparative observations with Calcutta, 1° Fahr. is the equivalent of every 362 feet of ascent; and twenty comparative observations with Dorjiling give 1° for every 340 feet.   The barometer rose and fell at the same hours as at lower

---

* This new species has been described for the "Flora Indica" of Dr. Thomson and myself: it is a remarkable plant, very closely resembling, and as it were representing, the *D. Brunonianum* of the Western Himalaya. The latter plant smells powerfully of musk, but not so disagreeably as this does.

elevations; the tide amounting to 0·060 inch, between 9·50 A.M. and 4 P.M.

I left Tungu on the 30th of July, and spent that night at Tallum, where a large party of men had just arrived, with loads of madder, rice, canes, bamboos, planks, &c., to be conveyed to Tibet on yaks and ponies.* On the following day I descended to Lamteng, gathering a profusion of fine plants by the way.

The flat on which I had encamped at this place in May and June, being now a marsh, I took up my abode for two days in one of the houses, and paid the usual penalty of communication with these filthy people; for which my only effectual remedy was boiling all my garments and bedding. Yet the house was high, airy, and light; the walls composed of bamboo, lath, and plaster.

Tropical Cicadas ascend to the pine-woods above Lamteng in this month, and chirp shrilly in the heat of the day; and glow-worms fly about at night. The common Bengal and Java toad, *Bufo scabra*, abounded in the marshes, a remarkable instance of wide geographical distribution, for a Batrachian which is common at the level of the sea under the tropics.

On the 3rd of August I descended to Choongtam, which I reached on the 5th. The lakes on the Chateng flat (alt. 8,750 feet) were very full, and contained many English water-plants :† the temperature of the water was 92° near

* About 300 loads of timber, each of six planks, are said to be taken across the Kongra Lama pass annually; and about 250 of rice, besides canes, madder, bamboos, cottons, cloths, and *Symplocos* leaves for dyeing. This is, no doubt, a considerably exaggerated statement, and may refer to both the Kongra Lama and Donkia passes.

† *Sparganium ramosum, Eleocharis palustris, Scirpus triqueter,* and *Callitriche verna?* Some very tropical genera ascend thus high; as *Paspalum* amongst grasses, and *Scleria,* a kind of sedge.

the edges, where a water-insect (*Notonecta*) was swimming about.

Below this I passed an extensive stalactitic deposit of lime, and a second occurred lower down, on the opposite side of the valley. The apparently total absence of limestone rocks in any part of Sikkim (for which I made careful search), renders these deposits, which are far from unfrequent, very curious. Can the limestone, which appears in Tibet, underlie the gneiss of Sikkim? We cannot venture to assume that these lime-charged streams, which in Sikkim burst from the steep flanks of narrow mountain spurs, at elevations between 1000 and 7000 feet, have any very remote or deep origin. If the limestone be not below the gneiss, it must either occur intercalated with it, or be the remains of a formation now all but denuded in Sikkim.

Terrific landslips had taken place along the valley, carrying down acres of rock, soil, and pine-forests, into the stream. I saw one from Kampo Samdong, on the opposite flank of the valley, which swept over 100 yards in breadth of forest. I looked in vain for any signs of scratching or scoring, at all comparable to that produced by glacial action. The bridge at the Tuktoong, mentioned at p. 31, being carried away, we had to ascend for 1000 feet (to a place where the river could be crossed) by a very precipitous path, and descend on the opposite side. In many places we had great difficulty in proceeding, the track being obliterated by the rains, torrents, and landslips. Along the flats, now covered with a dense rank vegetation, we waded ancle, and often knee, deep in mud, swarming with leeches; and instead of descending into the valley of the now too swollen Lachen, we made long detours, rounding spurs by canes and bamboos suspended from trees.

At Choongtam the rice-fields were flooded: and the whole flat was a marsh, covered with tropical grasses and weeds, and alive with insects, while the shrill cries of cicadas, frogs and birds, filled the air. Sand-flies, mosquitos, cockroaches, and enormous cockchafers,* *Mantis*, great locusts, grasshoppers, flying-bugs, crickets, ants, spiders, caterpillars, and leeches, were but a few of the pests that swarmed in my tent and made free with my bed. Great lazy butterflies floated through the air; *Thecla* and *Hesperides* skipped about, and the great *Nymphalidæ* darted around like swallows. The venomous black cobra was common, and we left the path with great caution, as it is a lazy reptile, and lies basking in the sun; many beautiful and harmless green snakes, four feet long, glided amongst the bushes. My dogs caught a "Rageu,"† a very remarkable animal, half goat and half deer; the flesh was good and tender, dark-coloured, and lean.

I remained here till the 15th of August,‡ arranging my

---

* *Eucerris Griffithii*, a magnificent species. Three very splendid insects of the outer ranges of Sikkim never occurred in the interior: these are a gigantic Curculio (*Calandra*) a wood-borer; a species of Goliath-beetle, *Cheirotonus Macleaii*, and a smaller species of the same rare family, *Trigonophorus nepalensis;* of these the former is very scarce, the latter extremely abundant, flying about at evenings; both are flower-feeders, eating honey and pollen. In the summer of 1848, the months at Dorjiling were well marked by the swarms of peculiar insects that appeared in inconceivable numbers; thus, April was marked by a great black *Passalus*, a beetle one-and-a-half inch long, that flies in the face and entangles itself in the hair; May, by stag-beetles and longicorns; June, by *Coccinella* (lady-birds), white moths, and flying-bugs; July, by a *Dryptis?* a long-necked carabideous insect; August, by myriads of earwigs, cockroaches, Goliath-beetles, and cicadas; September, by spiders.

† "Ragoah," according to Hodgson: but it is not the *Procapra picticaudata* of Tibet.

‡ Though 5° further north, and 5,268 feet above the level of Calcutta, the mean temperature at Choongtam this month was only 12° 5 cooler than at Calcutta; forty observations giving 1° Fahr. as equal to 690 feet of elevation; whereas in May the mean of twenty-seven observations gave 1° Fahr. as equal to 260 feet, the mean difference of temperature being then 25°. The mean maximum of the day was 80°, and was attained at 11 A. M., after which clouds formed, and the

Lachen valley collections previous to starting for the Lachoong, whence I hoped to reach Tibet again by a different route, crossing the Donkia pass, and thence exploring the sources of the Teesta at the Cholamoo lakes.

Whilst here I ascertained the velocity of the currents of the Lachen and Lachoong rivers. Both were torrents, than which none could be more rapid, short of becoming cataracts : the rains were at their height, and the melting of the snows at its maximum. I first measured several hundred yards along the banks of each river above the bridges, repeating this several times, as the rocks and jungle rendered it very difficult to do it accurately: then, sitting on the bridge, I timed floating masses of different materials and sizes that were thrown in at the upper point. I was surprised to find the velocity of the Lachen only nine miles per hour, for its waters seemed to shoot past with the speed of an arrow, but the floats showed the whole stream to be so troubled with local eddies and back-waters, that it took from forty-three to forty-eight seconds for each float to pass over 200 yards, as it was perpetually submerged by under-currents. The breadth of the river averaged sixty-eight feet, and the discharge was 4,420 cubic feet of water per second. The temperature was 57°.

At the Lachoong bridge the jungle was still denser, and the banks quite inaccessible in many places. The mean velocity was eight miles an hour, the breadth ninety five feet, the depth about the same as that of the Lachen, giving a discharge of 5,700 cubic feet of water per

thermometer fell to 66° at sunset, and 56° at night. In my blanket tent the heat rose to upwards of 100° in calm weather. The afternoons were generally squally and rainy.

second;* its temperature was also 57°. These streams retain an extraordinary velocity, for many miles upwards; the Lachen to its junction with the Zemu at 9000 feet, and the Zemu itself as far up as the Thlonok, at 10,000 feet, and the Lachoong to the village of that name, at 8000 feet: their united streams appear equally rapid till they become the Teesta at Singtam.†

On the 15th of August, having received supplies from Dorjiling, I started up the north bank of the Lachoong, following the Singtam Soubah, who accompanied me officially, and with a very bad grace; poor fellow, he expected me to have returned with him to Singtam, and thence gone back to Dorjiling, and many a sore struggle we had on this point. At Choongtam he had been laid up with ulcerated legs from the bites of leeches and sand-flies, which required my treatment.

The path was narrow, and ran through a jungle of mixed tropical and temperate plants,‡ many of which are not found at this elevation on the damp outer ranges of Dorjiling. We crossed to the south bank by a fine cane-bridge forty yards long, the river being twenty-eight across: and here I have to record the loss of my dog Kinchin; the companion of all my late journeyings, and to whom I had become really attached. He had a bad habit, of which I

* Hence it appears that the Lachoong, being so much the more copious stream, should in one sense be regarded as the continuation of the Teesta, rather than the Lachen, which, however, has by far the most distant source. Their united streams discharge upwards of 10,000 cubic feet of water per second in the height of the rains! which is, however, a mere fraction of the discharge of the Teesta when that river leaves the Himalaya. The Ganges at Hurdwar discharges 8000 feet per second during the dry season.

† The slope of the bed of the Lachen from below the confluence of the Zemu to the village of Singtam is 174 feet per mile, or 1 foot in 30; that of the Lachoong from the village of that name to Singtam is considerably less.

‡ As *Paris, Dipsacus, Circœa, Thalictrum, Saxifraga ciliaris, Spiranthes, Malva, Hypoxis, Anthericum, Passiflora, Drosera, Didymocarpus,* poplar, *Calamagrostis,* and *Eupatorium.*

had vainly tried to cure him, of running for a few yards on
the round bamboos by which the cane-bridges are crossed,
and on which it was impossible for a dog to retain his
footing : in this situation he used to get thoroughly
frightened, and lie down on the bamboos with his legs
hanging over the water, and having no hold whatever. I
had several times rescued him from this perilous position,
which was always rendered more imminent from the
shaking of the bridge as I approached him. On the
present occasion, I stopped at the foot of some rocks
below the bridge, botanizing, and Kinchin having
scrambled up the rocks, ran on to the bridge. I could
not see him, and was not thinking about him, when
suddenly his shrill, short barks of terror rang above the
roaring torrent. I hastened to the bridge, but before I
could get to it, he had lost his footing, and had disappeared. Holding on by the canes, I strained my eyes
till the bridge seemed to be swimming up the valley, and
the swift waters to be standing still, but to no purpose ;
he had been carried under at once, and swept away miles
below. For many days I missed him by my side on the
mountain, and by my feet in camp. He had become a
very handsome dog, with glossy black hair, pendent
triangular ears, short muzzle, high forehead, jet-black
eyes, straight limbs, arched neck, and a most glorious tail
curling over his back.*

A very bad road led to the village of Keadom, situated
on a flat terrace several hundred feet above the river, and
6,609 feet above the sea, where I spent the night. Here
are cultivated plantains and maize, although the elevation

* The woodcut at vol. i. p. 90, gives the character of the Tibet mastiff, to which
breed his father belonged; but it is not a portrait of himself, having been
sketched from a dog of the pure breed, in the Zoological Society's Gardens, by
C. Jenyns, Esq.

is equal to parts of Dorjiling, where these plants do
not ripen.

The river above Keadom is again crossed, by a plank
bridge, at a place where the contracted streams flow
between banks forty feet high, composed of obscurely
stratified gravel, sand, and water-worn boulders. Above
this the path ascends lofty flat-topped spurs, which
overhang the river, and command some of the most
beautiful scenery in Sikkim. The south-east slopes are
clothed with *Abies Brunoniana* at 8000 feet elevation, and
cleft by a deep ravine, from which projects what appears
to be an old moraine, fully 1500 or perhaps 2000 feet
high. Extensive landslips on its steep flank expose
(through the telescope) a mass of gravel and angular
blocks, while streams cut deep channels in it.

This valley is far more open and grassy than that of the
Lachen, and the vegetation also differs much.* In the afternoon we reached Lachoong, which is by far the most picturesque village in the temperate region of Sikkim. Grassy
flats of different levels, sprinkled with brushwood and scattered clumps of pine and maple, occupy the valley; whose
west flanks rise in steep, rocky, and scantily wooded grassy
slopes. About five miles to the north the valley forks; two
conspicuous domes of snow rising from the intermediate
mountains. The eastern valley leads to lofty snowed
regions, and is said to be impracticable; the Lachoong flows
down the western, which appeared rugged, and covered with
pine woods. On the east, Tunkra mountain† rises in a

---

* *Umbelliferæ* and *Compositæ* abound, and were then flowering; and an orchis
(*Satyrium Nepalense*), scented like our English *Gymnadenia*, covered the ground in
some places, with tall green *Habenariæ* and a yellow *Spathoglottis*, a genus with
pseudo-bulbs. Of shrubs, *Xanthoxylon, Rhus, Prinsepia, Cotoneaster, Pyrus*, poplar
and oak, formed thickets along the path; while there were as many as eight and
nine kinds of balsams, some eight feet high.

† This mountain is seen from Dorjiling; its elevation is about 18,700 fœt.

superb unbroken sweep of dark pine-wood and cliffs, surmounted by black rocks and white fingering peaks of snow.
South of this, the valley of the Tunkrachoo opens, backed
by sharp snowed pinnacles, which form the continuation of
the Chola range; over which a pass leads to the Phari
district of Tibet, which intervenes between Sikkim and
Bhotan.   Southwards the view is bounded by snowy
mountains, and the valley seems blocked up by the
remarkable moraine-like spur which I passed above
Keadóm.

Larch.

LACHOONG VALLEY AND VILLAGE, LOOKING SOUTH.

Stupendous moraines rise 1500 feet above the Lachoong
in several concentric series, curving downwards and outwards, so as to form a bell-shaped mouth to the valley of
the Tunkrachoo.   Those on the upper flank are much the

largest; and the loftiest of them terminates in a conical hill crowned with Boodhist flags, and its steep sides cut into horizontal roads or terraces, one of which is so broad and flat as to suggest the idea of its having been cleared by art.

Abies Smithiana.　　　*Larch.*　　　　　　　　　*Abies Brunoniana.*

LOFTY ANCIENT MORAINES IN THE LACHOONG VALLEY, LOOKING SOUTH-EAST.

On the south side of the Tunkrachoo river the moraines are also more or less terraced, as is the floor of the Lachoong valley, and its east slopes, 1000 feet up.*

---

* I have since been greatly struck with the similarity between the features of this valley, and those of Chamouni (though the latter is on a smaller scale) above the Lavanchi moraine. The spectator standing in the expanded part below the village of Argentière, and looking upwards, sees the valley closed above by the ancient moraine of the Argentière glacier, and below by that of Lavanchi; and on all sides the slopes are cut into terraces, strewed with boulders. I found traces

The river is fourteen yards broad, and neither deep nor rapid : the village is on the east bank, and is large for Sikkim ; it contains fully 100 good wooden houses, raised on posts, and clustered together without order. It was muddy and intolerably filthy, and intersected by some small streams, whose beds formed the roads, and, at the same time, the common sewers of the natives. There is some wretched cultivation in fields,* of wheat, barley, peas, radishes, and turnips. Rice was once cultivated at this elevation (8000 feet), but the crop was uncertain ; some very tropical grasses grow wild here, as *Eragrostis* and *Panicum*. In gardens the hollyhock is seen : it is said to be introduced through Tibet from China ; also *Pinus excelsa* from Bhotan, peaches, walnuts, and weeping willows. A tall poplar was pointed out to me as a great wonder ; it had two species of *Pyrus* growing on its boughs, evidently from seed ; one was a mountain ash, the other like *Pyrus Aria*.

Soon after camping, the Lachoong Phipun, a very tall, intelligent, and agreeable looking man, waited on me with the usual presents, and a request that I would visit his sick father. His house was lofty and airy : in the inner room the sick man was stretched on a board, covered with a blanket, and dying of pressure on the brain ; he was surrounded by a deputation of Lamas from Teshoo Loombo, sent for in this emergency. The principal one was a fat fellow, who sat cross-legged before a block-printed Tibetan

of stratified pebbles and sand on the north flank of the Lavanchi moraine however, which I failed to discover in those of Lachoong. The average slope of these pine-clad Sikkim valleys much approximates to that of Chamouni, and never approaches the precipitous character of the Bernese Alps' valleys, Kandersteg, Lauterbrunnen, and Grindelwald.

* Full of such English weeds as shepherd's purse, nettles, *Solanum nigrum*, and dock ; besides many Himalayan ones, as balsams, thistles, a beautiful geranium, mallow, *Haloragis* and Cucurbitaceous plants.

book, plates of raw meat, rice, and other offerings, and the bells, dorje, &c. of his profession. Others sat around, reading or chanting services, and filling the room with incense. At one end of the apartment was a good library in a beautifully carved book-case.

HEAD AND FEET OF TIBET MARMOT.

CHAPTER XXII.

THE Singtam Soubah being again laid up here from the consequences of leech-bites, I took the opportunity of visiting the Tunkra-lah pass, represented as the most snowy in Sikkim ; which I found to be the case. The route lay over the moraines on the north flank of the Tunkrachoo, which are divided by narrow dry gullies,* and composed of enormous blocks disintegrating into a deep layer of clay. All are clothed with luxuriant herbage and flowering shrubs,† besides small larches and pines,

---

* These ridges of the moraine, separated by gullies, indicate the progressive retirement of the ancient glacier, after periods of rest. The same phenomena may be seen, on a diminutive scale, in the Swiss Alps, by any one who carefully examines the lateral and often the terminal moraines of any retiring or diminishing glacier, at whose base or flanks are concentric ridges, which are successive deposits.

† *Ranunculus, Clematis, Thalictrum, Anemone, Aconitum variegatum* of Europe,

rhododendrons and maples; with *Enkianthus*, *Pyrus*, cherry, *Pieris*, laurel, and *Goughia*. The musk-deer inhabits these woods, and at this season I have never seen it higher. Large monkeys are also found on the skirts of the pine-forests, and the *Ailurus ochraceus* (Hodgs.), a curious long-tailed animal peculiar to the Himalaya, something between a diminutive bear and a squirrel. In the dense and gigantic forest of *Abies Brunoniana* and silver fir, I measured one of the former trees, and found it twenty-eight feet in girth, and above 120 feet in height. The *Abies Webbiana* attains thirty-five feet in girth, with a trunk unbranched for forty feet.

The path was narrow and difficult in the wood, and especially along the bed of the stream, where grew ugly trees of larch, eighty feet high, and abundance of a new species of alpine strawberry with oblong fruit. At 11,560 feet elevation, I arrived at an immense rock of gneiss, buried in the forest. Here currant-bushes were plentiful, generally growing on the pine-trunks, in strange association with a small species of *Begonia*, a hothouse tribe of plants in England. Emerging from the forest, vast old moraines are crossed, in a shallow mountain valley, several miles long and broad, 12,000 feet above the sea, choked with rhododendron shrubs, and nearly encircled by snowy mountains. Magnificent gentians grew here, also *Senecio*, *Corydalis*, and the *Aconitum luridum* (n. sp.), whose root is said to be as virulent as *A. ferox* and *A. Napellus*.* The

a scandent species, Berberry, *Deutzia*, *Philadelphus*, Rose, Honeysuckle, Thistles, Orchis, *Habenaria*, *Fritillaria*, *Aster*, *Calimeris*, *Verbascum thapsus*, *Pedicularis*, *Euphrasia*, *Senecio*, *Eupatorium*, *Dipsacus*, *Euphorbia*, Balsam, *Hypericum*, *Gentiana*, *Halenia*, *Codonopsis*, *Polygonum*.

* The result of Dr. Thomson's and my examination of the Himalayan aconites (of which there are seven species) is that the one generally known as *A. ferox*, and which supplies a great deal of the celebrated poison, is the common *A. Napellus* of Europe.

plants were all fully a month behind those of the Lachen valley at the same elevation. Heavy rain fell in the afternoon, and we halted under some rocks : as I had brought no tent, my bed was placed beneath the shelter of one, near which the rest of the party burrowed. I supped off half a yak's kidney, an enormous organ in this animal.

On the following morning we proceeded up the valley, towards a very steep rocky barrier, through which the river cut a narrow gorge, and beyond which rose lofty snowy mountains : the peak of Tunkra being to our left hand (north). Saxifrages grew here in profuse tufts of golden blossoms, and *Chrysosplenium*, rushes, mountain-sorrel (*Oxyria*), and the bladder-headed *Saussurea*, whose flowers are enclosed in inflated membranous bracts, and smell like putrid meat : there were also splendid primroses, the spikenard valerian, and golden Potentillas.

The ascent was steep and difficult, up a stony valley bounded by precipices ; in this the river flowed in a north-west direction, and we were obliged to wade along it, though its waters were bitterly cold, the temperature being 39° At 15,000 feet we passed from great snow beds to the surface of a glacier, partly an accumulation of snow, increased by lateral glaciers : its slope was very gentle for several miles ; the surface was eroded by rain, and very rough, whilst those of the lateral glaciers were ribboned, crevassed, and often conspicuously marked with dirt-bands.

A gently sloping saddle, bare of snow, which succeeds the glacier, forms the top of the Tunkra pass ; it unites two snowy mountains, and opens on the great valley of the Machoo, which flows in a part of Tibet between Sikkim and Bhotan ; its height is 16,083 feet above the sea by barometer, and 16,137 feet by boiling-point. Nothing can

be more different than the two slopes of this pass; that by
which I had come presented a gentle snowy acclivity,
bounded by precipitous mountains; while that which
opened before me was a steep, rocky, broad, grassy valley,
where not a particle of snow was to be seen, and yaks
were feeding near a small lake not 1000 feet down. Nor
were snowy mountains visible anywhere in this direction,
except far to the south-east, in Bhotan. This remarkable
difference of climate is due to the southerly wind which
ascends the Tibetan or Machoo valley being drained by
intervening mountains before reaching this pass, whilst the
Sikkim current brings abundant vapours up the Teesta
and Lachoong valleys.

Chumulari lies to the E.N.E. of the Tunkra pass, and is
only twenty six miles distant, but not seen; Phari is two
marches off, in an easterly direction, and Choombi one to
the south-east　Choombi is the general name given to a
large Tibetan province that embraces the head of the
Machoo river, and includes Phari, Eusa, Choombi, and
about thirteen other villages, corresponding to as many
districts, that contain from under a dozen to 300 houses
each, varying with the season and state of trade. The
latter is considerable, Phari being, next to Dorjiling, the
greatest Tibetan, Bhotan, Sikkim, and Indian entrepôt
along the whole Himalaya east of Nepal. The general
form of Choombi valley is triangular, the broader end
northwards: it is bounded by the Chola range on the
west from Donkia to Gipmoochi, and by the Kamphee
or Chakoong range to the east; which is, I believe,
continuous with Chumulari. These meridional ranges
approximate to the southward, so as to form a natural
boundary to Choombi. The Machoo river, rising from
Chumulari, flows through the Choombi district, and enters

Bhotan at a large mart called Rinchingoong, whence it
flows to the plains of India, where it is called at Cooch-
Behar, the Torsha, or, as some say, the Godadda, and
falls into the Burrampooter.

The Choombi district is elevated, for the only cultivation
is a summer or alpine one, neither rice, maize, nor millet
being grown there: it is also dry, for the great height
of the Bhotan mountains and the form of the Machoo
valley cut off the rains, and there is no dense forest. It is
very mountainous, all carriage being on men's and yaks'
backs, and is populous for this part of the country, the
inhabitants being estimated at 3000, in the trading season,
when many families from Tibet and Bhotan erect booths at
Phari.

A civil officer at Phari collects the revenue under the
Lhassan authorities, and there is also a Tibetan fort, an
officer, and guard. The inhabitants of this district more
resemble the Bhotanese than Tibetans, and are a thievish set,
finding a refuge under the Paro-Pilo of Bhotan,* who taxes
the refugees according to the estimate he forms of their
plunder. The Tibetans seldom pursue the culprits, as the
Lhassan government avoids all interference south of their
own frontier.

From Choombi to Lhassa is fifteen days' long journeys
for a man mounted on a stout mule; all the rice passing
through Phari is monopolised there for the Chinese troops

* There was once a large monastery, called Kazioo Goompa, at Choombi, with
upwards of one hundred Lamas. During a struggle between the Sikkim and
Bhotan monks for superiority in it, the abbot died. His avatar reappeared in two
places at once ! in Bhotan as a relative of the Paro-Pilo himself, and in Sikkim as
a brother of the powerful Gangtok Kajee. Their disputes were referred to the
Dalai Lama, who pronounced for Sikkim. This was not to be disputed by the
Pilo, who, however, plundered the Goompa of its silver, gold, and books, leaving
nothing but the bare walls for the successful Lama ! The Lhassan authorities
made no attempt to obtain restitution, and the monastery has been consequently
neglected.

at Lhassa. The grazing for yaks and small cattle is excellent
in Choombi, and the *Pinus excelsa* is said to grow
abundantly there, though unknown in Sikkim, but I have
not heard of any other peculiarity in its productions.

Very few plants grew amongst the stones at the top of the
Tunkra pass, and those few were mostly quite different from
those of Palung and Kongra Lama. A pink-floweerd
Arenaria, two kinds of *Corydalis,* the cottony *Saussurea,* and
diminutive primroses, were the most conspicuous.* The
wind was variable, blowing alternately up both valleys,
bringing much snow when it blew from the Teesta, though
deflected to a north-west breeze; when, on the contrary, it
blew from Tibet, it was, though southerly, dry. Clouds
obscured all distant view. The temperature varied
between noon and 1·30 p.m. from 39° to 40° 5, the air
being extremely damp.

Returning to the foot of the glacier, I took up my
quarters for two days under an enormous rock overlooking
the broad flat valley in which I had spent the previous
night, and directly fronting Tunkra mountain, which bore
north about five miles distant. This rock was sixty to
eighty feet high, and 15,250 feet above the sea; it was
of gneiss, and was placed on the top of a bleak ridge,
facing the north; no shrub or bush being near it. The
gentle slope outwards of the rock afforded the only shelter,
and a more utterly desolate place than Lacheepia, as it
is called, I never laid my unhoused head in. It com
manded an incomparable view due west across the
Lachoong and Lachen valleys, of the whole group of
Kinchinjunga snows, from Tibet southwards, and as such
was a most valuable position for geographical purposes.

---

* The only others were *Leontopodium, Sedum,* Saxifrage, *Ranunculus hyperboreus,*
Ligularia, two species of *Polygonum,* a *Trichostomum, Stereocaulon,* and *Lecidea
geographica*, not one grass or sedge.

The night was misty, and though the temperature was 35°, I was miserably cold; for my blankets being laid on the bare ground, the chill seemed to strike from the rock to the very marrow of my bones. In the morning the fog hung till sunrise, when it rose majestically from all the mountain tops; but the view obtained was transient, for in less than an hour the dense woolly banks of fog which choked the valleys ascended like a curtain to the warmed atmosphere above, and slowly threw a veil over the landscape. I waited till the last streak of snow was shut out from my view, when I descended, to breakfast on Himalayan grouse (*Tetrao-perdix nivicola*), a small gregarious bird which inhabits the loftiest stony mountains, and utters a short cry of " Quiok, quiok ; " in character and appearance it is intermediate between grouse and partridge, and is good eating, though tough.

Hoping to obtain another view, which might enable me to correct the bearings taken that morning, I was tempted to spend a second night in the open air at Lacheepia, passing the day botanizing * in the vicinity, and taking observations of the barometer and wet-bulb : I also boiled three thermometers by turns, noting the grave errors likely to attend observations of this instrument for elevation.† Little rain fell during the day, but it was heavy at night, though there was fortunately no wind ; and I made a more comfortable bed with tufts of juniper brought up from below. Our fire was principally of wet rhododendron wood,

* Scarcely a grass, and no *Astragali,* grow on these stony and snowy slopes : and the smallest heath-like *Andromeda,* a still smaller *Menziesia* (an arctic genus, previously unknown in the Himalaya) and a prostrate willow, are the only woody-stemmed plants above 15,000 feet.

† These will be more particularly alluded to in the Appendix, where will be found a comparison of elevations, deduced from boiling-point and from barometric observations.   The height of Lacheepia is 14,912. feet by boiling-point, and 15,262 feet by barometer.

with masses of the aromatic dwarf species, which, being full
of resinous glands, blazed with fury   Next day, after a
very transient glimpse of the Kinchinjunga snows, I
descended to Lachoong, where I remained for some days
botanizing.   During my stay I was several times awakened
by all the noises and accompaniments of a night-attack or
alarm; screaming voices, groans, shouts, and ejaculations,
the beating of drums and firing of guns, and flambeaux of
pine-wood gleaming amongst the trees, and flitting from
house to house.   The cause, I was informed, was the
presence of a demon, who required exorcisement, and who
generally managed to make the villagers remember his
visit, by their missing various articles after the turmoil
made to drive him away.   The custom of driving out
demons in the above manner is constantly practised by the
Lamas in Tibet: MM. Huc and Gabet give a graphic
account of such an operation during their stay at
Kounboum.

On the 29th of August I left Lachoong and proceeded
up the valley.   The road ran along a terrace, covered
with long grass, and bounded by lofty banks of unstratified gravel and sand, and passed through beautiful groves
of green pines, rich in plants.   No oak nor chesnut
ascends above 9000 feet here or elsewhere in the interior
of Sikkim, where they are replaced by a species of hazel
(*Corylus*); in the North Himalaya, on the other hand, an
oak (*Quercus semecarpifolia*, see vol. i. p. 187) is amongst
the most alpine trees, and the nut is a different species,
more resembling the European.   On the outer Sikkim ranges
oaks (*Q. annulata ?*) ascend to 10,000 feet, and there is
no hazel.   Above the fork, the valley contracts extremely,
and its bed is covered with moraines and landslips, which
often bury the larches and pines.   Marshes occur here and

there, full of the sweet-scented Hierochloe grass, the Scotch *Thalictrum alpinum*, and an *Eriocaulon*, which ascends to 10,000 feet.   The old moraines were very difficult to cross, and on one I found a barricade, which had been erected to deceive me regarding the frontier, had I chosen this route instead of the Lachen one, in May.

Broad flats clothed with rhododendron, alternate with others covered with mud, boulders, and gravel, which had flowed down from the gorges on the west, and which still contained trees, inclined in all directions, and buried up to their branches; some of these débâcles were 400 yards across, and sloped at an angle of 2° to 3°, bearing on their surfaces blocks fifteen yards in diameter.*   They seem to subside materially, as I perceived they had left marks many feet higher on the tree-trunks.   Such débâcles must often bury standing forests in a very favourable material, climate, and position for becoming fossilized.

On the 30th of August I arrived at Yeumtong, a small summer cattle-station, on a flat by the Lachoong, 11,920 feet above the sea; the general features of which closely resemble those of the narrow Swiss valleys.   The west flank is lofty and precipitous, with narrow gullies still retaining the winter's snow, at 12,500 feet ; the east gradually slopes up to the two snowy domes seen from Lachoong ; the bed of the valley is alternately a flat lake-bed, in which the river meanders at the rate of three and a half miles an hour, and sudden descents, cumbered with old moraines, over which it rushes in sheets of foam.   Silver-firs ascend nearly to 13,000 feet, where they are replaced by large junipers, sixty feet high : up the valley Chango Khang is seen, with a superb glacier descending to about 14,000 feet on its south

---

* None were to be compared in size and extent with that at Bex, at the mouth of the Rhone valley.

flank.   Enormous masses of rock were continually precipitated from the west side, close to the shed in which I had
taken up my quarters, keeping my people in constant
alarm, and causing a great commotion among the yaks,
dogs, and ponies.   On the opposite side of the river is
a deep gorge; in which an immense glacier descends
lower than any I have seen in Sikkim.   I made several
attempts to reach it by the gully of its discharging stream,
but was always foiled by the rocks and dense jungle of
pines, rhododendron, and dwarf holly.

The snow-banks on the face of the dome-shaped mountain appearing favourable for ascertaining the position of
the level of perpetual snow, I ascended to them on the 6th
of September, and found the mean elevation along an even,
continuous, and gradual slope, with a full south-west
exposure, to be 15,985 feet by barometer, and 15,816
feet by boiling-point. These beds of snow, however
broad and convex, cannot nevertheless be distinguished
from glaciers : they occupy, it is true, mountain slopes,
and do not fill hollows (like glaciers commonly so called),
but they display the ribboned structure of ice, and being
viscous fluids, descend at a rate and to a distance
depending on the slope, and on the amount of annual
accumulation behind   Their termination must therefore
be far below that point at which all the snow that falls
melts, which is the theoretical line of perpetual snow.
Before returning I attempted to proceed northwards to
the great glacier, hoping to descend by its lateral moraine,
but a heavy snow-storm drove me down to Yeumtong.

Some hot-springs burst from the bank of the Lachen a
mile or so below the village : they are used as baths, the
patient remaining three days at a time in them, only retiring
to eat in a little shed close by.   The discharge amounts to

a few gallons per minute; the temperature at the source is
112° 6, and 106° in the bath.* The water has a slightly saline
taste; it is colourless, but emits bubbles of sulphuretted
hydrogen gas, blackening silver. A cold spring (temperature 42°) emerged close by, and the Lachoong not ten yards
off, was 47° to 50°. A conferva grows in the hot water,
and the garnets are worn out of the gneiss rock exposed to
its action.

The Singtam Soubah had been very sulky since leaving
Choongtam, and I could scarcely get a drop of milk or a slice
of curd here. I had to take him to task severely for sanc-
tioning the flogging of one of my men; a huntsman, who
had offered me his services at Choongtam, and who was a
civil, industrious fellow, though he had procured me little
besides a huge monkey, which had nearly bitten off the head
of his best dog. I had made a point of consulting the
Soubah before hiring him, for fear of accidents; but this
did not screen him from the jealousy of the Choongtam
Lama, who twice flogged him in the Goompa with rattans
(with the Soubah's consent), alleging that he had quitted
his service for mine. My people knew of this, but were
afraid to tell me, which the poor fellow did himself.

The Lachoong Phipun visited me on the 7th of September: he had officiously been in Tibet to hear what the
Tibetan people would say to my going to Donkia, and
finding them supremely indifferent, returned to be my
guide. A month's provision for ten men having arrived
from Dorjiling, I left Yeumtong the following day for
Momay Samdong, the loftiest yak grazing station in
Sikkim (Palung being too cold for yaks), and within a
day's journey of the Donkia pass.

---

* This water boiled at 191° 6, the same at which snow-water and that of the
river did; giving an elevation of 11,730 feet. Observations on the mineral constituents of the water will be found in the Appendix.

The valley remains almost level for several miles, the road continuing along the east bank of the Lachen. Shoots of stones descend from the ravines, all of a white fine-grained granite, stained red with a minute conferva, which has been taken by Himalayan travellers for red snow ; * a phenomenon I never saw in Sikkim.

At a fork of the valley several miles above Yeumtong, and below the great glacier of Chango Khang, the ancient moraines are prodigious, much exceeding any I have else-where seen, both in extent, in the size of the boulders, and in the height to which the latter are piled on one another. Many boulders I measured were twenty yards across, and some even forty ; and the chaotic scene they presented baffles all description : they were scantily clothed with stunted silver firs.

Beyond this, the path crosses the river, and ascends rapidly over a mile of steeply sloping landslip, composed of angular fragments of granite, that are constantly falling from above, and are extremely dangerous.   At 14,000 feet, trees and shrubs cease, willow and honeysuckle being the last ; and thence onward the valley is bleak, open, and stony, with lofty rocky mountains on either side.   The south wind brought a cold drizzling rain, which numbed us, and two of the lads who had last come up from Dorjiling were seized with a remittent fever, originally contracted in the hot valleys ; luckily we found some cattle-sheds, in which I left them, with two men to attend on them.

Momay Samdong is situated in a broad part of the Lachoong valley, where three streams meet ; it is on the west of Chango Khang, and is six miles south-east of Kin-

* Red snow was never found in the Antarctic regions during Sir James Ross's South Polar voyage ; nor do I know any authentic record of its having been seen in the Himalaya.

chinjhow, and seven south-west of Donkia: it is in the same latitude as Palung, but scarcely so lofty.   The mean of fifty-six barometrical observations cotemporaneous with Calcutta makes it 15,362 feet above the sea; nearly the elevation of Lacheepia (near the Tunkra pass), from which, however, its scenery and vegetation entirely differ.

I pitched my tent close to a little shed, at the gently sloping base of a mountain that divided the Lachoong river from a western tributary.   It was a wild and most exposed spot: long stony mountains, grassy on the base near the river; distant snowy peaks, stupendous precipices, moraines, glaciers, transported boulders, and rocks rounded by glacial action, formed the dismal landscape which everywhere met the view.   There was not a bush six inches high, and the only approach to woody plants were minute creeping willows and dwarf rhododendrons, with a very few prostrate junipers and *Ephedra*.

The base of the spur was cut into broad flat terraces, composed of unstratified sand, pebbles, and boulders; the remains, doubtless, of an enormously thick glacial deposit. The terracing is as difficult to be accounted for in this valley as in that of Yangma (East Nepal); both valleys being far too broad, and descending too rapidly to admit of the hypothesis of their having been blocked up in the lower part, and the upper filled with large lakes.*   Another

---

* The formation of small lakes, however, between moraines and the sides of the valleys they occupy, or between two successively formed moraines (as I have elsewhere mentioned), will account for very extensive terraced areas of this kind; and it must be borne in mind that when the Momay valley was filled with ice, the breadth of its glacier at this point must have been twelve miles, and it must have extended east and west from Chango Khang across the main valley, to beyond Donkia.   Still the great moraines are wanting at this particular point, and though atmospheric action and the rivers have removed perhaps 200 feet of glacial shingle, they can hardly have destroyed a moraine of rocks, large enough to block up the valley.

tributary falls into the Lachoong at Momay, which leads
eastwards up to an enormous glacier that descends from
Donkia. Snowy mountains rise nearly all round it: those
on its south and east divide Sikkim from the Phari
province in Tibet; those on the north terminate in a
forked or cleft peak, which is a remarkable and conspicuous
feature from Momay. This, which I have called forked
Donkia,* is the termination of a magnificent amphitheatre
of stupendous snow-clad precipices, continuously upwards
of 20,000 feet high, that forms the east flank of the upper
Lachoong. From Donkia top again, the mountains sweep
round to the westward, rising into fingered peaks of extraordinary magnificence; and thence—still running west—
dip to 18,500 feet, forming the Donkia pass, and rise again
as the great mural mass of Kinchinjhow. This girdle of
mountains encloses the head waters of the Lachoong, which
rises in countless streams from its perpetual snows, glaciers,
and small lakes: its north drainage is to the Cholamoo
lakes in Tibet; in which is the source of the Lachen, which
flows round the north base of Kinchinjhow to Kongra Lama.

The bottom of the Lachoong valley at Momay is broad,
tolerably level, grassy, and covered with isolated mounds
and ridges that point down the valley, and are the remains
of glacial deposits. It dips suddenly below this, and some
gneiss rocks that rise in its centre are remarkably *mouton
néed* or rounded, and have boulders perched on their summits. Though manifestly rounded and grooved by ancient
glaciers, I failed to find scratches on these weather-worn
rocks.†

---

* Its elevation by my observations is about 21,870 feet.

† I have repeatedly, and equally in vain, sought for scratchings on many of the
most conspicuously moutonnéed gneiss rocks of Switzerland. The retention of
such markings depends on other circumstances than the mere hardness of the
rock, or amount of aqueous action. What can be more astonishing than to see

The Lachoong is here twelve or fifteen yards wide, and runs over a pebbly bed, cutting a shallow channel through the deposits, down to the subjacent rock, which is in some cases scooped out six or eight feet deep by its waters. I do not doubt that the flatness of the floor of the Momay valley is caused by the combined action of the streams that drained the three glaciers which met here; for the tendency of retiring glaciers is to level the floors of valleys, by giving an ever-shifting direction to the rivers which drain them, and which spread detritus in their course. Supposing these glaciers to have had no terminal moraines, they might still have forced immense beds of gravel into positions that would dam up lakes between the ice and the flanks of the valleys, and thus produce much terracing on the latter.*

On our arrival, we found that a party of buxom, good-natured looking girls who were tending yaks, were occupying the hut, which, however, they cheerfully gave up to my people, spreading a black tent close by for them-

---

these most delicate scratches retained in all their sharpness on rocks clothed with seaweed and shells, and exposed at every tide, in the bays of Western Scotland !

* We are still very ignorant of many details of ice action, and especially of the origin of many enormous deposits which are not true moraines. These, so con spicuous in the lofty Himalayan valleys, are not less so in those of the Swiss Alps witness that broad valley in which Grindelwald village is situated, and which is covered to an immense depth with an angular detritus, moulded into hills and valleys; also the whole broad open Upper Rhone valley, above the village of Munster, and below that of Obergestelen. The action of broad glaciers on gentle slopes is to raise their own beds by the accumulation of gravel which their lower surface carries and pushes forward. I have seen small glaciers thus raised 300 feet; leaving little doubt in my mind that the upper Himalayan valleys were thus choked with deposit 1000 feet thick, of which indeed the proofs remain along the flanks of the Yangma valley. The denuding and accumulating effects of ice thus give a contour to mountain valleys, and sculpture their flanks and floors far more rapidly than sea action, or the elements. After a very extensive experience of ice in the Antarctic ocean, and in mountainous countries, I cannot but conclude that very few of our geologists appreciate the power of ice as a mechanical agent, which can hardly be over-estimated, whether as glacier, iceberg, or pack ice, heaping shingle along coasts.

selves; and next morning they set off with all their effects
packed upon the yaks. The ground was marshy, and
covered with cowslips, *Ranunculus*, grasses and sedges,
Cyananthus, blue asters, gentians, &c. The spot appearing
highly favourable for observations, I determined to remain
here during the equinoctial month, and put my people on
"two-thirds allowance," *i. e*, four pounds of rice daily for
three men, allowing them to send down the valley to cater for
what more they could get. The Singtam Soubah was intensely
disgusted with my determination: he accompanied me next
day to the pass, and having exhausted his persuasions,
threats, and warnings about snow, wind, robbers, starvation,
and Cheen sepoys, departed on the 12th for Yeumtong,
leaving me truly happy for the first time since quitting
Dorjiling. I had now a prospect of uninterruptedly
following up my pursuits at an elevation little below that
of the summit of Mont Blanc, surrounded by the loftiest
mountains, and perhaps the vastest glaciers on the globe;
my instruments were in perfect order, and I saw around
me a curious and varied flora.

The morning of the 9th of September promised fair,
though billowy clouds were rapidly ascending the valley.
To the eastward my attention was directed to a double
rainbow; the upper was an arch of the usual form, and the
lower was the curved illuminated edge of a bank of cumulus,
with the orange hues below. We took the path to the
Donkia pass, fording the river, and ascending in a north-
east direction, along the foot of stony hills that rise at a
gradual slope of 12° to broad unsnowed ridges, 18,000 to
19,000 feet high. Shallow valleys, glacier-bound at their
upper extremities, descend from the still loftier rearward
mountains; and in these occur lakes. About five miles up, a
broad opening on the west leads to Tomo Chamo, as the

eastern summit of Kinchinjhow is called.*   Above this the valley expands very much, and is stony and desert: stupendous mountains, upwards of 21,000 feet high, rear themselves on all sides, and the desolation and grandeur of the scene are unequalled in my experience.   The path again crosses the river (which is split into many channels), and proceeds northwards, over gravelly terraces and rocks with patches of Scotch alpine grasses (*Festuca ovina* and *Poa laxa*), sedges, *Stipa*, dandelion, *Allardia*, gentians, *Saussurea*, and *Astragalus*, varied with hard hemispherical mounds of the alsineous plant mentioned at p. 89

I passed several shallow lakes at 17,500 feet; their banks were green and marshy, and supported thirty or forty kinds of plants.   At the head of the valley a steep rocky crest, 500 feet high, rises between two precipitous snowy peaks, and a very fatiguing ascent (at this elevation) leads to the sharp rocky summit of the Donkia pass, 18,466 feet above the sea by barometer, and 17,866 by boiling point.   The view on this occasion was obscured by clouds and fogs, except towards Tibet, in which direction it was magnificent; but as I afterwards twice ascended this pass, and also crossed it, I shall here bring together all the particulars I noted.

The Tibetan view, from its novelty, extent, and singularity,

---

* On one occasion I ascended this valley, which is very broad, flat, and full of lakes at different elevations; one, at about 17,000 feet elevation is three-quarters of a mile long, but not deep: no water-plants grew in it, but there were plenty of others round its margin.   I collected, in the dry bed of a stream near it, a curious white substance like thick felt, formed of felspathic silt (no doubt the product of glacial streams) and the siliceous cells of infusoriæ.   It much re sembles the fossil or meteoric paper of Germany, which is also formed of the lowest tribes of fresh-water plants, though considered by Ehrenberg as of animal origin.   A vein of granite in the bottom of the valley had completely altered the character of the gneiss, which contained veins of jasper and masses of amorphous garnet.   Much olivine is found in the fissures of the gneiss   this mineral is very rare in Sikkim, but I have also seen it in the fissures of the white gneissy granite of the surrounding heights.

demands the first notice: the Cholamoo lake lay 1500 feet
below me, at the bottom of a rapid and rocky descent; it
was a blue sheet of water, three or four miles from north
to south, and one and a half broad, hemmed in by rounded
spurs from Kinchinjhow on one side, and from Donkia
on the other: the Lachen flowed from its northern extremity, and turning westward, entered a broad barren valley,
bounded on the north by red stony mountains, called
Bhomtso, which I saw from Kongra Lama, and ascended
with Dr. Campbell in the October following: though
18,000 to 19,000 feet high, these mountains were wholly
unsnowed.   Beyond this range lay the broad valley of the
Arun, and in the extreme north-west distance, to the north
of Nepal, were some immense snowy mountains, reduced
to mere specks on the horizon.   The valley of the Arun
was bounded on the north by very precipitous black rocky
mountains, sprinkled with snow; beyond these again,
from north to north-west, snow topped range rose over
range in the clear purple distance.   The nearer of these
was the Kiang-lah, which forms the axis or water-shed of
this meridian; its south drainage being to the Arun river,
and its north to the Yaru-tsampu: it appeared forty to
fifty miles off, and of great mean elevation (20,000 feet):
the vast snowy mountains that rose beyond it were, I was
assured, beyond the Yaru, in the salt lake country.*   A
spur from Chomiomo cut off the view to the southward
of north-west, and one from Donkia concealed all to the
east of north.

The most remarkable features of this landscape were its

---

* This salt country was described to me as enormously lofty, perfectly
sterile, and fourteen days' march for loaded men and sheep from Jigatzi:
there is no pasture for yaks, whose feet are cut by the rocks. The salt is
dug (so they express it) from the margin of lakes; as is the carbonate of soda,
"Pieu" of the Tibetans.

J.D.H delt.

John Murray. Albe

Tibet and Cholamoo Lakes

from Donkia Pass, 18,500 ft.

enormous elevation, and its colours and contrast to the black, rugged, and snowy Himalaya of Sikkim. All the mountains between Donkia pass and the Arun were comparatively gently sloped, and of a yellow red colour, rising and falling in long undulations like dunes, 2000 to 3000 feet above the mean level of the Arun valley, and perfectly bare of perpetual snow or glaciers. Rocks everywhere broke out on their flanks, and often along their tops, but the general contour of that immense area was very open and undulating, like the great ranges of Central Asia, described by MM. Huc and Gabet. Beyond this again, the mountains were rugged, often rising into peaks which, from the angles I took here, and subsequently at Bhomtso, cannot be below 24,000 feet, and are probably much higher. The most lofty mountains were on the range north of Nepal, not less than 120 miles distant, and which, though heavily snowed, were below the horizon of Donkia pass.

Cholamoo lake lay in a broad, scantily grassed, sandy and stony valley; snow-beds, rocks, and glaciers dipped abruptly towards its head, but on its west bank a lofty brick-red spur sloped upwards from it, conspicuously cut into terraces for several hundred feet above its waters.

Kambajong, the chief Tibetan village near this, after Phari and Giantchi, is situated on the Arun (called in Tibet "Chomachoo"), on the road from Sikkim to Jigatzi * and

---

* I have adopted the simplest mode of spelling this name that I could find, and omitted the zong or jong, which means fort, and generally terminates it. I think it would not be difficult to enumerate fully a dozen ways of spelling the word, of which Shigatzi, Digarchi, and Djigatzi are the most common. The Tibetans tell me that they cross two passes after leaving Donkia, or Kongra Lama, *en route* for Jigatzi, on both of which they suffer from head aches and difficulty of breathing; ʳ ᵘ is over the Kambajong range; the other, much loftier, is over that of Kiang-lah: as they do not complain of Bhomtso, which is also crossed, and is 18,500 feet, the others may be very lofty indeed.   distance from Donkia pass to Jigatzi is said to be ten days' journey for loaded yaks. Now, according to Turner's observations (evidently

Teshoo Loombo. I did not see it, but a long, stony mountain range above the town is very conspicuous, its sides presenting an interrupted line of cliffs, resembling the portholes of a ship : some fresh fallen snow lay at the base, but none at the top, which was probably 18,500 feet high. The banks of the Arun are thence inhabited at intervals all the way to Tingré, where it enters Nepal.

Donkia rises to the eastward of the pass, but its top is not visible. I ascended (over loose rocks) to between 19,000 and 20,000 feet, and reached vast masses of blue ribboned ice, capping the ridges, but obtained no further prospect. To the west, the beetling east summit of Kinchinjhow rises at two miles distance, 3000 to 4000 feet above the pass. A little south of it, and north of Chango Khang, the view extends through a gap in the Sebolah range, across the valley of the Lachen, to Kinchinjunga, distant forty-two miles. The monarch of mountains looked quite small and low from this point, and it was difficult to believe it was 10,000 feet more lofty than my position I repeatedly looked from it to the high Tibetan mountains in the extreme north-west distance, and was more than ever struck with the apparently immense distance, and consequent altitude of the latter : I put, however, no reliance on such estimates.

To the south the eye wandered down the valley of the

taken with great care) that capital is in latitude 29° 4′ 20″ north, or only seventy miles north of Donkia; and as the yak travels at the rate of sixteen miles a day, the country must be extraordinarily rugged, or the valleys tortuous. Turner took eight or nine days on his journey from Phari to Teshoo Loombo, a distance of only eighty miles; yet he is quoted as an authority for the fact of Tibet being a plain ! he certainly crossed an undulating country, probably 16,000 to 17,000 feet high; a continuation eastwards of the Cholamoo features, and part of the same mountain range that connects Chumulari and Donkia : he had always lofty mountains in sight, and rugged ones on either side, after he had entered the Painomchoo valley. It is a remarkable and significant fact that Turner never appears to have seen Chumulari after having passed it, nor Donkia, Kinchinjhow, or Kinchinjunga at any time.

Lachoong to the mountains of the Chola range, which appear so lofty from Dorjiling, but from here are sunk far below the horizon : on comparing these with the northern landscape, the wonderful difference between their respective snow-levels, amounting to fully 5000 feet, was very apparent. South-east the stupendous snowy amphitheatre formed by the flank of Donkia was a magnificent spectacle.

This wonderful view forcibly impressed me with the fact, that all eye-estimates in mountainous countries are utterly fallacious, if not corrected by study and experience. I had been led to believe that from Donkia pass the whole country of Tibet sloped away in descending steppes to the Tsampu, and was more or less of a plain; and could I have trusted my eyes only, I should have confirmed this assertion so far as the slope was concerned. When, however, the levelled theodolite was directed to the distance, the reverse was found to be the case. Unsnowed and apparently low mountains touched the horizon line of the telescope; which proves that, if only 37 miles off, they must, from the dip of the horizon, be at least 1000 feet higher than the observer's position. The same infallible guide cuts off mountain-tops and deeply snowed ridges, which to the unaided eye appear far lower than the point from which they are viewed; but which, from the quantity of snow on them, must be many thousand feet higher, and, from the angle they subtend in the instrument, must be at an immense distance. The want of refraction to lift the horizon, the astonishing precision of the outlines, and the brilliancy of the images of mountains reduced by distance to mere specks, are all circumstances tending to depress them to appearance. The absence of trees, houses, and familiar objects to assist the eye in the appreciation of distance, throws back the whole landscape; which, seen through the rarified atmosphere of 18,500 feet,

looks as if diminished by being surveyed through the wrong
end of a telescope.

A few rude cairns were erected on the crest of the pass,
covered with wands, red banners, and votive offerings of
rags.   I found a fine slab of slate, inscribed with the
Tibetan characters, "Om Mani Padmi hom," which Meepo
allowed me to take away, as the reward of my exertions.
The ridge is wholly formed of angular blocks of white
gneissy granite, split by frost *   There was no snow on the
pass itself, but deep drifts and glaciers descended in hollows
on the north side, to 17,000 feet.   The rounded northern
red shoulder of Kinchinjhow by Cholamoo lake, apparently
19,000 feet high, was quite bare, and, as I have said, 1
ascended Donkia to upwards of 19,000 feet before I found
the rocks crusted with ice,† and the ground wholly frozen.
I assume, therefore, that 19,000 feet at this spot is not
below the mean level at which all the snow melts that falls
on a fair exposure to the south : this probably coincides
with a mean temperature of 20°.   Forty miles further
north (in Tibet) the same line is probably at 20,000 feet ;
for there much less snow falls, and much more melts in
proportion.‡   From the elevation of about 19,300 feet,

* It was not a proper granite, but a highly metamorphic felspathic gneiss, with
very little mica; being, I suspect, a gneiss which by metamorphic action was
almost remolten into granite : the lamination was obscure, and marked by faint
undulating lines of mica; it cleaves at all angles, but most generally along
fissures with highly polished undulated black surfaces.   The strike of the same
rock near at hand was north-west, and dip north-east, at various angles.

† Snow, transformed into ice throughout its whole mass : in short, glacial ice
in all physical characters.

‡ Two secondary considerations materially affecting the melting of snow, and
hence exerting a material influence on the elevation of the snow-line, appear to
me never to have been sufficiently dwelt upon.   Both, however, bear directly upon
the great elevation of the snow-line in Tibet.   From the imperfect transmission of
the heating rays of the sun through films of water, which transmit perfectly the
luminous rays, it follows that the direct effects of the rays, in clear sunshine, are very
different at equal elevations of the moist outer and dry inner Himalaya.   Secondly,

which I attained on Donkia, I saw a fine illustration of that atmospheric phenomenon called the " spectre of the Brocken," my own shadow being projected on a mass of thin mist that rose above the tremendous precipices over which I hung.   My head was surrounded with a brilliant circular glory or rainbow.*

The temperature of the Donkia pass is much higher than might be anticipated from its great elevation, and from the fact of its being always bitterly cold to the feelings.   This is no doubt due to the warmth of the ascending currents, and to the heat evolved during the condensation of their vapours.   I took the following observations :—

		Temp.	D. P.	Differ-ence.	Ten-sion.	Humidity.
Sept. 9	1·30—3·30 P.M.	41° 8	30° 3	11°·5—	0·1876	0·665
„ 27	1·15—3·15 P.M.	49° 2	32° 6	16°·6—	0·2037	0·560
Oct. 19	3·0 —3·30 P.M.	40° 1	25° 0	15°·1—	0·1551	0·585

The first and last of these temperatures were respectively 42° 3 and 46° 4 lower than Calcutta, which, with the proper deduction for latitude, allows 508 and 460 feet as equivalent to 1° Fahr.   I left a minimum thermometer on the summit on the 9th of September, and removed it on the 27th, but it had been lifted and turned over by the action of the frost and snow on the loose rocks amongst which I had placed it; the latter appearing to have been completely shifted.   Fortunately, the instrument escaped unhurt, with the index at 28°.

naked rock and soil absorb much more heat than surfaces covered with vegetation, and this heat again radiated is infinitely more rapidly absorbed by snow (or other white surfaces) than the direct heat of the sun's rays is.   Hence, at equal elevations the ground heats sooner, and the snow is more exposed to the heat thus radiated in arid Tibet, than in the wooded and grassed mountains of Sikkim.

* Probably caused by spiculæ of ice floating in the atmosphere, the lateral surfaces of which would then have an uniform inclination of 60° : this, according to the observations of Mariotte, Venturi, and Fraunhœfer being the angle necessary for the formation of halos.

A violent southerly wind, with a scud of mist, and sometimes snow, always blew over the pass: but we found shelter on the north face, where I twice kindled a fire, and boiled my thermometers.*   On one occasion I felt the pulses of my party several times during two hours' repose (without eating); the mean of eight persons was 105°, the extremes being 92° and 120°, and my own 108°

One flowering plant ascends to the summit; the alsinaceous one mentioned at p. 89.   The Fescue grass, a little fern (*Woodsia*), and a *Saussurea* † ascend very near the summit, and several lichens grow on the top, as *Cladonia vermicularis*, the yellow *Lecidea geographica*, and the orange *L. miniata;* ‡ also some barren mosses.   At 18,300 feet, I found on one stone only a fine Scotch lichen, a species of *Gyrophora*, the "*tripe de roche*" of Arctic voyagers, and the food of the Canadian hunters; it is also abundant on the Scotch alps.

Before leaving, I took one more long look at the boundless prospect; and, now that its important details were secured, I had leisure to reflect on the impression it produced.   There is no loftier country on the globe than that embraced by this view, and no more howling wilderness; well might the Singtam Soubah and every Tibetan describe it as the loftiest, coldest, windiest, and most barren country in the world.   Were it buried in everlasting snows, or

* On the 9th of September the boiling-point was 181° 3, and on the 27th, 181° 2. In both observations, I believe the kettle communicated a higher temperature to the thermometer than that of the water, for the elevations deduced are far too low.

† A pink-flowered woolly *Saussurea*, and *Delphinium glaciale*, are two of the most lofty plants; both being commonly found from 17,500 to 18,000 feet.

‡ This is one of the most Arctic, Antarctic, and universally diffused plants. The other lichens were *Lecidea atro-alba*, *oreina*, *elegans*, and *chlorophana*, all alpine European and Arctic species.   At 17,000 feet occur *Lecunora ventosa*, *physodes*, *candelaria*, *sórdida*, *atra*, and the beautiful Swiss *L. chrysoleuca*, also European species.

burnt by a tropical sun, it might still be as utterly sterile ;
but with such sterility I had long been familiar.   Here
the colourings are those of the fiery desert or volcanic
island, while the climate is that of the poles.   Never, in
the course of all my wanderings, had my eye rested on a
scene so dreary and inhospitable.   The "cities of the
plain" lie sunk in no more death-like sea than Cholamoo
lake, nor are the tombs of Petra hewn in more desolate
cliffs than those which flank the valley of the Tibetan Arun.

On our return my pony strained his shoulder amongst
the rocks ; as a remedy, the Lachoong Phipun plunged a
lancet into the muscle, and giving me his own animal,
rode mine down.*   It drizzled and sleeted all the way,
and was dark before we arrived at the tent.

At night the Tibetan dogs are let loose, when they howl
dismally : on one occasion they robbed me of all my meat, a
fine piece of yak's flesh.   The yaks are also troublesome,
and bad sleepers ; they used to try to effect an entrance
into my tent, pushing their muzzles under the flaps at the
bottom, and awakening me with a snort and moist hot
blast.   Before the second night I built a turf wall round
the tent, and in future slept with a heavy tripod by my
side, to poke at intruders.

Birds flock to the grass about Momay ; larks, finches,
warblers, abundance of sparrows, feeding on the yak-

---

* These animals, called Tanghan, are wonderfully strong and enduring ; they
are never shod, and the hoof often cracks, and they become pigeon-toed : they are
frequently blind of one eye, when they are called "zemik" (blind ones), but this
is thought no great defect.   They average 5*l.* to 10*l.* for a good animal in Tibet ;
and the best fetch 40*l.* to 50*l.* in the plains of India, where they become
acclimated and thrive well.   Giantchi (Jhansi-jeuug of Turner) is the best mart
for them in this part of Tibet, where some breeds fetch very high prices.   The
Tibetans give the foals of value messes of pig's blood and raw liver, which they
devour greedily, and it is said to strengthen them wonderfully ; the custom is, I
believe, general in central Asia.   Humboldt (Pers. Nar. iv. p. 320) describes the
horses of Caraccas as occasionally eating salt meat.

droppings, and occasionally the hoopoe; waders, cormorants, and wild ducks were sometimes seen in the streams, but most of them were migrating south. The yaks are driven out to pasture at sunrise, and home at sunset, till the middle of the month, when they return to Yeumtong All their droppings are removed from near the tents, and piled in heaps; as these animals, unlike their masters, will not sleep amid such dirt These heaps swarm with the maggots of two large flies, a yellow and black, affording abundant food to red-legged crows, ravens, and swallows. Butterflies are rare; the few are mostly *Colias*, *Hipparchia*, *Polyommatus*, and *Melitœa*, these I have seen feeding at 17,000 feet; when found higher, they have generally been carried up by currents. Of beetles, an *Aphodeus*, in yak-droppings, and an *Elaphrus*, a predaceous genus inhabiting swamps, are almost the only ones I saw. The wild quadrupeds are huge sheep, in flocks of fifty, the *Ovis Ammon* called "Gnow" I never shot one, not having time to pursue them for they were very seldom seen, and always at great elevations. The larger marmot is common, and I found the horns of the "Tchiru" antelope. Neither the wild horse, fox, hare, nor tailless rat, cross the Donkia pass. White clover, shepherd's purse, dock, plantain, and chickweed, are imported here by yaks; but the common *Prunella* of Europe is wild, and so is a groundsel like *Senecio Jacobœa*, *Ranunculus*, *Sibbaldia*, and 200 other plants. The grasses are numerous; they belong chiefly to *Poa*, *Festuca*, *Stipa*, and other European genera.

I repeatedly attempted to ascend both Kinchinjhow and Donkia from Momay, and generally reached from 18,000 to 19,000 feet, but never much higher.* The observations

---

* An elevation of 20,000, and perhaps 22,000 feet might, I should think, easily be attained by practice, in Tibet, north of Sikkim.

taken on these excursions are sufficiently illustrated by those of Donkia pass : they served chiefly to perfect my map, measure the surrounding peaks, and determine the elevation reached by plants; all of which were slow operations, the weather of this month being so bad that I rarely returned dry to my tent; fog and drizzle, if not sleet and snow, coming on during every day, without exception.

I made frequent excursions to the great glacier of Kinchinjhow. Its valley is about four miles long, broad and flat : Chango-khang * rears its blue and white cliffs 4,500 feet above its west flank, and throws down avalanches of stones and snow into the valley. Hot springs † burst from the ground near some granite rocks on its floor, about 16,000 feet above the sea, and only a mile below the glacier, and the water collects in pools : its temperature is 110°, and in places 116°, or 4° hotter than that of the Yeumtong hot-springs, though 4000 feet higher, and of precisely the same character. A *Barbarea* and some other plants make the neighbourhood of the hot-springs a little oasis, and the large marmot is common, uttering its sharp, chirping squeak.

The terminal moraine is about 500 feet high, quite

---

* The elevation of this mountain is about 20,560 feet, by the mean of several observations taken from surrounding localities.

† Supposing the mean temperature of the air at the elevation of the Momay springs to be 26° or 28°, which may be approximately assumed, and that, as some suppose, the heat of thermal springs is due to the internal temperature of the globe; then according to the law of increment of heat in descending (of 1° for fifty feet) we should find the temperature of 110° at a depth of 4,100 feet, or at 11,900 feet above the level of the sea. Direct experiment with internal heat has not, however, been carried beyond 2000 feet below the surface, and as the ratio of increment diminishes with the depth, that above assigned to the temperature of 110° is no doubt much too little. The Momay springs more probably owe their temperature to chemical decomposition of sulphurets of metals. I found pyrites in Tibet on the north flank of the mountain Kinchinjhow, in limestones asssociated with shales.

barren, and thrown obliquely across the valley, from north-
east to south-west, completely hiding the glacier. From
its top successive smaller parallel ridges (indicating the
periodic retirements of the glacier) lead down to the ice,
which must have sunk several hundred feet. This glacier
descends from Kinchinjhow, the huge cliff of whose eastern
extremity dips into it. The surface, less than half a mile
wide, is exceedingly undulated, and covered with large
pools of water, ninety feet deep, and beds of snow, and is
deeply corroded ; gigantic blocks are perched on pinnacles
of ice on its surface, and the gravel cones * are often twenty
feet high. The crevassing so conspicuous on the Swiss
glaciers is not so regular on this, and the surface appears
more like a troubled ocean ; due, no doubt, to the copious
rain and snow-falls throughout the summer, and the
corroding power of wet fogs. The substance of the ice
is ribboned, dirt-bands are seen from above to form long
loops on some parts, and the lateral moraines, like the
terminal, are high above the surface. These notes, made
previous to reading Professor Forbes's travels in the Alps,
sufficiently show that perpetual snow, whether as ice or
glacier, obeys the same laws in India as in Europe ; and I
have no remarks to offer on the structure of glaciers, that
are not well illustrated and explained in the above-
mentioned admirable work.

Its average slope for a mile above the terminal moraines
was less than 5°, and the height of its surface above the sea
16,500 feet by boiling-point ; the thickness of its ice
probably 400 feet. Between the moraine and the west
flank of the valley is a large lake, with terraced banks,
whose bottom (covered with fine felspathic silt) is several

---

* For a description of this curious phenomenon, which has been illustrated by
Agassiz, see "Forbes's Alps," p. 26 and 347.

hundred feet above that of the valley; it is half a mile long, and a quarter broad, and fed partly by glaciers of the second order on Chango-khang and Sebolah, and partly by filtration through the lateral moraine.

GNEISS-BLOCK WITH GRANITE BANDS, ON THE KINCHINJHOW GLACIER.

CHAPTER XXIII.

On the 20th of September I ascended to the great Donkia
glaciers, east of Momay; the valley is much longer than
that leading to the Kinchinjhow glacier, and at 16,000
or 17,000 feet elevation, containing four marshes or lakes,
alternating with as many transverse moraines that have
dammed the river. These moraines seem in some cases
to have been deposited where rocks in the bed of the
valley obstructed the downward progress of the ancient
glacier; hence, when this latter finally retired, it rested at
these obstructions, and accumulated there great deposits,
which do not cross the valley, but project from each side
obliquely into it. The rocks *in situ* on the floor of the
valley are all *moutonnéed* and polished on the top,
sides, and face looking up the valley, but are rugged on

that looking down it : gigantic blocks are poised on some. The lowest of the ancient moraines completely crosses the river, which finds its way between the boulders.

Under the red cliff of Forked Donkia the valley becomes very broad, bare, and gravelly, with a confusion of moraines, and turns more northwards. At the angle, the present terminal moraine rises like a mountain (I assumed it to be about 800 feet high),* and crosses the valley from N.N.E. to S.S.W. From the summit, which rises above the level of the glacier, and from which I assume its present retirement, a most striking scene opened. The ice filling an immense basin, several miles broad and long, formed a low dome,† with Forked Donkia on the west, and a serried range of rusty-red scarped mountains, 20,000 feet high on the north and east, separating large tributary glaciers. Other still loftier tops of Donkia appeared behind these, upwards of 22,000 feet high, but I could not recognise. the true summit (23,176 feet). The surface was very rugged, and so deeply honeycombed that the foot often sank from six to eight inches in

---

* This is the largest and longest terminal moraine backed by an existing glacier that I examined with care : I doubt its being so high as the moraine of the Allalein glacier below the Mat-maark sea in the Sachs valley (Valais, Switzerland) ; but it is impossible to compare such objects from memory : the Donkia one was much the most uniform in height.

† This convexity of the ice is particularly alluded to by Forbes ("Travels in the Alps," p. 386), as the "renflément" of Rendu and "surface bombée" of Agassiz, and is attributed to the effects of hydrostatic pressure tending to press the lower layers of ice upwards to the surface. My own impression at the time was, that the convexity of the surface of the Donkia glacier was due to a subjacent mountain spur running south from Donkia itself. I know, however, far too little of the topography of this glacier to advance such a conjecture with any confidence. In this case, as in all similar ones, broad expanses being covered to an enormous depth with ice, the surface of the latter must in some degree be modified by the ridges and valleys it conceals. The typical "surface bombée," which is conspicuous in the Himalaya glaciers, I was wont (in my ignorance of the mechanical laws of glaciers) to attribute to the more rapid melting of the edges of the glacier by the radiated heat of its lateral moraines and of the flanks of the valley that it occupies.

crisp wet ice. I proceeded a mile on it, with much more
difficulty than on any Swiss glacier: this was owing to the
elevation, and the corrosion of the surface into pits and
pools of water; the crevasses being but few and distant.
I saw no dirt-bands on looking down upon it from a point
I attained under the red cliff of Forked Donkia, at an
elevation of 18,307 feet by barometer, and 18,597 by
boiling-point. The weather was very cold, the thermometer
fell from 41° to 34°, and it snowed heavily after 3 P.M.

The strike of all the rocks (gneiss with granite veins)
seemed to be north-east, and dip north-west 30°. Such
also were the strike and dip on another spur from Donkia,
north of this, which I ascended to 19,000 feet, on the
26th of September: it abutted on the scarped precipices,
3000 feet high, of that mountain. I had been attracted
to the spot by its bright orange-red colour, which I found
to be caused by peroxide of iron. The highly crystalline
nature of the rocks, at these great elevations, is due to the
action of veins of fine-grained granite, which sometimes alter
the gneiss to such an extent that it appears as if fused
into a fine granite, with distinct crystals of quartz and
felspar; the most quartzy layers are then roughly crys-
tallized into prisms, or their particles are aggregated into
spheres composed of concentric layers of radiating crystals,
as is often seen in agates. The rearrangement of the mineral
constituents by heat goes on here just as in trap, cavities
filled with crystals being formed in rocks exposed to great
heat and pressure. Where mica abounds, it becomes
black and metallic; and the aluminous matter is crystal-
lised in the form of garnets.

At these great heights the weather was never fine for
more than an hour at a time, and a driving sleet followed
by thick snow drove me down on both these occasions.

Another time I ascended a third spur from this great mountain, and was overtaken by a heavy gale and thunder-storm, the latter is a rare phenomenon : it blew down my tripod and instruments which I had thought securely

SUMMIT OF FORKED DONKIA, AND "GOA." ANTELOPES.

propped with stones, and the thermometers were broken, but fortunately not the barometer. On picking up the latter, which lay with its top down the hill, a large bubble of air appeared, which I passed up and down the tube, and then allowed to escape; when I heard a rattling of broken glass in the cistern.   Having another barometer *

* This barometer (one of Newman's portable instruments) I have now at Kew : it was compared with the Royal Society's standard before leaving England ; and varied according to comparisons made with the Calcutta standard 0·012 during its travels ; on leaving Calcutta its error was 0 ; and on arriving in England, by

at my tent, I hastened to ascertain by comparison whether
the instrument which had travelled with me from England,
and taken so many thousand observations, was seriously
damaged: to my delight an error of 0·020 was all I could
detect at Momay and all other lower stations.   On my return
to Dorjiling in December, I took it to pieces, and found the
lower part of the bulb of the attached thermometer broken
off, and floating on the mercury.   Having quite expected

VIEW FROM AN ELEVATION OF 18,000 FEET OF THE EAST TOP OF KINCHINJHOW, AND OF TIBET,
OVER THE RIDGE THAT CONNECTS IT WITH DONKIA. WILD SHEEP (OVIS AMMON) IN THE
FOREGROUND.

this, I always checked the observations of the attached
thermometer by another, but—how, it is not easy to say—
the broken one invariably gave a correct temperature.

the standard of the Royal Society, + ·004. I have given in the Appendix some
remarks on the use of these barometers, which (though they have obvious defects),
are less liable to derangement, far more portable, and stand much heavier shocks
than those of any other construction with which I am familiar.

The Kinchinjhow spurs are not accessible to so great an elevation as those of Donkia, but they afford finer views over Tibet, across the ridge connecting Kinchinjow with Donkia.

Broad summits here, as on the opposite side of the valley, are quite bare of snow at 18,000 feet, though where they project as sloping hog-backed spurs from the parent mountain, the snows of the latter roll down on them and form glacial caps, the reverse of glaciers in valleys, but which overflow, as it were, on all sides of the slopes, and are ribboned * and crevassed.

On the 18th of September I ascended the range which divides the Lachen from the Lachoong valley, to the Sebolah pass, a very sharp ridge of gneiss, striking north-west and dipping north-east, which runs south from Kinchinjhow to Chango-khang. A yak-track led across the Kinchinjhow glacier, along the bank of the lake, and thence westward up a very steep spur, on which was much glacial ice and snow, but few plants above 16,000 feet. At nearly 17,000 feet I passed two small lakes, on the banks of one of which I found bees, a May-fly (*Ephemera*) and gnat; the two latter bred on stones in the water, which (the day being fine) had a temperature of 53°, while that of the large lake at the glacier, 1000 feet lower, was only 39°.

The view from the summit commands the whole castellated front of Kinchinjhow, the sweep of the Donkia cliffs to the east, Chango-khang's blunt cone of ribbed snow† over head, while to the west, across the grassy Palung dunes rise Chomiomo, the Thlonok mountains, and Kinchin-

---

* The convexity of the curves, however, seems to be upwards. Such reversed glaciers, ending abruptly on broad stony shoulders quite free of snow, should on no account be taken as indicating the lower limit of perpetual snow.

† This ridging or furrowing of steep snow-beds is explained at vol. i. p. 237.

junga in the distance.* The Palung plains, now yellow
with withered grass, were the most curious part of the view :
hemmed in by this range which rises 2000 feet above
them, and by the Lachen hills on the east, they appeared a
dead level, from which Kinchinjhow reared its head, like
an island from the ocean.† The black tents of the Tibetans
were still there, but the flocks were gone. The broad
fosse-like valley of the Chachoo was at my feet, with the
river winding along its bottom, and its flanks dotted with
black juniper bushes.

The temperature at this elevation, between 1 and 3 P.M.,
varied from 38° to 59°; the mean being 46°·5, with the
dew-point 34°·6. The height I made 17,585 feet by
barometer, and 17,517 by boiling-point. I tried the pulses
of eight persons after two hours' rest; they varied from
80 to 112, my own being 104. As usual at these heights,
all the party were suffering from giddiness and headaches.

Throughout September various parties passed my tent at
Momay, generally Lamas or traders : the former, wrapped
in blankets, wearing scarlet and gilt mitres, usually rode
grunting yaks, which were sometimes led by a slave-boy or a
mahogany-faced nun, with a broad yellow sheep-skin cap
with flaps over her ears, short petticoats, and striped boots.
The domestic utensils, pots, pans, and bamboos of butter,

---

* The latter bore 241° 30′; it was distant about thirty-four miles, and subtended
an angle of 3° 2′ 30.″ The rocks on its north flanks were all black, while those
forming the upper 10,000 feet of the south face were white: hence, the top is
probably granite, overlaid by the gneiss on the north.

† It is impossible to contemplate the abrupt flanks of all these lofty mountains,
without contrasting them with the sloping outlines that prevail in the southern
parts of Sikkim. All such precipices are, I have no doubt, the results of sea
action; and all posterior influence of sub-aërial action, aqueous or glacial, tends to
wear these precipices into slopes, to fill up valleys and to level mountains. Of all
such influences heavy rain-falls and a luxuriant vegetation are probably the most
active ; and these features are characteristic of the lower valleys of Sikkim, which
are consequently exposed to very different conditions of wear and tear from those
which prevail on these loftier rearward ranges.

tea-churn, bellows, stools, books, and sacred implements, usually hung rattling on all sides of his holiness, and a sumpter yak carried the tents and mats for sleeping.   On several occasions large parties of traders, with thirty or forty yaks* laden with planks, passed, and occasionally a shepherd with Tibet sheep, goats, and ponies. I questioned many of these travellers about the courses of the Tibetan rivers; they all agreed† in stating the Kambajong or Chomachoo river, north of the Lachen, to be the Arun of Nepal, and that it rose near the Ramchoo lake (of Turner's route).   The lake itself discharges either into the Arun, or into the Painomchoo (flowing to the Yaru); but this point I could never satisfactorily ascertain.

The weather at Momay, during September, was generally bad after 11 A.M.: little snow or rain fell, but thin mists and drizzle prevailed; less than one inch and a half of rain was collected, though upwards of eleven fell at Calcutta, and rather more at Dorjiling. The mornings were sometimes fine, cold, and sunny, with a north wind which had blown down the valley all night, and till 9 A.M., when the south-east wind, with fog, came on.   Throughout the day a north current blew above the southern; and when the mist was thin, the air sparkled with spiculæ of snow, caused by the cold dry upper current condensing the vapours of the lower.   This southern current passes over the tops of the loftiest mountains, ascending to 24,000 feet, and discharging frequent showers

---

* About 600 loaded yaks are said to cross the Donkia pass annually.

† One lad only, declared that the Kambajong river flowed north-west to Dobtah and Sarrh, and thence turned north to the Yaru; but all Campbell's itineraries, as well as mine, make the Dobtah lake drain into the Chomachoo, north of Wallanchoon; which latter river the Nepalese also affirm flows into Nepal, as the Arun.   The Lachen and Lachoong Phipuns both insisted on this, naming to me the principal towns on the way south-west from Kambajong along the river to Tingri Maidan, *viâ* Tashirukpa Chait, which is north of Wallanchoon pass.

in Tibet, as far north as Jigatzi, where, however, violent dry easterly gales are the most prevalent.

The equinoctial gales set in on the 21st, with a falling barometer, and sleet at night; on the 23rd and 24th it snowed heavily, and being unable to light a fire at the entrance of my tent, I spent two wretched days, taking observations; on the 25th it cleared, and the snow soon melted. Frosty nights succeeded, but the thermometer only fell to 31° once during the month, and the maximum once rose to 62°·5. The mean temperature from the 9th to the 30th September was 41°·6,* which coincided with that of 8 A.M. and 8 P.M.; the mean maximum, 52°.2, minimum, 34°·7, and consequent range, 17°·5.† On seven nights the radiating thermometer fell much below the temperature of the air, the mean being 10°·5 and maximum 14°·2; and on seven mornings the sun heated the black-bulb thermometer considerably, on the mean to 62°·6 above the air; maximum 75° 2, and minimum, 43°. The greatest heat of the day occurred at noon: the most rapid rise of temperature (5°) between 8 and 9 A.M., and the greatest fall (5°·5), between 3 and 4 P.M. A sunk thermometer fell from 52°·5 to 51°·5 between the 11th and 14th, when I was obliged to remove the thermometer owing to the accident mentioned above. The mercury in the barometer rose and fell contemporaneously with that at Calcutta and Dorjiling, but the amount of tide was considerably less, and, as is usual during the equinoctial month, on some days it scarcely moved, whilst on others it rose and fell rapidly. The tide amounted to 0·062 of an inch.

On the 28th of the month the Singtam Soubah came up

---

* The result of fifty-six comparative observations between Calcutta and Momay, give 40°·5 difference, which, after corrections, allows 1° Fahr. for every 438 feet of ascent.

† At Dorjiling the September range is only 9°·5; and at Calcutta 10°.

from Yeumtong, to request leave to depart for his home, on account of his wife's illness; and to inform me that Dr. Campbell had left Dorjiling, accompanied (in compliance with the Rajah's orders) by the Tchebu Lama. I therefore left Momay on the 30th, to meet him at Choongtam, arriving at Yeumtong the same night, amid heavy rain and sleet.

Autumnal tints reigned at Yeumtong, and the flowers had disappeared from its heath-like flat; a small eatable cherry with a wrinkled stone was ripe, and acceptable in a country so destitute of fruit.* Thence I descended to Lachoong, on the 1st of October, again through heavy rain, the snow lying on the Tunkra mountain at 14,000 feet. The larch was shedding its leaves, which turn red before they fall; but the annual vegetation was much behind that at 14,000 feet, and so many late flowerers, such as *Umbelliferæ* and *Compositæ*, had come into blossom, that the place still looked gay and green: the blue climbing gentian (*Crawfurdia*) now adorned the bushes; this plant would be a great acquisition in English gardens. A *Polygonum* still in flower here, was in ripe fruit near Momay, 6000 feet higher up the valley.

On the following day I made a long and very fatiguing march to Choongtam, but the coolies were not all able to accomplish it. The backwardness of the flora in descending was even more conspicuous than on the previous day: the jungles, at 7000 feet, being gay with a handsome Cucurbitaceous plant. Crossing the Lachoong cane-bridge, I paid the tribute of a sigh to the memory of my poor dog, and reached my old camping-ground at Choongtam by

---

* The absence of *Vaccinia* (whortleberries and cranberries) and eatable *Rubi* (brambles) in the alpine regions of the Himalaya is very remarkable, and they are not replaced by any substitute. With regard to Vaccinium, this is the more anomalous, as several species grow in the temperate regions of Sikkim.

10 P.M., having been marching rapidly for twelve hours. My bed and tent came up two hours later, and not before the leeches and mosquitos had taxed me severely. On the 4th of October I heard the nightingale for the first time this season.

Expecting Dr. Campbell on the following morning, I proceeded down the river to meet him: the whole valley was buried under a torrent or débacle of mud, shingle, and boulders, and for half a mile the stream was dammed up into a deep lake. Amongst the gneiss and granite boulders brought down by this débacle, I collected some actinolites; but all minerals are extremely rare in Sikkim and I never heard of a gem or crystal of any size or beauty, or of an ore of any consequence, being found in this country.

I met my friend on the other side of the mud torrent, and I was truly rejoiced to see him, though he was looking much the worse for his trying journey through the hot valleys at this season; in fact, I know no greater trial of the constitution than the exposure and hard exercise that is necessary in traversing these valleys, below 5000 feet, in the rainy season: delay is dangerous, and the heat, anxiety, and bodily suffering from fatigue, insects, and bruises, banish sleep, and urge the restless traveller onward to higher and more healthy regions. Dr. Campbell had, I found, in addition to the ordinary dangers of such a journey, met with an accident which might have proved serious; his pony having been dashed to pieces by falling over a precipice, a fate he barely escaped himself, by adroitly slipping from the saddle when he felt the animal's foot giving way.

On our way back to Choongtam, he detailed to me the motives that had led to his obtaining the authority of the Deputy-Governor of Bengal (Lord Dalhousie being absent)

for his visiting Sikkim. Foremost, was his earnest desire
to cultivate a better understanding with the Rajah and his
officers. He had always taken the Rajah's part, from a conviction that he was not to blame for the misunderstandings
which the Sikkim officers pretended to exist between
their country and Dorjiling; he had, whilst urgently
remonstrating with the Rajah, insisted on forbearance on
my part, and had long exercised it himself. In detailing
the treatment to which I was subjected, I had not hesitated
to express my opinion that the Rajah was more compro-
mised by it than his Dewan : Dr. Campbell, on the contrary, knew that the Dewan was the head and front of the
whole system of annoyance. In one point of view it
mattered little who was in the right; but the transaction
was a violation of good faith on the part of the Sikkim
government towards the British, for which the Rajah,
however helpless, was yet responsible. To act upon my
representations alone would have been unjust, and no course
remained but for Dr. Campbell to inquire personally into
the matter. The authority to do this gave him also the
opportunity of becoming acquainted with the country which
we were bound to protect, as well by our interest as by
treaty, but from which we were so jealously excluded, that
should any contingency occur, we were ignorant of what
steps to take for defence, and, indeed, of what we should
have to defend.

On the 6th of October we left Choongtam for my second
visit to the Kongra Lama pass, hoping to get round by the
Cholamoo lakes and the Donkia pass. As the country
beyond the frontier was uninhabited, the Tchebu Lama saw
no difficulty in this, provided the Lachen Phipun and the
Tibetans did not object. Our great obstacle was the
Singtam Soubah, who (by the Rajah's order) accompanied

us to clear the road, and give us every facility, but who was
very sulky, and undisguisedly rude to Campbell; he was in
fact extremely jealous of the Lama, who held higher authority
than he did, and who alone had the Rajah's confidence.

Our first day's march was of about ten miles to one of
the river-flats, which was covered with wild apple-trees,
whose fruit, when stewed with sugar, we found palatable.
The Lachen river, though still swollen, was comparatively
clear; the rains usually ceasing, or at least moderating, in
October: its water was about 5° colder than in the beginning
of August.

During the second day's march we were stopped at the
Taktoong river by the want of a bridge, which the Singtam
Soubah refused to exert himself to have repaired; its waters
were, however, so fallen, that our now large party soon
bridged it with admirable skill. We encamped the second
night on Chateng, and the following day made a long
march, crossing the Zemu, and ascending half-way to
Tallum Samdong. The alpine foliage was rapidly changing
colour; and that of the berberry turning scarlet, gave a
warm glow to the mountain above the forest. Lamteng
village was deserted: turnips were maturing near the
houses, and buckwheat on the slope behind; the latter is a
winter-crop at lower elevations, and harvested in April. At
Zemu Samdong the willow-leaves were becoming sear and
yellow, and the rose-bushes bore enormous scarlet hips, two
inches long, and covered with bristles; they were sweet,
and rather good eating. Near Tungu (where we arrived
on the 9th) the great Sikkim currant was in fruit; its
berries are much larger than the English, and of the
same beautiful red colour, but bitter and very acid;
they are, however, eaten by the Tibetans, who call them
"Kewdemah."

Near the village I found Dr. Campbell remonstrating with the Lachen Phipun on the delays and rude treatment I had received in June and July: the man, of course, answered every question with falsehoods, which is the custom of these people, and produced the Rajah's orders for my being treated with every civility, as a proof that he must have behaved as he ought! The Singtam Soubah, as was natural, hung back, for it was owing to him alone that the orders had been contravened, and the Phipun appealed to the bystanders for the truth of this.

The Phipun. (accompanied by his Larpun or subordinate officer) had prepared for us a sumptuous refreshment of tea-soup, which was brewing by the road, and in which all animosities were soon washed away. We took up our abode at Tungu in a wooden hut under the great rock, where we were detained for several days by bad weather. I was assured that during all August and September the weather had been uniformly gloomy, as at Momay, though little rain had fallen.

We had much difficulty in purchasing a sufficient number of blankets * for our people, and in arranging for our journey, to which the Lachen Phipun was favourable, promising us ponies for the expedition. The vegetation around was wholly changed since my July visit: the rhododendron scrub was verdigris-green from the young leaves which burst in autumn, and expose at the end of each branchlet a flower-bud covered with resinous scales, which are thrown off in the following spring. The jungle was spotted yellow with the withered birch, maple and mountain-ash, and scarlet with berberry bushes; while above, the pastures were yellow-brown with the dead grass, and streaked with snow.

---

* These were made of goat's wool, teazed into a satiny surface by little teazle-like brushes of bamboo.

Amongst other luxuries, we procured the flesh of yak calves, which is excellent veal : we always returned the foot for the mother to lick while being milked, without which she yields nothing. The yak goes nine months with calf, and drops one every two years, bearing altogether ten or twelve : the common Sikkim cow of lower elevations, at Dorjiling invariably goes from nine and a half to ten months, and calves annually : ponies go eleven months, and foal nearly every year. In Tibet the sheep are annually sheared ; the ewes drop their young in spring and autumn, but the lambs born at the latter period often die of cold and starvation, and double lambing is unknown ; whereas, in the plains of Bengal (where, however, sheep cannot be said to thrive without pulse fodder) twins are constantly born. At Dorjiling the sheep drop a lamb once in the season. The Tibetan mutton we generally found dry and stringy.

In these regions many of my goats and kids had died foaming at the mouth and grinding their teeth ; and I here discovered the cause to arise from their eating the leaves of *Rhododendron cinnabarinum** ("Kema Kechoong," Lepcha : Kema signifying Rhododendron) : this species alone is said to be poisonous ; and when used as fuel, it causes the face to swell and the eyes to inflame ; of which I observed several instances. As the subject of fire-wood is of every-day interest to the traveller in these regions, I may here mention that the rhododendron woods afford poor fires ; juniper burns the brightest, and with least smoke. *Abies Webbiana*, though emitting much smoke, gives a cheerful fire, far superior to larch,† spruce, or *Abies Brunoniana*. At Dorjiling, oak is the common

---

* The poisonous honey produced by other species is alluded to at vol. i., p. 201. An *Andromeda* and a *Gualtheria*, I have been assured are equally deleterious.

† The larch of northern Asia (*Larix Europæa*) is said to produce a pungent smoke, which I never observed to be the case with the Sikkim species.

fuel; alder is also good.   Chestnut is invariably used for
blacksmith's charcoal.   Magnolia has a disagreeable odour,
and laurel burns very badly.

The phenomenon of phosphorescence is most conspicuous
on stacks of fire-wood.   At Dorjiling, during the damp,
warm, summer months (May to October), at elevations of
5000 to 8000 feet, it may be witnessed every night by
penetrating a few yards into the forest—at least it was so
in 1848 and 1849; and during my stay there billets of
decayed wood were repeatedly sent to me by residents,
with inquiries as to the cause of their luminosity.   It is
no exaggeration to say that one does not need to move
from the fireside to see this phenomenon, for if there is
a partially decayed log amongst the fire-wood, it is almost
sure to glow with a pale phosphoric light.   A stack of
fire-wood, collected near my host's (Mr. Hodgson) cottage,
presented a beautiful spectacle for two months (in July and
August), and on passing it at night, I had to quiet my
pony, who was always alarmed by it.   The phenomenon
invariably accompanies decay, and is common on oak,
laurel (*Tetranthera*), birch, and probably other timbers; it
equally appears on cut wood and on stumps, but is most
frequent on branches lying close to the ground in the wet
forests.   I have reason to believe that it spreads with great
rapidity from old surfaces to freshly cut ones.   That it is a
vital phenomenon, and due to the mycelium of a fungus,
I do not in the least doubt, for I have observed it occasionally circumscribed by those black lines which are often
seen to bound mycelia on dead wood, and to precede a
more rapid decay.   I have often tried, but always in vain,
to coax these mycelia into developing some fungus, by
placing them in damp rooms, &c.   When camping in
the mountains, I frequently caused the natives to bring

phosphorescent wood into my tent, for the pleasure of watching its soft undulating light, which appears to pale and glow with every motion of the atmosphere; but except in this difference of intensity, it presents no change in appearance night after night. Alcohol, heat, and dryness soon dissipate it; electricity I never tried. It has no odour, and my dog, who had a fine sense of smell, paid no heed when it was laid under his nose.*

The weather continuing bad, and snow falling, the country people began to leave for their winter-quarters at Lamteng. In the evenings we enjoyed the company of the Phipun and Tchebu Lama, who relished a cup of sugarless tea more than any other refreshment we could offer. From them we collected much Tibetan information :—the former was an inveterate smoker, using a pale, mild tobacco, mixed largely with leaves of the small wild Tibetan Rhubarb, called "Chula." Snuff is little used, and is principally procured from the plains of India.

We visited Palung twice, chiefly in hopes that Dr. Campbell might see the magnificent prospect of Kinchinjhow from its plains : the first time we gained little beyond a ducking, but on the second (October the 15th) the view was superb; and I likewise caught a glimpse of Kinchinjunga from the neighbouring heights, bearing south 60° west and distant forty miles. I also measured barometrically the elevation at the great chait on the plains, and found it

---

* As far as my observations go, this phenomenon of light is confined to the lower orders of vegetable life, to the fungi alone, and is not dependent on irritability. I have never seen luminous flowers or roots, nor do I know of any authenticated instance of such, which may not be explained by the presence of mycelium or of animal life. In the animal kingdom, luminosity is confined, I believe, to the Invertebrata, and is especially common amongst the Radiata and Mollusca; it is also frequent in the Entromostracous Crustacea, and in various genera of most orders of insects. In all these, even in the Sertulariæ, I have invariably observed the light to be increased by irritation, in which respect the luminosity of animal life differs from that of vegetable.

15,620 feet, and by carefully boiled thermometers, 15,283, on the 13th October, and 15,566 on the 15th : the difference being due to the higher temperature on the latter day, and to a rise of 0° 3 on both boiling-point thermometers above what the same instruments stood at on the 13th. The elevation of Tungu from the October barometrical observations was only seven feet higher than that given by those of July ; the respective heights being 12,766 feet in July, and 12,773 in October.* The mean temperature had fallen from 50° in July to 41°, and that of the sunk thermometer from 57° to 51° 4. The mean range in July was 23° 3, and in October 13° 8 ; the weather during the latter period being, however, uniformly cold and misty, this was much below the mean monthly range, which probably exceeds 30°. Much more rain fell in October at Tungu than at Dorjiling, which is the opposite to what occurs during the rainy season.

October 15*th*. Having sent the coolies forward, with instructions to halt and camp on this side the Kongra Lama pass, we followed them, taking the route by Palung, and thence over the hills to the Lachen, to the east of which we descended, and further up its valley joined the advanced party in a rocky glen, called Sitong, an advantageous camping ground, from being sheltered by rocks which ward off the keen blasts : its elevation is 15,370 feet above the sea, and the magnificent west cliff of Kinchinjhow towers over it not a mile distant, bearing due east,

---

* The elevation of Tungu by boiling-point was 12,650 feet by a set of July observations, 12,818 by a set taken on the 11th of October, and 12,544 by a set on the 14th of October : the discrepancies were partly due to the temperature corrections, but mainly to the readings of the thermometers, which were —

July 28	sunset	189·5	air	47° 3	elev.	12,650
Oct. 11	noon	189·5	„	37° 6	„	12,818
Oct. 14	sunset	190·1	„	45° 3	„	12,544

and subtending an angle of 24° 3′. The afternoon was misty, but at 7 p.m. the south-east wind fell, and was immediately succeeded by the biting north return current, which dispelled the fog: hoar-frost sparkled on the ground, and the moon shone full on the snowy head of Kinchin-jhow, over which the milky-way and the broad flashing orbs of the stars formed a jewelled diadem. The night was very windy and cold, though the thermometer fell no lower than 22°, that placed in a polished parabolic reflector to 20°, and another laid on herbage to 17° 5.

On the 16th we were up early. I felt very anxious about the prospect of our getting round by Donkia pass and Cholamoo, which would enable me to complete the few remaining miles of my long survey of the Teesta river, and which promised immense results in the views I should obtain of the country, and of the geology and botany of these lofty snowless regions. Campbell, though extremely solicitous to obtain permission from the Tibetan guard, (who were waiting for us on the frontier), was never-theless bound by his own official position to yield at once to their wishes, should they refuse us a passage.

The sun rose on our camp at 7·30 a.m., when the north wind fell; and within an hour afterwards the temperature had risen to 45°. Having had our sticks * warmed and handed to us, we started on ponies, accompanied by the Lama only, to hold a parley with the Tibetans; ordering the rest of the party to follow at their leisure. We had not proceeded far when we were joined by two Tibetan Sepoys, who, on our reaching the pass, bellowed

---

* It was an invariable custom of our Lepcha and Tibetan attendants, to warm the handles of our sticks in cold weather, before starting on our daily marches. This is one of many little instances I could adduce, of their thought-fulness and attention to the smallest comforts of the stranger and wanderer in their lands.

lustily for their companions; when Campbell and the Lama drew up at the chait of Kongra Lama, and announced his wish to confer with their commandant.

My anxiety was now wound up to a pitch; I saw men with matchlocks emerging from amongst the rocks under Chomiomo, and despairing of permission being obtained, I goaded my pony with heels and stick, and dashed on up the Lachen valley, resolved to make the best of a splendid day, and not turn back till I had followed the river to the Cholamoo lakes. The Sepoys followed me a few paces, but running being difficult at 16,000 feet, they soon gave up the chase.

A few miles ride in a north-east direction over an open, undulating country, brought me to the Lachen, flowing westwards in a broad, open, stony valley, bounded by Kinchinjhow on the south, (its face being as precipitous as that on the opposite side), and on the north by the Peuka-thlo, a low range of rocky, sloping mountains, of which the summits were 18,000 to 19,000 feet above the sea. Enormous erratic blocks of gneiss strewed the ground, which was sandy or gravelly, and cut into terraces along the shallow, winding river, the green and sparkling waters of which rippled over pebbles, or expanded into lagoons. The already scanty vegetation diminished rapidly: it consisted chiefly of scattered bushes of a dwarf scrubby honey-suckle and tufts of nettle, both so brittle as to be trodden into powder, and the short leafless twiggy *Ephedra*, a few inches higher. The most alpine rhododendron (*R. nivale*) spread its small rigid branches close to the ground; the hemispherical *Arenaria*, another type of sterility, rose here and there, and tufts of *Myosotis, Artemisia, Astragali*, and *Androsace*, formed flat cushions level with the soil. Grass was very scarce, but a running wiry sedge (*Carex Moorcroftii*) bound the sand, like the *Carex arenaria* of our English coasts.

A more dismally barren country cannot well be conceived, nor one more strongly contrasting with the pastures of Palung at an equal elevation. The long lofty wall of Kinchinjhow and Donkia presents an effectual barrier to the transmission of moisture to the head of the Lachen valley, which therefore becomes a type of such elevations in Tibet. As I proceeded, the ground became still more sandy, chirping under the pony's feet; and where harder, it was burrowed by innumerable marmots, foxes, and the "Goomchen," or tail-less rat (*Lagomys badius*), sounding hollow to the tread, and at last becoming so dangerous that I was obliged to dismount and walk.

The geological features changed as rapidly as those of the climate and vegetation, for the strike of the rocks being north-west, and the dip north-east, I was rising over the strata that overlie the gneiss. The upper part of Kinchinjhow is composed of bold ice-capped cliffs of gneiss; but the long spurs that stretch northwards from it are of quartz, conglomerates, slates, and earthy red clays, forming the rounded terraced hills I had seen from Donkia pass. Between these spurs were narrow valleys, at whose mouths stupendous blocks of gneiss rest on rocks of a much later geological formation.

Opposite the most prominent of these spurs the river (16,800 feet above the sea) runs west, forming marshes, which were full of *Zannichellia palustris* and *Ranunculus aquatilis*, both English and Siberian plants: the waters contained many shells, of a species of *Lymnæa;* * and the soil

---

* This is the most alpine living shell in the world; my specimens being from nearly 17,000 feet elevation; it is the *Lymnæa Hookeri*, Reeve ("Proceedings of the Zoological Society," No. 204).

near the edge, which was covered with tufts of short grass, was whitened with effloresced carbonate of soda. Here were some square stone enclosures two feet high, used as pens, and for pitching tents in; within them I gathered some unripe barley.

Beyond this I recognised a hill of which I had taken bearings from Donkia pass, and a few miles further, on rounding a great spur of Kinchinjunga, I arrived in sight of Cholamoo lakes, with the Donkia mountain rearing its stupendous precipices of rock and ice on the east. My pony was knocked up, and I felt very giddy from the exertion and elevation; I had broken his bridle, and so led him on by my plaid for the last few miles to the banks of the lake; and there, with the pleasant sound of the waters rippling at my feet, I yielded for a few moments to those emotions of gratified ambition which, being unalloyed by selfish considerations for the future, become springs of happiness during the remainder of one's life.

The landscape about Cholamoo lakes was simple in its elements, stern and solemn; and though my solitary situation rendered it doubly impressive to me, I doubt whether the world contains any scene with more sublime associations than this calm sheet of water, 17,000 feet above the sea, with the shadows of mountains 22,000 to 24,000 feet high, sleeping on its bosom.

There was much short grass about the lake, on which large antelopes, "Chiru" (*Antilope Hodgsoni*),* and deer, "Goa" (*Procapra picticaudata*, Hodgson), were feeding. There were also many slate-coloured hares with white rumps

---

* I found the horns of this animal on the south side of the Donkia pass, but I never saw a live one except in Tibet. The *Procapra* is described by Mr. Hodgson, "Bengal As. Soc. Jour., 1846, p. 338," and is introduced into the cut at p. 139.

(*Lepus oiostolus*), with marmots and tail-less rats. The
abundance of animal life was wonderful, compared with the
want of it on the south side of Donkia pass, not five
miles distant in a straight line! it is partly due to the

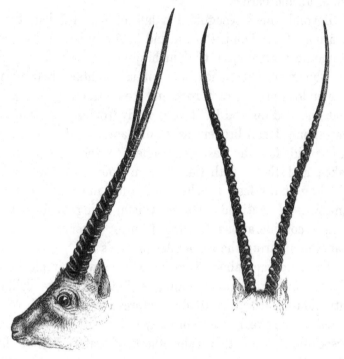

ANTELOPE'S HEAD.*

profusion of carbonate of soda, of which all ruminants are
fond, and partly to the dryness of the climate, which is
favourable to all burrowing quadrupeds. A flock of
common English teal were swimming in the lake, the
temperature of which was 55°.

* The accompanying figures of the heads of the Chiru (Antilope Hodgsoni),
were sketched by Lieut. Maxwell (of the Bengal Artillery), from a pair brought to
Dorjiling; it is the so-called unicorn of Tibet, and of MM. Huc and Gabet's
narrative,—a name which the profile no doubt suggested.

I had come about fifteen miles from the pass, and arrived at 1 P.M., remaining half an hour. I could not form an idea as to whether Campbell had followed or not, and began to speculate on the probability of passing the night in the open air, by the warm side of my steed. Though the sun shone brightly, the wind was bitterly cold, and I arrived at the stone dykes of Yeumtso at 3 .P.M., quite exhausted with fatigue and headache. I there found, to my great relief, the Tchebu Lama and Lachen Phipun: they were in some alarm at my absence, for they thought I was not aware of the extreme severity of the temperature on the north side of the snows, or of the risk of losing my way; they told me that after a long discourse with the Dingpun (or commander) of the Tibetan Sepoys, the latter had allowed all the party to pass; that the Sepoys had brought on the coolies, who were close behind, but that they themselves had seen nothing of Campbell; of whom the Lama then went in search.

The sun set behind Chomiomo at 5 P.M., and the wind at once dropped, so local are these violent atmospheric currents, which are caused by the heating of the upper extremities of these lofty valleys, and consequent rarefaction of the air. Intense terrestrial radiation immediately follows the withdrawal of the sun's rays, and the temperature sinks rapidly.

Soon after sunset the Lama returned, bringing Campbell; who, having mistaken some glacier-fed lakes at the back of Kinchinjhow for those of Cholamoo, was looking for me. He too had speculated on having to pass the night under a rock, with one plaid for himself and servant; in which case I am sure they would both have been frozen to death, having no pony to lie down beside. He told me that after I had quitted Kongra Lama, leaving him with

the Tchebu Lama and Phipun, the Dingpun and twenty
men came up, and very civilly but formally forbade their
crossing the frontier; but that upon explaining his motives,
and representing that it would save him ten days' journey,
the Dingpun had relented, and promised to conduct the
whole party to the Donkia pass.

We pitched our little tent in the corner of the cattle-pen,
and our coolies soon afterwards came up; mine were in
capital health, though suffering from headaches, but
Campbell's were in a distressing state of illness and fatigue,
with swollen faces and rapid pulses, and some were insen-
sible from symptoms like pressure on the brain; * these
were chiefly Ghorkas (Nepalese). The Tibetan Dingpun and
his guard arrived last of all, he was a droll little object, short,
fat, deeply marked with small-pox, swarthy, and greasy, he
was robed in a green woollen mantle, and was perched on
the back of a yak, which also carried his bedding, and
cooking utensils, the latter rattling about its flanks, horns,
neck, and every point of support: two other yaks bore the
tents of the party. His followers were tall savage looking
fellows, with broad swarthy faces, and their hair in short
pig-tails. They wore the long-sleeved cloak, short trousers,
and boots, all of thick woollen, and felt caps on their
heads. Each was armed with a long matchlock slung
over his back, with a moveable rest having two prongs
like a fork, and a hinge, so as to fold up along the barrel,
when the prongs project behind the shoulders like antelope
horns, giving the uncouth warrior a droll appearance.

* I have never experienced bleeding at the nose, ears, lips or eyelids, either in
my person or that of my companions, on these occasions; nor did I ever meet
with a recent traveller who has. Dr. Thomson has made the same remark, and
when in Switzerland together we were assured by Auguste Balmat, François
Coutet, and other experienced Mont Blanc guides, that they never witnessed
these symptoms nor the blackness of the sky, so frequently insisted upon by
alpine travellers.

A dozen cartridges, each in an iron case, were slung round the waist, and they also wore the long knife, flint, steel, and iron tobacco-pipe, pouch, and purse, suspended to a leathern girdle.

The night was fine, but intensely cold, and the vault of heaven was very dark, and blazing with stars; the air was electrical, and flash lightning illumined the sky; this was the reflection of a storm that was not felt at Dorjiling, but which raged on the plains of India, beyond the Terai, fully 120 miles, and perhaps 150, south of our position. No thunder was heard. The thermometer fell to 5°, and that in the reflector to 3° 5 ; at sunrise it rose to 10°, and soon after 8 A.M. to 33°; till this hour the humidity was great, and a thin mist hung over the frozen surface of the rocky ground; when this dispersed, the air became very dry, and the black-bulb thermometer in the sun rose 60° above the temperature in the shade. The light of the sun, though sometimes intercepted by vapours aloft, was very brilliant.[*]

This being the migrating season, swallows flitted through the air; finches, larches, and sparrows were hopping over the sterile soil, seeking food, though it was difficult to say what. The geese [†] which had roosted by the river, cackled; the wild ducks quacked and plumed themselves; ouzels and waders screamed or

---

[*] My black glass photometer shut out the sun's disc at 10·509 inches, from the mean of four sets of observations taken between 7 and 10 A.M.

[†] An enormous quantity of water-fowl breed in Tibet, including many Indian species that migrate no further north. The natives collect their eggs for the markets at Jigatzi, Giantchi, and Lhassa, along the banks of the Yaru river, Ramchoo, and Yarbru and Dochen lakes. Amongst other birds the Sara, or great crane of India (see "Turner's Tibet," p. 212), repairs to these enormous elevations to breed. The fact of birds characteristic of the tropics dwelling for months in such climates is a very instructive one, and should be borne in mind in our speculations upon the climate supposed to be indicated by the imbedded bones of birds.

chirped ; and all rejoiced as they prepared themselves for the last flight of the year, to the valleys of the southern Himalaya, to the Teesta, and other rivers of the Terai and plains of India.

The Dingpun paid his respects to us in the morning, wearing, besides his green cloak, a white cap with a green glass button, denoting his rank ; he informed us that he had written to his superior officer at Kambajong, explaining his motives for conducting us across the frontier, and he drew from his breast a long letter, written on *Daphne** paper, whose ends were tied with floss silk, with a large red seal ; this he pompously delivered, with whispered orders, to an attendant, and sent him off. He admired our clothes extremely,† and then my percussion gun, the first he had seen ; but above all he admired rum and water, which he drank with intense relish, leaving a mere sip for his comrades at the bottom of his little wooden cup, which they emptied, and afterwards licked clean, and replaced in his breast for him. We made a large basin full of very weak grog for his party, who were all friendly and polite ; and having made us the unexpected offer of allowing us to rest ourselves for the day at Yeumtso, he left us, and practised his men at firing at a mark, but they were very indifferent shots.

I ascended with Campbell to the lake he had visited

---

* Most of the paper used in Tibet is, as I have elsewhere noticed, made from the bark of various species of *Daphneæ*, and especially of *Edgeworthia Gardneri*, and is imported from Nepal and Bhotan ; but the Tibetans, as MM. Huc and Gabet correctly state, manufacture a paper from the root of a small shrub : this I have seen, and it is of a much thicker texture and more durable than Daphne paper. Dr. Thomson informs me that a species of *Astragalus* is used in western Tibet for this purpose, the whole shrub, which is dwarf, being reduced to pulp.

† All Tibetans admire and value English broad-cloth beyond any of our products. Woollen articles are very familiar to them, and warm clothing is one of the first requisites of life.

on the previous day, about 600 or 800 feet above Yeumtso, and 17,500 feet above the sea : it is a mile and a half long, and occupies a large depression between two rounded spurs, being fed by glaciers from Kinchinjhow. The rocks of these spurs were all of red quartz and slates, cut into broad terraces, covered with a thick glacial talus of gneiss and granite in angular pebbles, and evidently spread over the surface when the glacier, now occupying the upper end of the lake, extended over the valley.

The ice on the cliffs and summit of Kinchinjhow was much greener and clearer than that on the south face (opposite Palung) ; and rows of immense icicles hung from the cliffs. A conferva grew in the waters of the lake, and short, hard tufts of sedge on the banks, but no other plants were to be seen.   Brahminee geese, teal, and widgeon, were swimming in the waters, and a beetle (*Elaphrus*) was coursing over the wet banks ; finches and other small birds were numerous, eating the sedge-seeds, and picking up the insects.   No view was obtained to the north, owing to the height of the mountains on the north flank of the Lachen.

At noon the temperature rose to 52° 5, and the black-bulb to 104° 5 ; whilst the north-west dusty wind was so dry, that the dew-point fell to 24° 2.

In the afternoon we crossed the valley, and ascended
Bhomtso, fording the river, whose temperature was 48°
Some stupendous boulders of gneiss from Kinchinjhow are
deposited in a broad sandy track on the north bank, by
ancient glaciers, which once crossed this valley from Kinchinjhow.

The ascent was alternately over steep rocky slopes, and
broad shelf-like flats ; many more plants grew here than I
had expected, in inconspicuous scattered tufts.*   The rocks

* Besides those before mentioned, there were Fescue-grass (*Festuca ovina* of
Scotland), a strong-scented silky wormwood (*Artemisia*), and round tufts of
Oxytropis chiliophylla, a kind of *Astragalus* that inhabits eastern and western Tibet ;
this alone was green : it formed great circles on the ground, the centre decaying,
and the annual shoots growing outwards, and thus constantly enlarging the circle.
A woolly *Leontopodium*, *Androsace*, and some other plants assumed nearly the
same mode of growth. The rest of the vegetation consisted of a *Sedum*, *Nardo-
stachys Jatamansi*, *Meconopsis horridula*, a slender *Androsace*, *Gnaphalium*, *Stipa*,
Salvia, *Draba*, *Pedicularis*, *Potentilla* or *Sibbaldia*, *Gentiana* and *Erigeron alpinus* of
Scotland. All these grow nearly up to 18,000 feet.

were nearly vertical strata of quartz, hornstone, and conglomerate, striking north-west, and dipping south-west 80°
The broad top of the hill was also of quartz, but covered
with angular pebbles of the rocks transported from
Kinchinjhow. Some clay-stone fragments were stained
red with oxide of iron, and covered with *Parmelia
miniata*; * this, with *Borrera*, another lichen, which forms
stringy masses blown along by the wind, were the only
plants, and they are among the most alpine in the world.

Bhomtso is 18,590 feet above the sea by barometer, and
18,305 by boiling-point: it presented an infinitely more
extensive prospect than I had ventured to anticipate, commanding all the most important Sikkim, North Bhotan, and
Tibetan mountains, including Kinchinjunga thirty-seven
miles to the south-west, and Chumulari thirty-nine miles
south-east. Due south, across the sandy valley of the
Lachen, Kinchinjhow reared its long wall of glaciers and
rugged precipices, 22,000 feet high, and under its cliffs lay
the lake to which we had walked in the morning: beyond
Kongra Lama were the Thlonok mountains, where I had
spent the month of June, with Kinchinjunga in the distance.
Westward Chomiomo rose abruptly from the rounded hills
we were on, to 22,000 feet elevation, ten miles distant. To
the east of Kinchinjhow were the Cholamoo lakes, with the
rugged mass of Donkia stretching in cliffs of ice and snow
continuously southwards to forked Donkia, which overhung
Momay Samdong.

A long sloping spur sweeps from the north of Donkia first
north, and then west to Bhomtso, rising to a height of more

---

* This minute lichen, mentioned at p. 130, is the most Arctic, Antarctic, and
Alpine in the world; often occurring so abundantly as to colour the rocks of an
orange red. This was the case at Bhomtso, and is so also in Cockburn Island in
the Antarctic ocean, which it covers so profusely that the rocks look as if brightly
painted. See "Ross's Voyage," vol. ii. p. 339.

than 20,000 feet without snow. Over this spur the celebrated Chumulari * peeps, bearing south-east, and from its
isolated position and sharpness looking low and small; it
appeared quite near, though thirty-nine miles distant.

North-east of Chumulari, and far beyond it, are several
meridional ranges of very much loftier snowy mountains,
which terminated the view of the snowy Himalaya; the
distance embraced being fully 150 miles, and perhaps much
more. Of one of these eastern masses † I afterwards took

---

* Some doubt still hangs over the identity of this mountain, chiefly owing to
Turner's having neglected to observe his geographical positions. I saw a much
loftier mountain than this, bearing from Bhomtso north 87° east, and it was called
Chumulari by the Tibetan Sepoys; but it does not answer to Turner's description
of an isolated snowy peak, such as he approached within three miles; and though
in the latitude he assigned to it, is fully sixty miles to the east of his route. A
peak, similar to the one he describes, is seen from Tonglo and Sinchul (see vol. i.
pp. 125 and 185); this is the one alluded to above, and it is identified by both
Tibetans and Lepchas at Dorjiling as the true Chumulari, and was measured by
Colonel Waugh, who placed it in lat. 27° 49′ north, long. 89° 18′ east. The latter
position, though fifteen miles south of what Turner gives it, is probably correct;
as Pemberton found that Turner had put other places in Bhotan twenty miles too
far north. Moreover, in saying that it is visible from Purnea in the plains of
Bengal, Turner refers to Kinchinjunga, whose elevation was then unknown.
Dr. Campbell ("Bengal As. Soc. Jour.," 1848), describes Chumulari from oral
information, as an isolated mountain encircled by twenty-one goompas, and perambulated by pilgrims in five days; the Lachoong Phipun, on the other hand,
who was a Lama, and well acquainted with the country, affirmed that Chumulari
has many tops, and cannot be perambulated; but that detached peaks near it may
be, and that it is to a temple near one of these that pilgrims resort. Again, the
natives use these names very vaguely, and as that of Kinchinjunga is often applied
equally to all or any part of the group of snows between the Lachen and Tambur
rivers, so may the term Chumulari have been used vaguely to Captain Turner or
to me. I have been told that an isolated, snow-topped, venerated mountain rises
about twenty miles south of the true Chumulari, and is called "Sakya-khang"
(Sakya's snowy mountain), which may be that seen from Dorjiling; but I incline to
consider Campbell's and Waugh's mountain as the one alluded to by Turner, and
it is to it that I here refer as bearing north 115° 30′ east from Bhomtso.

† These are probably the Ghassa mountains of Turner's narrative: bearings
which I took of one of the loftiest of them, from the Khasia mountains,
together with those from Bhomtso, would appear to place it in latitude
28° 10′ and longitude 90°, and 200 miles from the former station, and 90° east of
the latter. Its elevation from Bhomtso angles is 24,160 feet. I presume I also
saw Chumulari from the Khasia;· the most western peak seen thence being
in the direction of that mountain. Captain R. Strachey has most kindly

Pl IX.

J D.H.del.ᵗ

John Murray Albemarle Street, 1854.

W.L.Walton, lith.

Cholamo Lake & Lachen River, from Bomtso.

bearings and angular heights from the Khasia mountains, in Bengal, upwards of 200 miles south-east of its position.

Turning to the northward, a singular contrast in the view was presented : the broad sandy valley of the Arun lay a few miles off, and perhaps 1,500 feet below me; low brown and red ridges, 18,000 to 19,000 feet high, of stony sloping mountains with rocky tops, divided its feeders, which appeared to be dry, and to occupy flat sandy valleys. For thirty miles north no mountain was above the level of the theodolite, and not a particle of snow was to be seen : beyond that, rugged purple-flanked and snowy-topped mountains girdled the horizon, appearing no nearer than they did from the Donkia pass, and their angular heights and bearings being almost the same as from that point of view. The nearer of these are said to form the Kiang-lah chain, the furthest I was told by different authorities are in the salt districts north of Jigatzi.

To the north-east was the lofty region traversed by Turner on his route by the Ramchoo lakes to Teshoo Loombo; its elevation may be 17,000 feet * above the sea. Beyond it a gorge led through rugged mountains, by which I was told the Painom river flows north-west to the Yaru; and at an immense distance to the north-east were the Khamba mountains, a long blue range, which it is said

paid close attention to these bearings and distances, and recalculated the distances and heights : no confidence is, however, to be placed in the results of such minute angles, taken from immense distances. Owing in part no doubt to extraordinary refraction, the angles of the Ghassa mountain taken from the Khasia give it an elevation of 26,500 feet ! which is very much over the truth; and make that of Chumulari still higher : the distance from my position in the Khasia being 210 miles from Chumulari ! which is probably the utmost limit at which the human eye has ever discerned a terrestrial object.

* It is somewhat remarkable that Turner nowhere alludes to difficulty of breathing, and in one place only to head-ache (p. 209) when at these great elevations. This is in a great measure accounted for by his having been constantly mounted. I never suffered either in my breathing, head, or stomach when riding, even when at 18,300 feet.

divides the Lhassan or "U" from the "Tsang" (or
Jigatzi) province of Tibet; it appeared fully 100 miles off,
and was probably much more; it bore from N. 57° E. to
N. 70° E., and though so lofty as to be heavily snowed
throughout, was much below the horizon-line of Bhomtso;
it is crossed on the route from Jigatzi, and from Sikkim
to Lhassa,* and is considered very lofty, from affecting
the breathing. About twenty miles to the north-east are
some curious red conical mountains, said to be on the west
side of the Ramchoo lakes; they were unsnowed, and bore
N. 45° 30′ E. and N. 60° 30′ E. A sparingly-snowed group
bore N. 26° 30′ E., and another N. 79° E., the latter
being probably that mentioned by Turner as seen by him
from near Giantchi.

But the mountains which appeared both the highest and
the most distant on the northern landscape, were those I
described when at Donkia, as being north of Nepal and
beyond the Arun river, and the culminant peak of which
bore N. 55° W. Both Dr. Campbell and I made repeated
estimates of its height and distance by the eye; comparing
its size and snow-level with those of the mountains near us;
and assuming 4000 to 5000 feet as the minimum height
of its snowy cap; this would give it an elevation of 23,000
to 25,000 feet. An excellent telescope brought out no
features on its flanks not visible to the naked eye, and by
the most careful levellings with the theodolite, it was
depressed more than 0° 7′ below the horizon of Bhomtso,
whence the distance must be above 100 miles.

The transparency of the pale-blue atmosphere of these

---

* Lhassa, which lies north-east, may be reached in ten days from this, with
relays of ponies; many mountains are crossed, where the breath is affected, and
few villages are passed after leaving Giantchi, the "Jhansi jeung" of Turner's
narrative. See Campbell's "Routes from Dorjiling to Lhassa." ("Bengal As.
Soc. Journal.")

lofty regions can hardly be described, nor the clearness and precision with which the most distant objects are projected against the sky. From having afterwards measured peaks 200 and 210 miles distant from the Khasia mountains, I feel sure that I underrated the estimates made at Bhomtso, and I have no hesitation in saying, that the mean elevation of the sparingly-snowed * watershed between the Yaru and the Arun will be found to be greater than that of the snowy Himalaya south of it, and to follow the chain running from Donkia, north of the Arun, along the Kiang-lah mountains, towards the Nepal frontier, at Tingri Maidan. No part of that watershed perhaps rises so high as 24,000 feet, but its lowest elevation is probably nowhere under 18,000 feet.

This broad belt of lofty country, north of the snowy Himalaya, is the Dingcham province of Tibet, and runs along the frontier of Sikkim, Bhotan, and Nepal. It gives rise to all the Himalayan rivers, and its mean elevation is probably 15,000 to 15,500 feet: its general appearance, as seen from greater heights, is that of a much less mountainous country than the snowy and wet Himalayan regions; this is because its mean elevation is so enormous, that ranges of 20,000 to 22,000 feet appear low and insignificant upon it. The absence of forest and other obstructions to the view, the breadth and flatness of the valleys, and the undulating character of the lower ranges that

---

* Were the snow-level in Dingcham, as low as it is in Sikkim, the whole of Tibet from Donkia almost to the Yaru-Tsampu river would be everywhere intersected by glaciers and other impassable barriers of snow and ice, for a breadth of fifty miles; and the country would have no parallel for amount of snow beyond the Polar circles. It is impossible to conjecture what would have been the effects on the climate of northern India and central Asia under these conditions. When, however, we reflect upon the evidences of glacial phenomena that abound in all the Himalayan valleys at and above 9000 feet elevation, it is difficult to avoid the conclusion that such a state of things once existed, and that at a comparatively very recent period.

traverse its surface, give it a comparatively level appearance, and suggest the term "maidan" or "plains" to the Tibetan, when comparing his country with the complicated ridges of the deep Sikkim valleys. Here one may travel for many miles without rising or falling 3000 feet, yet never descending below 14,000 feet, partly because the flat winding valleys are followed in preference to exhausting ascents, and partly because the passes are seldom more than that elevation above the valleys; whereas, in Sikkim, rises and descents of 6000, and even 9000 feet, are common in passing from valley to valley, sometimes in one day's march.

The swarthy races of Dingcham have been elsewhere described; they are an honest, hospitable, and very hardy people, differing from the northern Tibetans chiefly in colour, and in invariably wearing the pigtail, which MM. Huc and Gabet assure us is not usual in Lhassa.* They are a pastoral race, and Campbell saw a flock of 400 hornless sheep, grazing on short sedges (*Carex*) and fescue-grass, in the middle of October, at 18,000 feet above the sea. An enormous ram attended the flock, whose long hair hung down to the ground; its back was painted red.

There is neither tree nor shrub in this country; and a very little wheat (which seldom ripens), barley, turnips, and radishes are, I believe, the only crops, except occasionally

---

* Amongst Lhassan customs alluded to by these travellers, is that of the women smearing their faces with a black pigment, the object of which they affirm to be that they may render themselves odious to the male sex, and thus avoid temptation. The custom is common enough, but the real object is to preserve the skin, which the dry cold wind peels from the face. The pigment is mutton-fat, blackened, according to Tchebu Lama, with catechu and other ingredients; but I believe more frequently by the dirt of the face itself. I fear I do not slander the Tibetan damsels in saying that personal cleanliness and chastity are both lightly esteemed amongst them; and as the Lama naïvely remarked, when questioned on the subject, "the Tibetan women are not so different from those of other countries as to wish to conceal what charms they possess."

peas. Other legumes, cabbages, &c., are cultivated in the sheltered valleys of the Yaru feeders, where great heat is reflected from the rocks; and there also stunted trees grow, as willows, walnuts, poplars, and perhaps ashes ; all of which, however, are said to be planted and scarce. Even at Teshoo Loombo and Jigatzi * buckwheat is a rare crop, and only a prostrate very hardy kind is grown. Clay teapots and pipkins are the most valuable exports to Sikkim from the latter city, after salt and soda. Jewels and woollen cloaks are also exported, the latter especially from Giantchi, which is famous for its woollen fabrics and mart of ponies.

Of the Yaru river at Jigatzi, which all affirm becomes the Burrampooter in Assam, I have little information to add to Turner's description : it is sixty miles north of Bhomtso, and I assume its elevation to be 13—14,000 feet ;† it takes an immense bend to the northward after

---

* Digarchi, Jigatzi, or Shigatzi jong (the fort of Shigatzi) is the capital of the "Tsang" province, and Teshoo Loombo is the neighbouring city of temples and monasteries, the ecclesiastical capital of Tibet, and the abode of the grand (Teshoo) Lama, or ever-living Boodh. Whether we estimate this man by the number of his devotees, or the perfect sincerity of their worship, he is without exception one of the most honoured beings living in the world. I have assumed the elevation of Jigatzi to be 13—14,000 feet, using as data Turner's October mean temperature of Teshoo Loombo, and the decrement for elevation of 400 feet to 1° Fahr.; which my own observations indicate as an approximation to the truth. Humboldt ("Asie Centrale," iii., p. 223) uses a much smaller multiplier, and infers the elevation of Teshoo Loombo to be between 9,500 and 10,000 feet. Our data are far too imperfect to warrant any satisfactory conclusions on this interesting subject ; but the accounts I have received of the vegetation of the Yaru valley at Jigatzi seem to indicate an elevation of at least 13,000 feet for the bed of that river. Of the elevation of Lhassa itself we have no idea: if MM. Huc and Gabet's statement of the rivers not being frozen there in March be correct, the climate must be very different from what we suppose.

† The Yaru, which approaches the Nepal frontier west of Tingri, and beyond the great mountain described at vol. i. p. 265, makes a sweep to the northward, and turns south to Jigatzi, whence it makes another and greater bend to the north, and again turning south flows west of Lhassa, receiving the Kechoo river from that holy city. From Jigatzi it is said to be navigable to near Lhassa by skin and plank-built boats. Thence it flows south-east to the Assam frontier, and while still in Tibet, is said to enter a warm climate, where tea, silk, cotton, and rice, are grown. Of its course after entering the Assam Himalaya little is known, and in

passing Jigatzi, and again turns south, flowing to the west of Lhassa, and at some distance from that capital. Lhassa, as all agree, is at a much lower elevation than Jigatzi ; and apricots (whose ripe stones Dr. Campbell procured for me) and walnuts are said to ripen there, and the Dama or Himalayan furze (*Caragana*), is said to grow there. The Bactrian camel also thrives and breeds at Lhassa, together with a small variety of cow (not the yak), both signs of a much more temperate climate than Jigatzi enjoys. It is, however, a remarkable fact that there are two tame elephants near the latter city, kept by the Teshoo Lama. They were taken to Jigatzi, through Bhotan, by Phari ; and I have been informed that they have become clothed with long hair, owing to the cold of the climate; but Tchebu Lama contradicted this, adding, that his countrymen were so credulous, that they would believe blankets grew on the elephants' backs, if the Lamas told them so.

No village or house is seen throughout the extensive area over which the eye roams from Bhomtso, and the general character of the desolate landscape was similar to that which I have described as seen from Donkia Pass (p. 124). The wild ass * grazing with its foal on the sloping downs,

---

answer to my enquiries why it had not been followed, I was always told that the country through which it flowed was inhabited by tribes of savages, who live on snakes and vermin, and are fierce and warlike. These are no doubt the Singpho, Bor and Bor-abor tribes who inhabit the mountains of upper Assam. A travelling mendicant was once sent to follow up the Dihong to the Burrampooter, under the joint auspices of Mr. Hodgson and Major Jenkins, the Commissioner of Assam; but the poor fellow was speared on the frontier by these savages. The concurrent testimony of the Assamese, that the Dihong is the Yaru, on its southern course to become the Burrampooter, renders this point as conclusively settled as any, resting on mere oral evidence, is likely to be.

* This, the *Equus Hemionus* of Pallas, the untameable Kiang of Tibet, abounds in Dingcham, and we saw several. It resembles the ass more than the horse, from its size, heavy head, small limbs, thin tail, and the stripe over the shoulder. The flesh is eaten and much liked. The Kiang-lah mountains are so named from their being a great resort of this creature. It differs widely from the wild ass of Persia, Sind, and Beloochistan, but is undoubtedly the same as the Siberian animal.

the hare bounding over the stony soil, the antelope scouring the sandy flats, and the fox stealing along to his burrow, are all desert and Tartarian types of the animal creation. The shrill whistle of the marmot alone breaks the silence of the scene, recalling the snows of Lapland to the mind; the kite and raven wheel through the air, 1000 feet over head, with as strong and steady a pinion as if that atmosphere possessed the same power of resistance that it does at the level of the sea.   Still higher in the heavens, long black V-shaped trains of wild geese cleave the air, shooting over the glacier-crowned top of Kinchinjhow, and winging their flight in one day, perhaps, from the Yaru to the Ganges, over 500 miles of space, and through 22,000 feet of elevation.   One plant alone, the yellow lichen (*Borrera*), is found at this height, and only as a visitor; for, Tartar-like, it emigrates over these lofty slopes and ridges, blown about by the violent winds.   I found a small beetle on the very top,* probably blown up also, for it was a flower-feeder, and seemed benumbed with cold.

Every night that we spent in Tibet, we enjoyed a magnificent display of sunbeams converging to the east, and making a false sunset.   I detailed this phenomenon when seen from the Kymore mountains, and I repeatedly saw it again in the Khasia, but never in the Sikkim Himalaya, whence I assume that it is most frequent in mountain plateaus.   As the sun set, broad purple beams rose from a dark, low, leaden bank on the eastern horizon, and spreading up to the zenith, covered the intervening space : they lasted through the twilight, from fifteen to twenty minutes, fading gradually into the blackness of

---

* I observed a small red *Acarus* (mite) at this elevation, both on Donkia and Kinchinjhow, which reminds me that I found a species of the same genus at Cockburn Island (in latitude 64° south, longitude 64° 49 west). This genus hence inhabits a higher southern latitude than any other land animal attains.

night.   I looked in vain for the beautiful lancet beam of
the zodiacal light; its position was obscured by Chomiomo.

On the 18th of October we had another brilliant
morning, after a cold night, the temperature having fallen
to 4°.   I took the altitude of Yeumtso by carefully boiling
two thermometers, and the result was 16,279 feet, the
barometrical observations giving 16,808 feet.   I removed a
thermometer sunk three feet in the gravelly soil, which
showed a temperature of 43°,* which is 12° 7 above the
mean temperature of the two days we camped here.

Our fires were made of dry yak droppings which soon
burn out with a fierce flame, and much black smoke; they
give a disagreeable taste to whatever is cooked with them.

Having sent the coolies forward to Cholamoo lake, we
re-ascended Bhomtso to verify my observations.   As on
the previous occasion a violent dry north-west wind blew,
peeling the skin from our faces, loading the air with grains
of sand, and rendering theodolite observations very uncertain; besides injuring all my instruments, and exposing
them to great risk of breakage.

The Tibetan Sepoys did not at all understand our
ascending Bhomtso a second time; they ran after
Campbell, who was ahead on a stout pony, girding up
their long garments, bracing their matchlocks tight over
their shoulders, and gasping for breath at every step, the
long horns of their muskets bobbing up and down as they
toiled amongst the rocks.   When I reached the top I
found Campbell seated behind a little stone wall which
he had raised to keep off the violent wind, and the
uncouth warriors in a circle round him, puzzled beyond
measure at his admiration of the view.   My instruments
perplexed them extremely, and in crowding round me, they

---

* It had risen to 43° 5 during the previous day.

broke my azimuth compass. They left us to ourselves when the fire I made to boil the thermometers went out, the wind being intensely cold. I had given my barometer to one of Campbell's men to carry, who not coming up, the latter kindly went to search for him, and found him on the ground quite knocked up and stupified by the cold, and there, if left alone, he would have lain till overtaken by death.

The barometer on the summit of Bhomtso stood at 15·548 inches;* the temperature between 11·30 A.M. and 2·30 P.M. fluctuated between 44° and 56°: this was very high for so great an elevation, and no doubt due to the power of the sun on the sterile soil, and consequent radiated heat. The tension of vapour was ·0763, and the dew-point was 5° 8, or 43° 5 below the temperature of the air. Such extraordinary dryness † and consequent evaporation, increased by the violent wind, sufficiently accounts for the height of the snow line; in further evidence of which, I may add that a piece of ice or snow laid on the ground here, does not melt, but disappears by evaporation.

The difference between the dry cold air of this elevation and that of the heated plains of India, is very great. During the driest winds of the Terai, in spring, the temperature is 80° to 90°, the tension of vapour is ·400 to ·500, with a dew-point 22° below the temperature, and upwards of six grains of vapour are suspended in the cubic foot of air; a thick haze obscures the heavens, and clouds of dust rise high in the air; here on the other hand (probably

---

* The elevation of Bhomtso, worked by Bessel's tables, and using corrected observations of the Calcutta barometer for the lower station, is 18,590 feet. The corresponding dew-point 4° 4 (49° 6 below that of the air at the time of observation). By Oltmann's tables the elevation is 18,540 feet. The elevation by boiling water is 18,305.

† The weight of vapour in a cubic foot of air was no more than $\frac{87}{1000}$ of a grain, and the saturation-point ·208.

owing to the rarity of the atmosphere and the low tension of its vapours), the drought is accompanied by perfect transparency, and the atmosphere is too attenuated to support the dust raised by the wind.

We descended in the afternoon, and on our way up the Lachen valley examined a narrow gulley in a lofty red spur from Kinchinjhow, where black shales were *in situ*, striking north-east, and dipping north-west 45°. These shales were interposed between beds of yellow quartz conglomerate, upon the latter of which rested a talus of earthy rocks, angular fragments of which were strewed about opposite this spur, but were not seen elsewhere.

It became dark before we reached the Cholamoo lake, where we lost our way amongst glaciers, moraines, and marshes. We expected to have seen the lights of the camp, but were disappointed, and as it was freezing hard, we began to be anxious, and shouted till the echos of our voices against the opposite bank were heard by Tchebu Lama, who met us in great alarm for our safety. Our camp was pitched some way from the shore, on a broad plain, 16,900 feet above the sea.* A cold wind descended from Donkia; yet, though more elevated than Yeumtso, the climate of Cholamoo, from being damper and misty, was milder. The minimum thermometer fell to 14.°

Before starting for Donkia pass on the following morning, we visited some black rocks which rose from the flat to the east of the lake. They proved to be of fossiliferous limestone, the strata of which were much disturbed : the strike

---

* This, which is about the level of the lake, gives the Lachen river a fall of about 1500 feet between its source and Kongra Lama, or sixty feet per mile following its windings. From Kongra Lama to Tallum it is 140 feet per mile; from Tallum to Singtam 160 feet; and from Singtam to the plains of India 50 feet per mile. The total fall from Cholamoo lake to its exit on the plains of India is eighty-five feet per mile. Its length, following its windings, is 195 miles, upwards of double the direct distance.

appeared in one part north-west, and the dip north east 45°: a large fault passed east by north through the cliff, and it was further cleft by joints running northwards. The cliff was not 100 yards long, and was about 70 thick; its surface was shivered by frost into cubical masses, and glacial boulders of gneiss lay on the top. The limestone rock was chiefly a blue pisolite conglomerate, with veins and crystals of white carbonate of lime, seams of shale, and iron pyrites. A part was compact and blue, very crystalline, and full of encrinitic fossils, and probably nummulites, but all were too much altered for determination.

This, from its mineral characters, appears to be the same limestone formation which occurs throughout the Himalaya and Western Tibet; but the fossils I collected are in too imperfect a state to warrant any conclusions on this subject. Its occurrence immediately to the northward of the snowy mountains, and in such very small quantities, are very remarkable facts. The neighbouring rocks of Donkia were gneiss with granite veins, also striking north-west and dipping north-east 10°, as if they overlay the limestone, but here as in all similar situations there was great confusion of the strata, and variation in direction and strike.

And here I may once for all confess that though I believe the general strike of the rocks on this frontier to be north-west, and the dip north-east, I am unable to affirm it positively; for though I took every opportunity of studying the subject, and devoted many hours to the careful measuring and recording of dips and strikes, on both faces of Kinchinjhow, Donkia, Bhomtso, and Kongra Lama, I am unable to reduce these to any intelligible system.*

The coolies of Dr. Campbell's party were completely

---

* North-west is the prevalent strike in Kumaon, the north-west Himalaya generally, and throughout Western Tibet, Kashmir, &c., according to Dr. Thomson.

knocked up by the rarified air; they had taken a whole day to march here from Yeumtso, scarcely six miles, and could eat no food at night. A Lama of our party offered up prayers * to Kinchinjhow for the recovery of a stout Lepcha lad (called Nurko), who showed no signs of animation, and had all the symptoms of serous apoplexy. The Lama perched a saddle on a stone, and burning incense before it, scattered rice to the winds, invoking Kinchin, Donkia, and all the neighbouring peaks. A strong dose of calomel and jalap, which we poured down the sick lad's throat, contributed materially to the success of these incantations.

The Tibetan Sepoys were getting tired of our delays, which so much favoured my operations; but though showing signs of impatience and sulkiness, they behaved well to the last; taking the sick man to the top of the pass on their yaks, and assisting all the party : nothing, however, would induce them to cross into Sikkim, which they considered as " Company's territory."

Before proceeding to the pass, I turned off to the east, and reascended Donkia to upwards of 19,000 feet, vainly hoping to get a more distant view, and other bearings of the Tibetan mountains. The ascent was over enormous piles of loose rocks split by the frost, and was extremely fatiguing. I reached a peak overhanging a steep precipice, at whose base were small lakes and glaciers, from which flowed several sources of the Lachen, afterwards swelled by the great affluent from Cholamoo lake. A few rocks striking north-east and dipping north-west, projected

---

* All diseases are attributed by the Tibetans to the four elements, who are propitiated accordingly in cases of severe illness. The winds are invoked in cases of affections of the breathing; fire in fevers and inflammations; water in dropsy, and diseases whereby the fluids are affected; and the God of earth when solid organs are diseased, as in liver-complaints, rheumatism, &c. Propitiatory offerings are made to the deities of these elements, but never sacrifices.

at the very summit, with frozen snow amongst them, beyond which the ice and precipices rendered it impossible to proceed : but though exposed to the north, there was no perpetual snow in the ordinary acceptation of the term, and an arctic European lichen (*Lecidea oreina*) grew on the top, so faintly discolouring the rocks as hardly to be detected without a magnifying-glass.

I descended obliquely, down a very steep slope of 35°, over upwards of a thousand feet of débris, the blocks on which were so loosely poised on one another, that it was necessary to proceed with the utmost circumspection, for I was alone, and a false step would almost certainly have been followed by breaking a leg. The alternate freezing and thawing of rain amongst these masses, must produce a constant downward motion in the whole pile of débris (which was upwards of 2000 feet high), and may account for the otherwise unexplained phenomenon of continuous shoots of angular rocks reposing on very gentle slopes in other places.*

The north ascent to the Donkia pass is by a path well selected amongst immense angular masses of rock, and over vast piles of débris : the strike on this, the north face, was again north-east, and dip north-west. I arrived at the top at 3 P.M., throughly fatigued, and found my faithful Lepcha lads (Cheytoong and Bassebo) nestling under a rock with my theodolite and barometers, having been awaiting my arrival in the biting wind for three hours. My pony stood there too, the picture of patience, and laden with

---

* May not the origin of the streams of quartz blocks that fill gently sloping broad valleys several miles long, in the Falkland Islands, be thus explained ? (See "Darwin's Journal," in Murray's Home and Col. Lib.) The extraordinary shifting in the position of my thermometer left among the rocks of the Donkia pass (see p. 129), and the mobile state of the slopes I descended on this occasion, first suggested this explanation to me. When in the Falkland Islands I was wholly unable to offer any explanation of the phenomenon there, to which my attention had been drawn by Mr. Darwin's narrative.

minerals. After repeating my observations, I proceeded to Momay Samdong, where I arrived after dusk. I left a small bottle of brandy and some biscuits with the lads, and it was well I did so, for the pony knocked up before reaching Momay, and rather than leave my bags of stones, they passed the night by the warm flank of the beast, under a rock at 18,000 feet elevation, without other food, fire, or shelter.

I found my companion encamped at Momay, on the spot I had occupied in September; he had had the utmost difficulty in getting his coolies on, as they threw down their light loads in despair, and lying with their faces to the ground, had to be roused from a lethargy that would soon have been followed by death.

We rested for a day at Momay, and on the 20th, attempted to ascend to the Donkia glacier, but were driven back by a heavy snow-storm. The scenery on arriving here, presented a wide difference to that we had left; snow lying at 16,500 feet, whereas immediately to the north of the same mountain there was none at 19,000 feet. Before leaving Momay, I sealed two small glass flasks containing the air of this elevation, by closing with a spirit lamp a very fine capillary tube, which formed the opening to each; avoiding the possibility of heating the contents by the hand or otherwise. The result of its analysis by Mr. Muller (who sent me the prepared flasks), was that it contained 36·538 per cent. in volume of oxygen; whereas his repeated analysis of the air of Calcutta gives 21 per cent. Such a result is too anomalous to be considered satisfactory.

I again visited the Kinchinjhow glacier and hot springs; the water had exactly the same temperature as in the previous month, though the mean temperature of the air was 8° or 9° lower. The minimum thermometer fell to 22°, being 10° lower than it ever fell in September.

We descended to Yeumtong in a cold drizzle, arriving by sunset; we remained through the following day, hoping to explore the lower glacier on the opposite side of the valley: which, however, the weather entirely prevented. I have before mentioned (p. 140) that in descending in autumn from the drier and more sunny rearward Sikkim valleys, the vegetation is found to be most backward in the lowest and dampest regions. On this occasion, I found asters, grasses, polygonums, and other plants that were withered, brown, and seeding at Momay (14,000 to 15,000 feet), at Yeumtong (12,000 feet) green and unripe; and 2000 feet lower still, at Lachoong, the contrast was even more marked. Thus the short backward spring and summer of the Arctic zone is overtaken by an early and forward seed-time and winter: so far as regards the effects of mean temperature, the warmer station is in autumn more backward than the colder. This is everywhere obvious in the prevalent plants of each, and is especially recognisable in the rhododendrons; as the following table shows:—

16,000 to 17,000 feet, *R. nivale* flowers in July; fruits in September =2 months.
13,000 „ 14,000 feet, *R. anthopogon* flowers in June; fruits in Oct. =4 months.
11,000 „ 12,000 feet, *R. campanulatum* flowers in May; fruits in Nov.=6 months.
8,000 „ 9,000 feet, *R. argenteum* flowers in April; fruits in Dec.=8 months.

And so it is with many species of *Compositæ* and *Umbelliferæ*, and indeed of all natural orders, some of which I have on the same day gathered in ripe fruit at 13,000 to 14,000 feet, and found still in flower at 9000 to 10,000 feet. The brighter skies and more powerful and frequent solar radiation at the greater elevations, account for this apparent inversion of the order of nature.*

* The distribution of the seasons at different elevations in the Himalaya gives rise to some anomalies that have puzzled naturalists. From the middle of

I was disappointed at finding the rhododendron seeds
still immature at Yeumtong, for I was doubtful whether the
same kinds might be met with at the Chola pass, which I
had yet to visit; besides which, their tardy maturation
threatened to delay me for an indefinite period in the
country. *Viburnum* and *Lonicera*, however, were ripe
and abundant; the fruits of both are considered poisonous
in Europe, but here the black berries of a species of the
former (called " Nalum ") are eatable and agreeable;
as are those of a *Gualtheria*, which are pale blue, and
called "Kalumbo." Except these, and the cherry mentioned above, there are no other autumnal fruits above
10,000 feet: brambles, strange as it may appear, do not
ascend beyond that elevation in the Sikkim Himalaya,
though so abundant below it, both in species and individuals, and though so typical of northern Europe.

At Lachoong we found all the yaks that had been
grazing till the end of September at the higher elevations,
and the Phipun presented our men with one of a gigantic
size, and proportionally old and tough. The Lepchas

October to that of May, vegetation is torpid above 14,000 feet, and indeed almost
uniformly covered with snow. From November till the middle of April, vegetation is also torpid above 10,000 feet, except that a few trees and bushes do not
ripen all their seeds till December. The three winter months (December, January,
and February) are all but dead above 6000 feet, the earliest appearance of spring
at Dorjiling (7000 feet) being at the sudden accession of heat in March. From
May till August the vegetation at each elevation is (in ascending order) a month
behind that below it; 4000 feet being about equal to a month of summer weather
in one sense. I mean by this, that the genera and natural orders (and sometimes
the species) which flower at 8000 feet in May, are not so forward at 12,000
feet till June, nor at 16,000 feet till July. After August, however, the reverse
holds good; then the vegetation is as forward at 16,000 feet as at 8000
feet. By the end of September most of the natural orders and genera have
ripened their fruit in the upper zone, though they have flowered as late as July;
whereas October is the fruiting month at 12,000, and November below 10,000
feet. Dr. Thomson does not consider that the more sunny climate of the loftier
elevations sufficiently accounts for this, and adds the stimulus of cold, which
must act by checking the vegetative organs and hastening maturation.

barbarously slaughtered it with arrows, and feasted on the flesh and entrails, singed and fried the skin, and made soup of the bones, leaving nothing but the horns and hoofs. Having a fine day, they prepared some as jerked meat, cutting it into thin strips, which they dried on the rocks. This (called "Schat-chew," dried meat) is a very common and favourite food in Tibet, I found it palatable; but on the other hand, the dried saddles of mutton, of which they boast so much, taste so strongly of tallow, that I found it impossible to swallow a morsel of them.*

We staid two days at Lachoong, two of my lads being again laid up with fever; one of them had been similarly attacked at the same place nearly two months before: the other lad had been repeatedly ill since June, and at all elevations. Both cases were returns of a fever caught in the low unhealthy valleys some months previously, and excited by exposure and hardship.

The vegetation at Lachoong was still beautiful, and the weather mild, though snow had descended to 14,000 feet on Tunkra. *Compositæ* were abundantly in flower, apples

---

* Raw dried split fish are abundantly cured (without salt) in Tibet; they are caught in the Yaru and great lakes of Ramchoo, Dobtah, and Yarbru, and are chiefly carp, and allied fish, which attain a large size. It is one of the most remarkable facts in the zoology of Asia, that no trout or salmon inhabits any of the rivers that débouche into the Indian Ocean (the so-called Himalayan trout is a species of carp). This widely distributed natural order of fish (*Salmonidæ*) is however, found in the Oxus, and in all the rivers of central Asia that flow north and west, and the *Salmo orientalis*, M'Clelland ("Calcutta Journ. Nat. Hist." iii., p. 283), was caught by Mr. Griffith (Journals, p. 404) in the Bamean river (north of the Hindo Koosh) which flows into the Oxus, and whose waters are separated by one narrow mountain ridge from those of the feeders of the Indus. The central Himalayan rivers often rise in Tibet from lakes full of fish, but have none (at least during the rains) in that rapid part of their course from 10,000 to 14,000 feet elevation: below that fish abound, but I believe invariably of different species from those found at the sources of the same rivers. The nature of the tropical ocean into which all the Himalayan rivers débouche, is no doubt the proximate cause of the absence of *Salmonidæ.* Sir John Richardson (Fishes of China Seas, &c., " in Brit. Ass. Rep. &c."), says that no species of the order has been found in the Chinese or eastern Asiatic seas.

in young fruit, bushes of *Cotoneaster* covered with scarlet
berries, and the brushwood silvery with the feathery heads
of *Clematis.*

I here found that I had lost a thermometer for high
temperatures, owing to a hole in the bag in which Cheytoong
carried those of my instruments which were in constant
use. It had been last used at the hot springs of the Kinchinjhow glacier; and the poor lad was so concerned at his
mishap, that he came to me soon afterwards, with his
blanket on his back, and a few handfuls of rice in a bag, to
make his salaam before setting out to search for it.   There
was not now a single inhabitant between Lachoong and
that dreary spot, and strongly against my wish he started,
without a companion.   Three days afterwards he overtook
us at Keadom, radiant with joy at having found the instrument: he had gone up to the hot springs, and vainly
sought around them that evening; then rather than lose the
chance of a day-light search on his way back, he had spent
the cold October night *in the hot water*, without fire or
shelter, at 16,000 feet above the sea.   Next morning his
search was again fruitless; and he was returning disconsolate,
when he descried the brass case glistening between two
planks of the bridge crossing the river at Momay, over which
torrent the instrument was suspended.   The Lepchas have
generally been considered timorous of evil spirits, and
especially averse to travelling at night, even in company.
However little this gallant lad may have been given to
superstition, he was nevertheless a Lepcha, born in a warm
region, and had never faced the cold till he became my
servant; and it required a stout heart and an honest one,
to spend a night in so awful a solitude as that which reigns
around the foot of the Kinchinjhow glacier.*

* The fondness of natives for hot springs wherever they occur is very natural,

The villagers at Keadom, where we slept on the 26th, were busy cutting the crops of millet, maize, and *Amaranthus*. A girl who, on my way down the previous month, had observed my curiosity about a singular variety of the maize, had preserved the heads on their ripening, and now brought them to me. The peaches were all gathered, and though only half ripe, were better than Dorjiling produce. A magnificent tree of *Bucklandia*, one of the most beautiful evergreens in Sikkim, grew near this village; it had a trunk twenty-one feet seven inches in girth, at five feet from the ground, and was unbranched for forty feet.* Ferns and the beautiful air-plant *Cœlogyne Wallichii* grew on its branches, with other orchids, while *Clematis* and *Stauntonia* climbed the trunk. Such great names (Buckland, Staunton, and Wallich) thus brought before the traveller's notice, never failed to excite lively and pleasing emotions: it is the ignorant and unfeeling alone who can ridicule the association of the names of travellers and naturalists with those of animals and plants.

We arrived at Choongtam (for the fourth time) at noon, and took up our quarters in a good house near the temple. The autumn and winter flowering plants now prevailed here, such as *Labiatæ*, which are generally late at this

and has been noticed by Humboldt, "Pers. Narr." iv. 195, who states that on Christianity being introduced into Iceland, the natives refused to be baptised in any but the water of the Geysers. I have mentioned at p. 117 the uses to which the Yeumtong hot springs are put; and the custom of using artificial hot baths is noticed at vol. i., p. 305.

* This superb tree is a great desideratum in our gardens; I believe it would thrive in the warm west of England. Its wood is brown, and not valuable as timber, but the thick, bright, glossy, evergreen foliage is particularly handsome, and so is the form of the crown. It is also interesting in a physiological point of view, from the woody fibre being studded with those curious microscopic discs so characteristic of pines, and which when occurring on fossil wood are considered conclusive as to the natural family to which such woods belong. Geologists should bear in mind that not only does the whole natural order to which *Bucklandia* belongs, possess this character, but also various species of *Magnoliaceæ* found in India, Australia, Borneo, and South America.

elevation; and grasses, which, though rare in the damp forest regions, are so common on these slopes that I here gathered twenty-six kinds. I spent a day here in order to collect seeds of the superb rhododendrons * which I had discovered in May, growing on the hills behind. The ascent was now difficult, from the length of the wiry grass, which rendered the slopes so slippery that it was impossible to ascend without holding on by the tussocks.

A ragged Tibetan mendicant (Phud) was amusing the people: he put on a black mask with cowrie shells for eyes, and danced uncouth figures with a kind of heel and toe shuffle, in excellent time, to rude Tibetan songs of his own: for this he received ample alms, which a little boy collected in a wallet. These vagrants live well upon charity; they bless, curse, and transact little affairs of all kinds up and down the valleys of Sikkim and Tibet; this one dealt in red clay teapots, sheep and puppies.

We found Meepo at Choongtam : I had given him leave (when here last) to go back to the Rajah, and to visit his wife; and he had returned with instructions to conduct me to the Chola and Yakla passes, in Eastern Sikkim. These passes, like that of Tunkra (p. 110), lead over the Chola range to that part of Tibet which is interposed between Sikkim and Bhotan. My road lay past the Rajah's residence, which we considered very fortunate, as apparently affording Campbell an opportunity of a conference with his highness, for which both he and the Tchebu Lama were most anxious.

On the way down the Lachen-Lachoong, we found the valley still flooded (as described at p. 20 and 146), and the alders standing with their trunks twelve feet under water;

* These Rhododendrons are now all flourishing at Kew and elsewhere: they are *R. Dalhousiæ, arboreum, Maddeni, Edgeworthii, Aucklandii* and *virgatum.*

but the shingle dam was now dry and hard: it would probably soften, and be carried away by the first rains of the following year. I left here the temperate flora of northern Sikkim, tropical forms commencing to appear: of these the nettle tribe were most numerous in the woods. A large grape,

TIBETAN PHUD.

with beautiful clusters of round purple berries, was very fair eating; it is not the common vine of Europe, which nevertheless is probably an Himalayan plant, the *Vitis Indica.**

* The origin of the common grape being unknown, it becomes a curious question to decide whether the Himalayan *Vitis Indica* is the wild state of that plant: an hypothesis strengthened by the fact of Bacchus, &c., having come from the East.

At Chakoong the temperature of the river, which in
May was 54°, was now 51° 5 at 3 P.M. We did not halt
here, but proceeded to Namgah, a very long and fatiguing
march. Thence a short march took us to Singtam, which
we reached on the 30th of October. The road by which I
had come up was for half the distance obliterated in most
parts by landslips,* but they were hard and dry, and the
leeches were gone.

Bad weather, and Campbell's correspondence with the
Durbar, who prevented all communication with the Rajah,
detained us here two days, after which we crossed to the
Teesta valley, and continued along its east bank to
Tucheam, 2000 feet above the river. We obtained a
magnificent view of the east face of Kinchinjunga, its
tops bearing respectively N. 62° W., and N. 63° W.: the
south slope of the snowed portion in profile was 34°, and
of the north 40°; but both appeared much steeper to the
eye, when unaided by an instrument.

The great shrubby nettle (*Urtica crenulata*) is common
here: this plant, called "Mealum-ma," attains fifteen
feet in height; it has broad glossy leaves, and though
apparently without stings, is held in so great dread,† that

---

* I took a number of dips and strikes of the micaceous rocks: the strike of
these was as often north-east as north-west; it was ever varying, and the strata
were so disturbed, as materially to increase the number and vast dimensions of
the landslips.

† The stinging hairs are microscopic, and confined to the young shoots, leaf
and flower-stalks. Leschenault de la Tour describes being stung by this nettle
on three fingers of his hand only at the Calcutta Botanical Gardens, and the
subsequent sneezing and running at the nose, followed by tetanic symptoms and
two days' suffering, nor did the effects disappear for nine days. It is a remarkable
fact that the plant stings violently only at this season. I frequently gathered it
with impunity on subsequent occasions, and suspected some inaccuracy in my
observations; but in Silhet both Dr. Thomson and I experienced the same effects
in autumn. Endlicher ("Lindley's Vegetable Kingdom") attributes the causticity
of nettle-juice to bicarbonate of ammonia, which Dr. Thomson and I ascertained
was certainly not present in this species.

I had difficulty in getting help to cut it down. I gathered many specimens without allowing any part to touch my skin; still the scentless effluvium was so powerful, that mucous matter poured from my eyes and nose all the rest of the afternoon, in such abundance, that I had to hold my head over a basin for an hour. The sting is very virulent, producing inflammation; and to punish a child with "Mealum-ma" is the severest Lepcha threat. Violent fevers and death have been said to ensue from its sting; but this I very much doubt.

TIBETAN IMPLEMENTS.

TEA-POT, CUP, AND BRICK OF TEA; KNIFE, TOBACCO-PIPE (ACROSS CHOP-STICKS), POUCH, AND FLINT-AND-STEEL.

CHAPTER XXV.

WE started on the 3rd of November for Tumloong (or Sikkim Durbar), Dr. Campbell sending Tchebu Lama forward with letters to announce his approach. A steep ascent, through large trees of *Rhododendron arboreum,* led over a sharp spur of mica-schist (strike north-west and dip north-east), beyond which the whole bay-like valley of the Ryott opened before us, presenting one of the most lovely and fertile landscapes in Sikkim. It is ten miles long, and three or four broad, flanked by lofty mountains, and its head girt by the beautiful snowy range of Chola, from which silvery rills descend through black pine-woods, dividing innumerable converging cultivated spurs, and uniting about 2000 feet below us, in a profound gorge. Everywhere were scattered houses, purple crops of buck-

wheat, green fields of young wheat, yellow millet, broad
green plantains, and orange groves.

We crossed spur after spur, often under or over
precipices about fifteen hundred feet above the river,
proceeding eastwards to the village of Rangang, whence
we caught sight of the Rajah's house. It was an irregular low stone building of Tibetan architecture, with
slanting walls and small windows high up under the
broad thatched roof, above which, in the middle, was a
Chinese-looking square copper-gilt canopy, with projecting
eaves and bells at the corners, surmounted by a ball
and square spire. On either gable of the roof was a
round-topped cylinder of gilded copper, something like a
closed umbrella; this is a very frequent and characteristic Boodhist ornament, and is represented in Turner's
plate of the mausoleum of Teshoo Lama ("Tibet"
plate xi.); indeed the Rajah's canopy at Tumloong is
probably a copy of the upper part of the building there
represented, having been built by architects from Teshoo
Loombo. It was surrounded by chaits, mendongs, poles
with banners, and other religious erections; and though
beautifully situated on a flat terrace overlooking the
valley, we were much disappointed with its size and
appearance.

On the brow of the hill behind was the large red
goompa of the Tupgain Lama, the late heir-apparent to
the temporal and spiritual authority in Sikkim; and near
it a nunnery called Lagong, the lady abbess of which
is a daughter of the Rajah, who, with the assistance of
sisters, keeps an enormous Mani, or praying-cylinder,
revolving perpetually to the prayer of "Om Mani
Padmi hom." On this side was a similar spur, on
which the gilded pinnacles and copper canopy of the

Phadong* goompa gleamed through the trees. At a considerable distance across the head of the valley was still a
third goompa, that of Phenzong.

We were met by a large party of armed Lepchas, dressed
in blue and white striped kirtles, broad loose scarlet jackets,
and the little bamboo wattle hat lined with talc, and surmounted by a peacock's feather; they escorted us to the
village, and then retired.

We encamped a few hundred feet below the Rajah's
house, and close by those of Meepo and the Tchebu
Lama's family, who are among the oldest and most
respectable of Tibetan origin in Sikkim. The population
on this, the north side of the Ryott, consists principally of
Sikkim Bhoteeas and Tibetans, while the opposite is
peopled by Lepchas. Crowds came to see us, and many
brought presents, with which we were overwhelmed; but
we could not help remarking that our cordial greetings
were wholly from the older families attached to the Rajah,
and from the Lamas; none proceeded from the Dewan's
relatives or friends, nor therefore any in the name of the
Rajah himself, or of the Sikkim government.

Tchebu Lama vainly used every endeavour to procure
for us an audience with his highness; who was surrounded by his councillors, or Amlah, all of whom were
adherents of the Dewan, who was in Tibet. My man
Meepo, and the Tchebu Lama, who were ordered to
continue in official attendance upon us, shrugged their
shoulders, but could suggest no remedy. On the following
morning Campbell was visited by many parties, amongst
whom were the Lama's family, and that of the late Dewan
(Ilam Sing), who implored us to send again to announce

* Phadong means Royal, and this temple answers to a chapel royal for the
Rajah.

our presence, and not to dismiss at once the moonshie
and his office,* who had accompanied us for the purpose of a conference with the Rajah.   Their wishes
were complied with, and we waited till noon before proceeding.

TCHEBU LAMA.

A gay and animated scene was produced by the con-
course of women, dressed in their pretty striped and crossed
cloaks, who brought tokens of good-will.   Amongst
them Meepo's wife appeared conspicuous from the large

* It is usual in India for Government officers when about to transact business,
to travel with a staff (called office) of native interpreters, clerks, &c., of whom the
chief is commonly called moonshie.

necklaces * and amulets, corals, and silver filagree work, with which her neck and shoulders were loaded : she wore on her head a red tiara ("Patuk") bedizened with seed pearls and large turquoises, and a gold fillet of filagree bosses united by a web of slender chains; her long tails were elaborately plaited, and woven with beads, and her cloak hooked in front by a chain of broad silver links studded with turquoises. White silk scarfs, the emblem of peace and friendship, were thrown over our hands by each party ; and rice, eggs, fowls, kids, goats, and Murwa beer, poured in apace, to the great delight of our servants.

We returned two visits of ceremony, one to Meepo's house, a poor cottage, to which we carried presents of chintz dresses for his two little girls, who were busy teazing their hair with cylindrical combs, formed of a single slender joint of bamboo slit all round half-way up into innumerable teeth. Our other visit was paid to the Lama's family, who inhabited a large house not far from the Rajah's. The lower story was an area enclosed by stone walls, into which the cattle, &c., were driven. An outside stone stair led to the upper story, where we were received by the head of the family, accompanied by a great con-course of Lamas. He conducted us to a beautiful little oratory at one end of the building, fitted up like a square temple, and lighted with latticed windows, covered with brilliant and tasteful paintings by Lhassan artists. The beams of the ceiling were supported by octagonal columns painted red, with broad capitals. Everywhere the lotus, the mani, and the chirki (or wheel with three rays, emblematic of the Boodhist Trinity), were introduced ; " Om Mani Padmi hom" in gilt letters, adorned the pro-

---

* The lumps of amber forming these (called " Poshea ") were larger than the fist : they are procured in East Tibet, probably from Birmah.

jecting end of every beam;* and the Chinese "cloud messenger," or winged dragon, floated in azure and gold along the capitals and beams, amongst scrolls and groups of flowers.   At one end was a sitting figure of Gorucknath in Lama robes, surrounded by a glory, with

CLASP OF A WOMAN'S CLOAK.

mitre and beads; the right hand holding the Dorje, and the forefinger raised in prayer.   Around was a good library of books.   More presents were brought here, and tea served.

The route to Chola pass, which crosses the range of that name south of the Chola peak (17,320 feet) at the head of this valley, is across the Ryott, and then eastwards along a

---

* A mythical animal with a dog's head and blood-red spot over the forehead was not uncommon in this chapel, and is also seen in the Sikkim temples and throughout Tibet. Ermann, in his Siberian Travels, mentions it as occurring in the Khampa Lama's temple at Maimao chin ; he conjectures it to have been the Cyclops of the Greeks, which according to the Homeric myth had a mark on the forehead, instead of an eye. The glory surrounding the heads of Tibetan deities is also alluded to by Ermann, who recognises in it the Nimbus of the ancients, used to protect the heads of statues from the weather, and from being soiled by birds ; and adds that the glory of the ancient masters in painting was no doubt introduced into the Byzantine school from the Boodhists.

lofty ridge. Campbell started at noon, and I waited
behind with Meepo, who wished me to see the Rajah's
dwelling, to which we therefore ascended; but, to my
guide's chagrin, we were met and turned back by a scribe,
or clerk, of the Amlah. We were followed by a messenger,
apologising and begging me to return; but I had already
descended 1000 feet, and felt no inclination to reascend
the hill, especially as there did not appear to be
anything worth seeing. Soon after I had overtaken
Campbell, he was accosted by an excessively dirty fellow,
who desired him to return for a conference with the
Amlah; this was of course declined, but, at the same
time, Campbell expressed his readiness to receive the
Amlah at our halting place.

The Ryott flows in a very tropical gorge 2000 feet above
the sea; from the proximity of the snowy mountains, its
temperature was only 64° 3. Thence the ascent is very
steep to Tumloong, where we took up our quarters. at a
rest-house called Rungpo (alt. 6008 feet). This road is
well kept, and hence onwards is traversed yearly by the
Rajah on his way to his summer residence of Choombi, two
marches beyond the Chola pass; whither he is taken to
avoid the Sikkim rains, which are peculiarly disagreeable
to Tibetans. Rungpo commands a most beautiful view
northwards, across the valley, of the royal residence,
temples, goompas, hamlets, and cultivation, scattered over
spurs that emerge from the forest, studded below with
tree-ferns and plantains, and backed by black pine-woods
and snowy mountains. In the evening the Amlah arrived
to confer with Campbell; at first there was a proposal of
turning us out of the house, in which there was plenty of
room besides, but as we declined to move, except by his
Highness's order, they put up in houses close by.

On the following morning they met us as we were departing for Chola pass, bringing large presents in the name of the Rajah, and excuses on their and his part for having paid us no respect at Tumloong, saying, that it was not the custom to receive strangers till after they had rested a day, that they were busy preparing a suitable reception, &c.; this was all false, and contrary to etiquette, but there was no use in telling them so. Campbell spoke firmly and kindly to them, and pointed out their incivility and the unfriendly tone of their whole conduct. They then desired Campbell to wait and discuss business affairs with them; this was out of the question, and he assured them that he was ever ready to do so with the Rajah, that he was now (as he had informed his Highness) on his way with me to the Chola and Yakla passes, and that we had, for want of coolies, left some loads behind us, which, if they were really friendly, they would forward. This they did, and so we parted; they (contrary to expectation) making no objection to Campbell's proceeding with me.

A long march up a very steep, narrow ridge took us by a good road to Laghep, a stone resting-house (alt., 10,475 feet) on a very narrow flat. I had abundance of occupation in gathering rhododendron-seeds, of which I procured twenty-four kinds* on this and the following day.

A very remarkable plant, which I had seen in flower in the Lachen valley, called "Loodoo-ma" by the Bhoteeas, and "Nomorchi" by Lepchas, grew on the ridge at 7000

---

* These occurred in the following order in ascending, commencing at 6000 feet.—1. *R. Dalhousiæ;* 2. *R. vaccinioides;* 3. *R. camelliæflorum;* 4. *R. arboreum.* Above 8000 feet:—5. *R. argenteum;* 6. *R. Falconeri;* 7. *R. barbatum;* 8. *R. Campbelliæ;* 9. *R. Edgeworthii;* 10. *R. niveum;* 11. *R. Thomsoni;* 12. *R. cinnabarinum;* 13. *R. glaucum.* Above 10,500 feet:—14. *R. lanatum;* 15. *R. virgatum;* 16. *R. campylocarpum;* 17. *R. ciliatum;* 18. *R. Hodgsoni;* 19. *R. campanulatum.* Above 12,000 feet:—20. *R. lepidotum;* 21. *R. fulgens;* 22. *R. Wightianum;* 23. *R. anthopogon;* 24. *R. setosum.*

feet; it bears a yellow fruit like short cucumbers, full of a soft, sweet, milky pulp, and large black seeds; it belongs to a new genus,* allied to *Stauntonia,* of which two Himalayan kinds produce similar, but less agreeable edible fruits ("Kole-pot," Lepcha). At Laghep, iris was abundant, and a small bushy berberry (*B. concinna*) with oval eatable berries. The north wall of the house (which was in a very exposed spot) was quite bare, while the south was completely clothed with moss and weeds.

The rocks above Laghep were gneiss; below it, micaschist, striking north-west, and dipping north-east, at a high angle. A beautiful yellow poppy-like plant grew in clefts at 10,000 feet; it has flowered in England, from seeds which I sent home, and bears the name of *Cathcartia.*†

We continued, on the following morning, in an easterly direction, up the same narrow steep ridge, to a lofty eminence called Phieung-goong (alt. 12,422 feet), from being covered with the Phieung, or small bamboo. *Abies Webbiana* begins here, and continues onwards, but, as on Tonglo, Mainom, and the other outer wetter Sikkim ranges, there is neither larch, *Pinus excelsa, Abies Smithiana,* or *A. Brunoniana.*

---

* This genus, for which Dr. Thomson and I, in our "Flora Indica," have proposed the name *Decaisnea* (in honour of my friend Professor J. Decaisne, the eminent French botanist), has several straight, stick-like, erect branches from the root, which bear spreading pinnated leaves, two feet long, standing out horizontally. The flowers are uni-sexual, green, and in racemes, and the fruits, of which two or three grow together, are about four inches long, and one in diameter. All the other plants of the natural order to which it belongs, are climbers.

† See "Botanical Magazine," for 1852. The name was given in honour of the memory of my friend, the late J. F. Cathcart, Esq., of the Bengal Civil Service. This gentleman was devoted to the pursuit of botany, and caused a magnificent series of drawings of Dorjiling plants to be made by native artists during his residence there. This collection is now deposited at Kew, through the liberality of his family, and it is proposed to publish a selection from the plates, as a tribute to his memory. Mr. Cathcart, after the expiration of his Indian service, returned to Europe, and died at Lausanne on his way to England.

Hence we followed an oblique descent of 1,500 feet, to the bed of the Rutto river, through thick woods of pines and *Rhododendron Hodgsoni*, which latter, on our again ascending, was succeeded by the various alpine kinds.   We halted at Barfonchen (alt. 11,233 feet), a stone-hut in the silver-fir forest.   Some yaks were grazing in the vicinity, and from their herdsman we learnt that the Dewan was at Choombi, on the road to Yakla; he had kept wholly out of the way during the summer, directing every unfriendly action to be pursued towards myself and the government by the Amlah, consisting of his brothers and relatives, whom he left at Tumloong.

The night was brilliant and starlight : the minimum thermometer fell to 27°, a strong north-east wind blew down the valley, and there was a thick hoar-frost, with which the black yaks were drolly powdered.   The broad leaves of *R. Hodgsoni* were curled, from the expansion of the frozen fluid in the layer of cells on the upper surface of the leaf, which is exposed to the greatest cold of radiation.   The sun restores them a little, but as winter advances, they become irrecoverably curled, and droop at the ends of the branches.

We left Barfonchen on the 7th November, and ascended the river, near which we put up a woodcock.   Emerging from the woods at Chumanako (alt. 12,590 feet), where there is another stone hut, the mountains become bleak, bare, and stony, and the rocks are all moutonnéed by ancient glaciers.   At 13,000 feet the ground was covered with ice, and all the streams were frozen.   Crossing several rocky ledges, behind which were small lakes, a gradual ascent led to the summit of the Chola pass, a broad low depression, 14,925 feet above the sea, wholly bare of snow.

Campbell had preceded me, and I found him conversing with some Tibetans, who told him that there was no road hence to Yakla, and that we should not be permitted to go to Choombi. As the Chinese guard was posted in the neighbourhood, he accompanied one of the Tibetans to see the commandant, whilst I remained taking observations. The temperature was 33°, with a violent, biting, dry east wind. The rocks were gneiss, striking north-east, and horizontal, or dipping north-west. The scanty vegetation consisted chiefly of grass and *Sibbaldia*.

In about an hour Meepo and some of my people came up and asked for Campbell, for whom the Tchebu Lama was waiting below: the Lama had remained at Rungpo, endeavouring to put matters on a better footing with the Amlah. Wishing to see the Tibet guard myself, I accompanied the two remaining Tibetans down a steep valley with cliffs on either hand, for several hundred feet, when I was overtaken by some Sikkim sepoys in red jackets, who wanted to turn me back forcibly: I was at a loss to understand their conduct, and appealed to the Tibetan sepoys, who caused them to desist. About 1000 feet down I found Campbell, with a body of about ninety Tibetans, a few of whom were armed with matchlocks, and the rest with bows and arrows. They were commanded by a Dingpun, a short swarthy man, with a flat-crowned cap with floss-silk hanging all round, and a green glass button in front; he wore a loose scarlet jacket, broadly edged with black velvet, and having great brass buttons of the Indian naval uniform; his subaltern was similarly dressed, but his buttons were those of the 44th Bengal Infantry. The commandant having heard of our wish to go round by Choombi, told Campbell that he had come purposely to inform him that there was no road that way to

Yakla; he was very polite, ordering his party to rise and salute me when I arrived, and doing the same when we both left.

On our return we were accompanied by the Dingpun of the Tibetans and a few of his people, and were soon met by more Sikkim sepoys, who said they were sent from the Durbar, to bring Campbell back to transact business; they behaved very rudely, and when still half a mile from the Sikkim frontier, jostled him and feigned to draw their knives, and one of them pointed a spear-headed bow to his breast. Campbell defended himself with a stick, and remonstrated with them on their rudeness; and I, who had nothing but a barometer in my hand, called up the Tibetans. The Dingpun came instantly, and driving the Sikkim people forward, escorted us to the frontier, where he took an inscribed board from the chait, and showing us the great vermilion seal of the Emperor of China (or more probably of the Lhassan authorities) on one side, and two small brown ones of the Sikkim Rajah on the other; and giving us to understand that here his jurisdiction ceased, he again saluted and left us.

On descending, I was surprised to meet the Singtam Soubah, whom I had not seen since leaving Tungu; he was seated on a rock, and I remarked that he looked ashy pale and haggard, and that he salaamed to me only, and not to Campbell; and that Tchebu Lama, who was with him, seemed very uncomfortable. The Soubah wanted Campbell to stop for a conference, which at such a time, and in such a wind, was impossible, so he followed us to Chumanako, where we proposed to pass the night.

A great party of Sikkim Bhoteeas had assembled here, all strangers to me: I certainly thought the concourse unusually large, and the previous conduct to Campbell, strange, rude, and quite unintelligible, especially before the Tibetans. But the Bhoteeas were always a queer, and often insolent people,* whom I was long ago tired of trying to understand, and they might have wanted to show off before their neighbours; and such was the confidence with which my long travels amongst them had inspired me, that the possibility of danger or violence never entered my head.

We went into the hut, and were resting ourselves on a log at one end of it, when, the evening being very cold, the people crowded in; on which Campbell went out, saying, that we had better leave the hut to them, and that he would see the tents pitched. He had scarcely left, when I heard him calling loudly to me, " Hooker! Hooker! the savages are murdering me!" I rushed to the door, and caught sight of him striking out with his fists, and struggling violently; being tall and powerful, he had already prostrated a few, but, a host of men bore him down, and appeared to be trampling on him; at the same moment I was myself seized by eight men, who forced me back into the hut, and down on the log, where they held me in a sitting posture, pressing me against the wall; here I spent a few moments of agony, as I heard my friend's stifled cries grow fainter and fainter. I struggled but little, and that only at first, for at least five-and-twenty men crowded round and laid their hands upon me, rendering any effort to move useless; they were, however, neither angry nor

* Captain Pemberton during his mission to Bhotan was repeatedly treated with the utmost insolence by the officials in that country (see Griffith's Journal). My Sirdar, Nimbo, himself a native of Bhotan, saw a good deal of the embassy when there, and told me many particulars as to the treatment to which it had been subjected, and the consequent low estimation in which both the ambassadors themselves and the Government whom they represented were held in Bhotan.

violent, and signed to me to keep quiet. I retained my
presence of mind, and felt comfort in remembering that I
saw no knives used by the party who fell on Campbell, and
that if their intentions had been murderous, an arrow
would have been the more sure and less troublesome
weapon. It was evident that the whole animus was
directed against Campbell, and though at first alarmed
on my own account, all the inferences which, with the
rapidity of lightning my mind involuntarily drew, were
favourable.

After a few minutes, three persons came into the hut,
and seated themselves opposite to me : I only recognised
two of them; namely, the Singtam Soubah, pale, trembling like a leaf, and with great drops of sweat trickling
from his greasy brow; and the Tchebu Lama, stolid,
but evidently under restraint, and frightened. The
former ordered the men to leave hold of me, and to stand
guard on either side, and, in a violently agitated manner,
he endeavoured to explain that Campbell was a prisoner
by the orders of the Rajah, who was dissatisfied with
his conduct as a government officer, during the past
twelve years; and that he was to be taken to the
Durbar and confined till the supreme government at
Calcutta should confirm such articles as he should be compelled to subscribe to ; he also wanted to know from me
how Campbell would be likely to behave. I refused to
answer any questions till I should be informed why I was
myself made prisoner ; on which he went away, leaving me
still guarded. My own Sirdar then explained that Campbell
had been knocked down, tied hand and foot, and taken
to his tent, and that all his coolies were also bound,
our captors claiming them as Sikkimites, and subjects
of the Rajah.

Shortly afterwards the three returned, the Soubah looking more spectral than ever, and still more violently agitated, and I thought I perceived that whatever were his plans, he had failed in them. He asked me what view the Governor-General would take of this proceeding? and receiving no answer, he went off with the Tchebu Lama, and left me with the third individual. The latter looked steadily at me for some time, and then asked if I did not know him. I said I did not, when he gave his name as Dingpun Tinli, and I recognised in him one of the men whom the Dewan had sent to conduct us to the top of Mainom the previous year (see vol. i. p. 305). This opened my eyes a good deal, for he was known to be a right-hand man of the Dewan's, and had within a few months been convicted of kidnapping two Brahmin girls from Nepal,* and had vowed vengeance against Campbell for the duty he performed in bringing him to punishment.

I was soon asked to go to my tent, which I found pitched close by; they refused me permission to see my fellow-prisoner, or to be near him, but allowed me to hang up my instruments, and arrange my collections. My guards were frequently changed during the night, Lepchas often taking a turn; they repeatedly assured me that there was no complaint or ill-feeling against me, that the better classes in Sikkim would be greatly ashamed of the whole affair, that Tchebu Lama was equally a prisoner, and that the grievances against Campbell were of a political nature, but what they were they did not know.

---

* This act as I have mentioned at v. i. p. 341, was not only a violation of the British treaty, but an outrage on the religion of Nepal. Jung Bahadoor demanded instant restitution, which Campbell effected; thus incurring the Dingpun's wrath, who lost, besides his prize, a good deal of money which the escapade cost him.

The night was very cold (thermometer 26°), and two inches of snow fell.   I took as many of my party as I could into my tent, they having no shelter fit for such an elevation (12,590 feet) at this season   Through the connivance of some of the people, I managed to correspond with Campbell, who afterwards gave me the following account of the treatment he had received.   He stated that on leaving the hut, he had been met by Meepo, who told him the Soubah had ordered his being turned out.   A crowd of sepoys then fell on him and brought him to the ground, knocked him on the head, trampled on him, and pressed his neck down to his chest as he lay, as if endeavouring to break it.   His feet were tied, and his arms pinioned behind, the wrist of the right hand being bound to the left arm above the elbow ;   the cords were then doubled, and he was violently shaken.   The Singtam Soubah directed all this, which was performed chiefly by the Dingpun Tinli and Jongpun Sangabadoo.*   After this the Soubah came to me, as I have related ;   and returning, had Campbell brought bound before him, and asked him, through Tchebu Lama, if he would write from dictation.   The Soubah was violent, excited, and nervous ; Tchebu Lama scared.   Campbell answered, that if they continued torturing him (which was done by twisting the cords round his wrists by a bamboo-wrench), he might say or do anything, but that his government would not confirm any acts thus extorted.   The Soubah became still more violent, shook his bow in Campbell's face, and drawing his hand significantly across his throat, repeated his questions, adding others, enquiring why he had refused to receive the Lassoo Kajee as Vakeel, &c. (see p. 2).

* This was the other man sent with us to Mainom, by the Dewan, in the previous December.

The Soubah's people, meanwhile, gradually slunk away, seeing which he left Campbell, who was taken to his tent.

Early next morning Meepo was sent by the Soubah, to ask whether I would go to Yakla pass, or return to Dorjiling, and to say that the Rajah's orders had been very strict that I was not to be molested, and that I might proceed to whatever passes I wished to visit, whilst Campbell was to be taken back to the Durbar, to transact business. I was obliged to call upon the Soubah and Dingpun to explain their conduct of the previous day, which they declared arose from no ill-feeling, but simply from their fear of my interfering in Campbell's behalf; they could not see what reason I had to complain, so long as I was neither hurt nor bound. I tried in vain to explain to them that they could not so play fast and loose with a British subject, and insisted that if they really considered me free, they should place me with Campbell, under whose protection I considered myself, he being still the Governor-General's agent.

Much discussion followed this : Meepo urged me to go on to Yakla, and leave these bad people; and the Soubah and Dingpun, who had exceeded their orders in laying hands on me, both wished me away. My course was, however, clear as to the propriety of keeping as close to Campbell as I was allowed, so they reluctantly agreed to take me with him to the Durbar.

Tchebu Lama came to me soon afterwards, looking as stolid as ever, but with a gulping in his throat; he alone was glad I was going with them, and implored me to counsel Campbell not to irritate the Amlah by a refusal to accede to their dictates, in which case his life might be the forfeit. As to himself, the opposite faction had now got the mastery, there was nothing for it but to succumb, and his throat would surely be cut. I endeavoured to comfort him with the assurance that they dared not hurt Campbell, and that this conduct of a party of ruffians, influenced by the Dewan and their own private pique, did not represent his Rajah's feelings and wishes, as he himself knew; but the poor fellow was utterly unnerved, and shaking hands warmly, with his eyes full of tears, he took his leave.

We were summoned by the Dingpun to march at 10 A.M.: I demanded an interview with Campbell first, which was refused; but I felt myself pretty safe, and insisting upon it, he was brought to me. He was sadly bruised about the head, arms, and wrists, walked very lame, and had a black eye to boot, but was looking stout and confident.

I may here mention that seizing the representative of a neighbouring power and confining him till he shall have become amenable to terms, is a common practice along the Tibet, Sikkim, and Bhotan frontiers. It had been resorted to in 1847, by the Bhotanese, under the instructions of the Paro Pilo, who waylaid the Sikkim Rajah when still in Tibet, on his return from Jigatzi, and beleagured him for two months, endeavouring to bring him to their terms about some border dispute; on this occasion the Rajah applied to the British government for assistance, which was refused; and he was ultimately rescued by a Tibetan force.

In the present case the Dewan issued orders that Campbell was to be confined at Tumloong till he himself should arrive there; and the Rajah was kept in ignorance of the affair. The Sepoys who met us on our approach to Tumloong on the 3rd of November, were, I suspect, originally sent for the purpose; and I think that the Amlah

also had followed us to Rungpo with the same object.
Their own extreme timidity, and the general good-feeling
in the country towards Campbell prevented its execution
before, and, as a last resource, they selected the Singtam
Soubah and Dingpun Tinli for the office, as being personally hostile to him. The Dewan meanwhile being in
Tibet, and knowing that we were about to visit the frontier,
for which I had full permission and escort, sent up the
Tibetan guard, hoping to embroil them in the affair; in
this he failed, and it drew upon him the anger of the
Lhassan authorities.* The Soubah, in endeavouring to
extort the new treaty by force, and the Dingpun, who
had his own revenge to gratify, exceeded their instructions in using violence towards Campbell, whom the
Dewan ordered should be simply taken and confined;
they were consequently disgraced, long before we were
released, and the failure of the stratagem thrown upon
their shoulders.

During the march down to Laghep, Campbell was
treated by the Dingpun's men with great rudeness: I kept

---

* In the following summer (1850), when the Rajah, Dewan, and Soubah, repaired
to Choombi, the Lhassan authorities sent a Commissioner to inquire into the affair,
understanding that the Dewan had attempted to embroil the Tibetans in it. The
commissioner asked the Rajah why he had committed such an outrage on the
representative of the British government, under whose protection he was; thus
losing his territory, and bringing English troops so near the Tibet frontier. The
Rajah answered that he never did anything of the kind; that he was old and
infirm, and unable to transact all his affairs; that the mischief had arisen out of the
acts and ignorance of others, and finally begged the Commissioner to investigate
the whole affair, and satisfy himself about it. During the inquiry that followed,
the Dewan threw all the blame on the Tibetans, who he said, were alone
implicated: this assertion was easily disproved, and on the conclusion of the
inquiry the Commissioner railed vehemently at the Dewan, saying:—"You
tried to put this business on the people of my country; it is an abominable lie.
You did it yourselves, and no one else. The Company is a great monarchy; you
insulted it, and it has taken its revenge. If you, or any other Tibetan, ever
again cause a rupture with the English, you shall be taken with ropes round your
necks to Pekin, there to undergo the just punishment of your offence under the
sentence of the mighty Emperor."

as near as I was allowed, quietly gathering rhododendron-seeds by the way.　At the camping-ground we were again separated, at which I remonstrated with the Dingpun, also complaining of his people's insolent behaviour towards their prisoner, which he promised should be discontinued.

The next day we reached Rungpo, where we halted for further instructions: our tents were placed apart, but we managed to correspond by stealth.　On the 10th of November we were conducted to Tumloong: a pony was brought for me, but I refused it, on seeing that Campbell was treated with great indignity, and obliged to follow at the tail of the mule ridden by the Dingpun, who thus marched him in triumph up to the village.

I was taken to a house at Phadong, and my fellow tra-veller was confined in another at some distance to the east-ward, a stone's throw below the Rajah's; and thrust into a little cage-like room.　I was soon visited by an old Lama, who assured me that we were both perfectly safe, but that there were many grievances against Campbell. The Soubah arrived shortly after, bringing me compli-ments, nominally in the Rajah's name, and a substantial present, consisting of a large cow, sheep, fowls, a brick of tea, bags of rice, flour, butter, eggs, and a profusion of vegetables.　I refused to take them on the friendly terms on which they were brought, and only accepted them as provisions during my detention.　I remonstrated again about our separation, and warned the Soubah of the inevitable consequence of this outrage upon the repre-sentative of a friendly power, travelling under the authority of his own government, unarmed and without escort: he was greatly perplexed, and assured me that Campbell's detention was only temporary, because he had not given

satisfaction to the Rajah, and as the latter could not get
answers to his demands from Calcutta in less than a month,
it was determined to keep him till then; but to send
me to Dorjiling.   He returned in the evening to tell me
that Campbell's men (with the exception only of the
Ghorkas *) had been seized, because they were runaway
slaves from Sikkim; but that I need not alarm myself,
for mine should be untouched.

The hut being small, and intolerably dirty, I pitched my
tent close by, and lived in it for seven days : I was not
guarded, but so closely watched, that I could not go out
for the most trifling purpose, except under surveillance.
They were evidently afraid of my escaping; I was however
treated with civility, but forbidden to communicate either
with Campbell or with Dorjiling.

The Soubah frequently visited me, always protesting I
was no prisoner, that Campbell's seizure was a very
trifling affair, and the violence employed all a mistake.
He always brought presents, and tried to sound me about
the government at Calcutta.   On the 12th he paid his
last visit, looking wofully dejected, being out of favour at
court, and dismissed to his home : he referred me to
Meepo for all future communications to the Rajah, and
bade me a most cordial farewell, which I regretted being
unable to return with any show of kind feeling.   Poor
fellow! he had staked his last, and lost it, when he undertook to seize the agent of the most powerful government
in the east, and to reduce him to the condition of a tool of
the Dewan.   Despite the many obstructions he had placed
in my way, we had not fallen out since July; we had been

---

* These people stood in far greater fear of the Nepalese than of the English,
and the reason is obvious: the former allow no infraction of their rights
to pass unnoticed, whereas we had permitted every article of our treaty to be
contravened.

constant companions, and though at issue, never at enmity. I had impeached him, and my grievances had been forwarded to the Rajah with a demand for his punishment, but he never seemed to owe me a grudge for that, knowing the Rajah's impotence as compared with the power of the Dewan whom he served; and, in common with all his party, presuming on the unwillingness of the British government to punish.

On the 13th of November I was hurriedly summoned by Meepo to the Phadong temple, where I was interrogated by the Amlah, as the Rajah's councillors (in this instance the Dewan's adherents) are called. I found four China mats placed on a stone bench, on one of which I was requested to seat myself, the others being occupied by the Dewan's elder brother, a younger brother of the Gangtok Kajee (a man of some wealth), and an old Lama: the conference took place in the open air and amongst an immense crowd of Lamas, men, women, and children.

I took the initiative (as I made a point of doing on all such occasions) and demanded proper interpreters, which were refused.; and the Amlah began a rambling interrogatory in Tibetan, through my Lepcha Sirdar Pakshok, who spoke very little Tibetan or Hindostanee, and my half-caste servant, who spoke as little English. The Dewan's brother was very nervously counting his beads, and never raised his eyes while I kept mine steadily upon him.

He suggested most of the queries, every one of which took several minutes, as he was constantly interrupted by the Kajee, who was very fat and stupid: the Lama scarcely spoke, and the bystanders never. My connection with the Indian government was first enquired into; next they came to political matters, upon which I declined entering; but I

gathered that their object was to oblige Campbell to accept
the Lassoo Kajee as Vakeel, to alter the slavery laws, to
draw a new boundary line with Nepal, to institute direct
communication between themselves and the Governor-
General,* and to engage that there should be no trade or
communication between Sikkim and India, except through
the Dewan : all of these subjects related to the terms of
the original treaty between the Rajah and the Indian
government. They told me they had sent these proposals
to the government through Dorjiling,† but had received no
acknowledgment from the latter place, and they wanted to
know the probable result at Calcutta. As the only answer
I could give might irritate them, I again declined giving
any. Lastly, they assured me that no blame was imputed
to myself, that on the contrary I had been travelling under
the Rajah's protection, who rejoiced in my success, that I
might have visited Yakla pass as I had intended doing, but
that preferring to accompany my friend, they had allowed
me to do so, and that I might now either join him, or continue to live in my tent : of course I joyfully accepted the
former proposal. After being refused permission to send
a letter to Dorjiling, except I would write in a character
which they could read, I asked if they had anything more
to say, and being answered in the negative, I was taken by

* They were prompted to demand this by an unfortunate oversight that
occurred at Calcutta some years before. Vakeels from the Sikkim Durbar repaired
to that capital, and though unaccredited by the Governor-General's agent at Dorjiling, were (in the absence of the Governor-General) received by the president of
the council in open Durbar. The effect was of course to reduce the Governor-
General's agent at Dorjiling to a cipher.

† These letters, which concluded with a line stating that Campbell was detained
at Tumloong till favourable answers should be received, had arrived at Dorjiling;
but being written in Tibetan, and containing matters into which no one but
Campbell could enter, they were laid on one side till his return. The interpreter
did not read the last line, which stated that Dr. Campbell was *detained* till
answers were received, and the fact of our capture and imprisonment therefore
remained unknown for several weeks.

Meepo to Campbell, heartily glad to end a parley which had lasted for an hour and a half.

I found my friend in good health and spirits, strictly guarded in a small thatched hut, of bamboo wattle and clay : the situation was pretty, and commanded a view of the Ryott valley and the snowy mountains; there were some picturesque chaits hard by, and a blacksmith's forge. Our walks were confined to a few steps in front of the hut, and included a puddle and a spring of water. We had one black room with a small window, and a fire in the middle on a stone; we slept in the narrow apartment behind it, which was the cage in which Campbell had been at first confined, and which exactly admitted us both, lying on the floor. Two or three Sepoys occupied an adjoining room, and had a peep-hole through the partition-wall.

My gratification at our being placed together was damped by the seizure of all my faithful attendants except my own servant, and one who was a Nepalese : the rest were bound, and placed in the stocks and close confinement, charged with being Sikkim people who had no authority to take service in Dorjiling. On the contrary they were all registered as British subjects, and had during my travels been recognised as such by the Rajah and all his authorities. Three times the Soubah and others had voluntarily assured me that my person and people were inviolate; nor was there any cause for this outrage but the fear of their escaping with news to Dorjiling, and possibly a feeling of irritation amongst the authorities at the failure of their schemes. Meanwhile we were not allowed to write, and we heard that the bag of letters which we had sent before our capture had been seized and burnt. Campbell greatly feared that they would threaten

Dorjiling with a night attack,* as we heard that the Lassoo Kajee was stationed at Namtchi with a party for that purpose, and all communication cut off, except through him.

* Threats of sacking Dorjiling had on several previous occasions been made by the Dewan, to the too great alarm of the inhabitants, who were ignorant of the timid and pacific disposition of the Lepchas, and of the fact that there are not fifty muskets in the country, nor twenty men able to use them. On this occasion the threats were coupled with the report that we were murdered, and that the Rajah had asked for 50,000 Tibetan soldiers, who were being marched twenty-five days' journey over passes 15,000 feet high, and deep in snow, and were coming to drive the English out of Sikkim! I need hardly observe that the Tibetans (who have repeatedly refused to interfere on this side the snows) had no hand in the matter, or that, supposing they could collect that number of men in all Tibet, it would be impossible to feed them for a week, there or in Sikkim. Such reports unfortunately spread a panic in Dorjiling: the guards were called in from all the outposts, and the ladies huddled into one house, whilst the males stood on the defensive; to the great amusement of the Amlah at Tumloong, whose insolence to us increased proportionally.

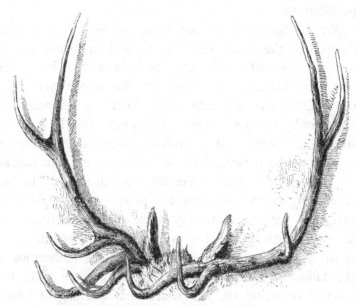

HORNS OF THE SHOWA STAG (*Cervus Wallichii*), A NATIVE OF CHOOMBI IN TIBET.
Length of antler, 4 feet 6 in.

CHAPTER XXVI.

SINCE his confinement, Dr. Campbell had been desired to
attend the Durbar for the purpose of transacting business,
but had refused to go, except by compulsion, considering
that in the excited state of the authorities, amongst whom
there was not one person of responsibility or judgment,
his presence would not only be useless, but he might be
exposed to further insult or possibly violence.

On the 15th of November we were informed that the
Dewan was on his way from Tibet: of this we were glad,
for knave as he was, we had hitherto considered him to
possess sense and understanding. His agents were
beginning to find out their mistake, and summoned to

council the principal Lamas and Kajees of the country,
who, to a man, repudiated the proceedings, and refused to
attend.   Our captors were extremely anxious to induce us
to write letters to Dorjiling, and sent spies of all kinds to
offer us facilities for secret correspondence.   The simplicity
and clumsiness with which these artifices were attempted
would have been ludicrous under other circumstances;
while the threat of murdering Campbell only alarmed us,
inasmuch as it came from people too stupid to be trusted.
We made out that all Sikkim people were excluded from
Dorjiling, and the Amlah consequently could not conceal
their anxiety to know what had befallen their letters to
government.

Meanwhile we were but scantily fed, and our imprisoned
coolies got nothing at all.   Our guards were supplied with
a handful of rice or meal as the day's allowance; they were
consequently grumbling,* and were daily reduced in
number.   The supplies of rice from the Terai, beyond
Dorjiling, were cut off by the interruption of communication, and the authorities evidently could not hold us long
at this rate : we sent up complaints, but of course received
no answer.

The Dewan arrived in the afternoon in great state,
carried in an English chair given him by Campbell some
years before, habited in a blue silk cloak lined with lamb-
skin, and wearing an enormous straw hat with a red tassel,

* The Rajah has no standing army; not even a body-guard, and these men
were summoned to Tumloong before our arrival : they had no arms and received
no pay, but were fed when called out on duty.   There is no store for grain,
no bazaar or market, in any part of the country, each family growing little
enough for its own wants and no more; consequently Sikkim could not stand
on the defensive for a week.   The Rajah receives his supply of grain in annual
contributions from the peasantry, who thus pay a rent in kind, which varies
from little to nothing, according to the year, &c.   He had also property of
his own in the Terai, but the slender proceeds only enabled him to trade
with Tibet for tea, &c.

and black velvet butterflies on the flapping brim.　He was accompanied by a household of women, who were laden with ornaments, and wore boots, and sat astride on ponies; many Lamas were also with him, one of whom wore a broad Chinese-like hat covered with polished copper foil.　Half a dozen Sepoys with matchlocks preceded him, and on approaching Tumloong, bawled out his titles, dignities, &c., as was formerly the custom in England.

RAJAH'S RESIDENCE, AND THE HUT ASSIGNED TO US.　ARRIVAL OF THE DEWAN.

At Dorjiling our seizure was still unknown: our letters were brought to us, but we were not allowed to answer them.　Now that the Dewan had arrived, we hoped to come to a speedy explanation with him, but he shammed sickness, and sent no answer to our messages, if indeed he

received them.   Our guards were reduced to one Sepoy
with a knife, who was friendly; and a dirty, cross-eyed
fellow named Thoba-sing, who, with the exception of
Tchebu Lama, was the only Bhoteea about the Durbar
who could speak Hindostanee, and who did it very imperfectly : he was our attendant and spy, the most bare-
faced liar I ever met with, even in the east; and as
cringing and obsequious when alone with us, as he was to
his masters on other occasions, when he never failed to
show off his authority over us in an offensive manner.
Though he was the most disagreeable fellow we were ever
thrown in contact with, I do not think that he was therefore selected, but solely from his possessing a few words of
Hindostanee, and his presumed capability of playing the spy.

The weather was generally drizzling or rainy, and we
were getting very tired of our captivity; but I beguiled
the time by carefully keeping my meteorological register,*
and by reducing many of my previous observations.
Each morning we were awakened at daybreak by the
prolonged echos of the conchs, trumpets, and cymbals,
beaten by the priests before the many temples in the
valley: wild and pleasing sounds, often followed by

---

* During the thirty days spent at Tumloong, the temperature was mild and
equable, with much cloud and drizzle, but little hard rain ; and we experienced
violent thunder-storms, followed by transient sunshine.   Unlike 1848, the rains
did not cease this year before the middle of December ; nor had there been one
fine month since April.   The mean temperature, computed from 150 observations,
was 50° 2, and from the maximum and minimum thermometer 49° 6, which is a
fair approximation to the theoretical temperature calculated for the elevation
and month, and allows a fall of 1° for 320 feet of ascent.   The temperature during
the spring (from 50 observations) varied during the day from 2° 4 to 5° 8 higher
than that of the air, the greatest differences occurring morning and evening.   The
barometric tide amounted to 0·091 between 9·50 A.M., and 4 P.M., which is less than
at the level of the plains of India, and more than at any greater elevation than
Tumloong.   The air was always damp, nearly saturated at night, and the mean
amount of humidity for ninety-eight observations taken during the day was only
0·850, corresponding to a dew-point of 49° 6, or 5° 2 below that of the air.

their choral chants.　After dark we sat over the fire, generally in company with a little Lepcha girl, who was appointed to keep us in fire-wood, and who sat watching our movements with childish curiosity.　Dolly, as we christened her, was a quick child and a kind one, intolerably dirty, but very entertaining from her powers of mimicry.　She was fond of hearing me whistle airs, and procured me a Tibetan Jews'-harp,* with which, and coarse tobacco, which I smoked out of a Tibetan brass pipe, I wiled away the dark evenings, whilst my cheerful companion amused himself with an old harmonicon, to the enchantment of Dolly and our guards and neighbours.

TIBET PIPE, AND TINDER-POUCH WITH STEEL ATTACHED.

The messengers from Dorjiling were kept in utter ignorance of our confinement till their arrival at Tumloong, when they were cross-questioned, and finally sent to us. They gradually became too numerous, there being only one apartment for ourselves, and such of our servants as

---

* This instrument (which is common in Tibet) is identical with the European, except that the tongue is produced behind the bow, in a strong steel spike, by which the instrument is held firmer to the mouth.

were not imprisoned elsewhere. Some of them were frightened out of their senses, and the state of abject fear and trembling in which one Limboo arrived, and continued for nearly a week, was quite distressing * to every one except Dolly, who mimicked him in a manner that was irresistibly ludicrous. Whether he had been beaten or threatened we could not make out, nor whether he had heard of some dark fate impending over ourselves—a suspicion which would force itself on our minds ; especially as Thoba-sing had coolly suggested to the Amlah the dispatching of Campbell, as the shortest way of getting out of the scrape ! We were also ignorant whether any steps were being taken at Dorjiling for our release, which we felt satisfied must follow any active measures against these bullying cowards, though they themselves frequently warned us that we should be thrown into the Teesta if any such were pursued.

So long as our money lasted, we bought food, for the Durbar had none to give; and latterly my ever charitable companion fed our guards, including Dolly and Thoba-sing, in pity to their pinched condition. Several families sent us small presents, especially that of the late estimable Dewan, Ilam-sing, whose widow and daughters lived close by, and never failed to express in secret their sympathy and good feeling.

Tchebu Lama's and Meepo's families were equally forward in their desire to serve us ; but they were marked men, and could only communicate by stealth.

---

* It amounted to a complete prostration of bodily and mental powers : the man trembled and started when spoken to, or at any noise, a cold sweat constantly bedewed his forehead, and he continued in this state for eight days. No kindness on Campbell's part could rouse him to give any intelligible account of his fears or their cause. His companions said he had lost his goroo, i.e., his charm, which the priest gives him while yet a child, and which he renews or gets re-sanctified as occasion requires. To us the circumstance was extremely painful.

Our coolies were released on the 18th, more than half starved, but the Sirdars were still kept in chains or the stocks: some were sent back to Dorjiling, and the British subjects billetted off amongst the villagers, and variously employed by the Dewan: my lad, Cheytoong, was set to collect the long leaves of a *Tupistra*, called " Purphiok," which yield a sweet juice, and were chopped up and mixed with tobacco for the Dewan's hookah.

November 20th.—The Dewan, we heard this day, ignored all the late proceedings, professing to be enraged with his brother and the Amlah, and refusing to meddle in the matter. This was no doubt a pretence: we had sent repeatedly for an explanation with himself or the Rajah, from which he excused himself on the plea of ill-health, till this day, when he apprized us that he would meet Campbell, and a cotton tent was pitched for the purpose.

We went about noon, and were received with great politeness and shaking of hands by the Dewan, the young Gangtok Kajee, and the old monk who had been present at my examination at Phadong. Tchebu Lama's brother was also there, as a member of the Amlah, lately taken into favour; while Tchebu himself acted as interpreter, the Dewan speaking only Tibetan. They all sat cross-legged on a bamboo bench on one side, and we on chairs opposite them: walnuts and sweetmeats were brought us, and a small present in the Rajah's name, consisting of rice, flour, and butter.

The Dewan opened the conversation both in this and another conference, which took place on the 22nd, by requesting Campbell to state his reasons for having desired these interviews. Neither he nor the Amlah seemed to have the smallest idea of the nature and consequences of the acts they had committed, and they therefore anxiously

sought information as to the view that would be taken
of them by the British Government. They could not see
why Campbell should not transact business with them
in his present condition, and wanted him to be the medium
of communication between themselves and Calcutta. The
latter confined himself to pointing out his own views of
the following subjects :—1. The seizing and imprisoning
of the agent of a friendly power, travelling unarmed and
without escort, under the formal protection of the Rajah, and
with the authority of his own government. 2. The aggra-
vation of this act of the Amlah, by our present detention
under the Dewan's authority. 3. The chance of collision,
and the disastrous consequences of a war, for which they
had no preparation of any kind. 4. The impossibility of
the supreme government paying any attention to their
letters so long as we were illegally detained.

All this sank deep into the Dewan's heart : he answered,
"You have spoken truth, and I will submit it all to the
Rajah;" but at the same time he urged that there was
nothing dishonourable in the imprisonment, and that the
original violence being all a mistake, it should be overlooked by both parties. We parted on good terms, and
heard shortly after the second conference that our release
was promised and arranged : when a communication *
from Dorjiling changed their plans, the Dewan conveniently
fell sick on the spot, and we were thrown back again.

In the meantime, however, we were allowed to write
to our friends, and to receive money and food, of which

* I need scarcely say that every step was taken at Dorjiling for our release, that
the most anxious solicitude for our safety could suggest. But the first communication to the Rajah, though it pointed out the heinous nature of his offence, was,
through a natural fear of exasperating our captors, couched in very moderate
language. The particulars of our seizure, and the reasons for it, and for our
further detention, were unknown at Dorjiling, or a very different line of policy
would have been pursued.

we stood in great need. I transmitted a private account
of the whole affair to the Governor-General, who was
unfortunately at Bombay, but to whose prompt and vigorous
measures we were finally indebted for our release. His
lordship expedited a despatch to the Rajah, such as the
latter was accustomed to receive from Nepal, Bhotan, or
Lhassa, and such as alone commands attention from these
half-civilized Indo-Chinese, who measure power by the
firmness of the tone adopted towards them; and who,
whether in Sikkim, Birmah, Siam, Bhotan, or China, have
too long been accustomed to see every article of our
treaties contravened, with no worse consequences than a
protest or a threat, which is never carried into execution
till some fatal step calls forth the dormant power of the
British Government.*

The end of the month arrived without bringing any
prospect of our release, whilst we were harassed by false
reports of all kinds. The Dewan went on the 25th to
a hot bath, a few hundred feet down the hill; he was led
past our hut, his burly frame tottering as if in great
weakness, but a more transparent fraud could not have
been practised : he was, in fact, lying on his oars, pending
further negociations. The Amlah proposed that Campbell
should sign a bond, granting immunity for all past offences
on their part, whilst they were to withdraw the letter of
grievances against him. The Lamas cast horoscopes for the

---

* We forget that all our concessions to these people are interpreted into weakness; that they who cannot live on an amicable equality with one another, cannot
be expected to do so with us; that all our talk of power and resources are mere
boasts to habitual bullies, so long as we do not exert ourselves in the correction of
premeditated insults. No Government can be more tolerant, more sincerely desirous
of peace, and more anxious to confine its sway within its own limits than that of
India, but it can only continue at peace by demanding respect, and the punctilious
enforcement of even the most trifling terms in the treaties it makes with Indo-
Chinese.

future, little presents continually arrived for us, and the Ranee sent me some tobacco, and to Campbell brown sugar and Murwa beer. The blacksmiths, who had been ostentatiously making long knives at the forge hard by, were dismissed; troops were said to be arriving at Dorjiling, and a letter sternly demanding our release had been received.

The Lamas of Pemiongchi, Changachelling, Tassiding, &c., and the Dewan's enemies, and Tchebu Lama's friends, began to flock from all quarters to Tumloong, demanding audience of the Rajah, and our instant liberation. The Dewan's game was evidently up; but the timidity of his opponents, his own craft, and the habitual dilatoriness of all, contributed to cause endless delays. The young Gangtok Kajee tried to curry favour with us, sending word that he was urging our release, and adding that he had some capital ponies for us to see on our way to Dorjiling! Many similar trifles showed that these people had not a conception of the nature of their position, or of that of an officer of the British Government.

The Tchebu Lama visited us only once, and then under surveillance; he renewed his professions of good faith, and we had every reason to know that he had suffered severely for his adherence to us, and consistent repudiation of the Amlah's conduct; he was in great favour with his brother Lamas, but was not allowed to see the Rajah, who was said to trust to him alone of all his counsellors. He told us that peremptory orders had arrived from Calcutta for our release, but that the Amlah had replied that they would not acknowledge the despatch, from its not bearing the Governor-General's great seal! The country-people refusing to be saddled with the keep of our coolies, they were sent to Dorjiling in small parties, charged to say that we were free, and following them.

The weather continued rainy and bad, with occasionally a few hours of sunshine, which, however, always rendered the ditch before our door offensive: we were still prevented leaving the hut, but as a great annual festival was going on, we were less disagreeably watched. Campbell was very unwell, and we had no medicine; and as the Dewan, accustomed to such duplicity himself, naturally took this for a *ruse*, and refused to allow us to send to Dorjiling for any, we were more than ever convinced that his own sickness was simulated.

On the 2nd and 3rd December we had further conferences with the Dewan, who said that we were to be taken to Dorjiling in six days, with two Vakeels from the Rajah. The Pemiongchi Lama, as the oldest and most venerated in Sikkim, attended, and addressed Campbell in a speech of great feeling and truth. Having heard, he said, of these unfortunate circumstances a few days ago, he had come on feeble limbs, and though upwards of seventy winters old, as the representative of his holy brotherhood, to tender advice to his Rajah, which he hoped would be followed. Since Sikkim had been connected with the British rule, it had experienced continued peace and protection; whereas before they were in constant dread of their lives and properties, which, as well as their most sacred temples, were violated by the Nepalese and Bhotanese. He then dwelt upon Campbell's invariable kindness and good feeling, and his exertions for the benefit of their country, and for the cementing of friendship, and hoped he would not let these untoward events induce an opposite course in future; but that he would continue to exert his influence with the Governor-General in their favour.

The Dewan listened attentively; he was anxious and

perplexed, and evidently losing his presence of mind : he
talked to us of Lhassa and its gaieties, dromedaries, Lamas,
and everything Tibetan; offered to sell us ponies cheap,
and altogether behaved in a most undignified manner;
ever and anon calling attention to his pretended sick leg,
which he nursed on his knee.   He gave us the acceptable
news that the government at Calcutta had sent up an
officer to carry on Campbell's duties, which had alarmed him
exceedingly.   The Rajah, we were told, was very angry at
our seizure and detention ; he had no fault to find with
the Governor-General's agent, and hoped he would be continued as such.   In fact, all the blame was thrown on the
brothers of the Dewan, and of the Gangtok Kajee, and
more irresponsible stupid boors could not have been found
on whom to lay it, or who would have felt less inclined
to commit such folly if it had not been put on them by
the Dewan.   On leaving, white silk scarfs were thrown
over our shoulders, and we went away, still doubtful, after
so many disappointments, whether we should really be
set at liberty at the stated period.

Although there was so much talk about our leaving, our
confinement continued as rigorous as ever. The Dewan
curried favour in every other way, sending us Tibetan
wares for purchase, with absurd prices attached, he being
an arrant pedlar. All the principal families waited on us,
desiring peace and friendship. The coolies who had not
been dismissed were allowed to run away, except my
Bhotan Sirdar, Nimbo, against whom the Dewan was
inveterate : * he, however, managed soon afterwards to
break a great chain with which his legs were shackled, and

* The Sikkim people are always at issue with the Bhotanese. Nimbo was a
runaway slave of the latter country, who had been received into Sikkim, and
retained there until he took up his quarters at Dorjiling.

marching at night, eluded a hot pursuit, and proceeded to the Teesta, swam the river, and reached Dorjiling in eight days; arriving with a large iron ring on each leg, and a link of several pounds weight attached to one.

Parting presents arrived from the Rajah on the 7th, consisting of ponies, cloths, silks, woollens, immense squares of butter, tea, and the usual et ceteras, to the utter impoverishment of his stores : these he offered to the two Sahibs, "in token of his amity with the British government, his desire for peace, and deprecation of angry discussions." The Ranee sent silk purses, fans, and such Tibetan paraphernalia, with an equally amicable message, that "she was most anxious to avert the consequences of whatever complaints had gone forth against Dr. Campbell, who might depend on her strenuous exertions to persuade the Rajah to do whatever he wished !" These friendly messages were probably evoked by the information that an English regiment, with three guns, was on its way to Sikkim, and that 300 of the Bhaugulpore Rangers had already arrived there. The government of Bengal sending another agent * to Dorjiling, was also a contingency they had not anticipated, having fully expected to get rid of any such obstacle to direct communication with the Governor-General.

A present from the whole population followed that of the Ranee, coupled with earnest entreaties that Campbell would resume his position at Dorjiling; and on the following day forty coolies mustered to arrange the baggage. Before we left, the Ranee sent three rupees to buy a

---

* Mr. Lushington, the gentleman sent to conduct Sikkim affairs during Dr. Campbell's detention : to whom I shall ever feel grateful for his activity in our cause, and his unremitting attention to every little arrangement that could alleviate the discomforts and anxieties of our position.

yard of chalé and some gloves, accompanying them with
a present of white silk, &c., for Mrs. Campbell, to whom
the commission was intrusted : a singular instance of the
insouciant simplicity of these odd people.

The 9th of December was a splendid and hot day, one
of the very few we had had during our captivity. We
left at noon, descending the hill through an enormous
crowd of people, who brought farewell presents, all wishing
us well. We were still under escort as prisoners of the
Dewan, who was coolly marching a troop of forty unloaded
mules and ponies, and double that number of men's loads
of merchandize, purchased during the summer in Tibet, to
trade with at Dorjiling and the Titalya fair! His impu-
dence or stupidity was thus quite inexplicable ; treating
us as prisoners, ignoring every demand of the authorities
at Dorjiling, of the Supreme Council of Calcutta, and of
the Governor-General himself ; and at the same time acting
as if he were to enter the British territories on the most
friendly and advantageous footing for himself and his
property, and incurring so great an expense in all this as
to prove that he was in earnest in thinking so.

Tchebu Lama accompanied us, but we were not allowed
to converse with him. We halted at the bottom of the
valley, where the Dewan invited us to partake of tea ; from
this place he gave us mules * or ponies to ride, and we
ascended to Yankoong, a village 3,867 feet above the sea.
On the following day we crossed a high ridge from the
Ryott valley to that of the Rungmi ; where we camped at
Tikbotang (alt. 3,763 feet), and on the 11th at Gangtok
Sampoo, a few miles lower down the same valley.

We were now in the Soubahship of the Gangtok Kajee, a

---

* The Tibet mules are often as fine as the Spanish : I rode one which had performed a journey from Choombi to Lhassa in fifteen days, with a man and load.

member of the oldest and most wealthy family in Sikkim;
he had from the first repudiated the late acts of the Amlah,
in which his brother had taken part, and had always
been hostile to the Dewan. The latter conducted himself
with disagreeable familiarity towards us, and *hauteur*
towards the people; he was preceded by immense kettle-
drums, carried on men's backs, and great hand-bells, which
were beaten and rung on approaching villages; on which
occasions he changed his dress of sky-blue for yellow silk
robes worked with Chinese dragons, to the indignation of
Tchebu Lama, an amber robe in polite Tibetan society
being sacred to royalty and the Lamas. We everywhere
perceived unequivocal symptoms of the dislike with
which he was regarded. Cattle were driven away, villages
deserted, and no one came to pay respects, or bring
presents, except the Kajees, who were ordered to attend,
and his elder brother, for whom he had usurped an
estate near Gangtok.

On the 13th, he marched us a few miles, and then
halted for a day at Serriomsa (alt. 2,820 feet), at the
bottom of a hot valley full of irrigated rice-crops and
plantain and orange-groves. Here the Gangtok Kajee
waited on us with a handsome present, and informed us
privately of his cordial hatred of the "upstart Dewan," and
hopes for his overthrow; a demonstration of which we took
no notice.* The Dewan's brother (one of the Amlah) also
sent a large present, but was ashamed to appear. Another
letter reached the Dewan here, directed to the Rajah; it
was from the Governor-General at Bombay, and had
been sent across the country by special messengers:

* Nothing would have been easier than for the Gangtok Kajee, or any other
respectable man in Sikkim, to have overthrown the Dewan and his party; but
these people are intolerably apathetic, and prefer being tyrannized over to the
trouble of shaking off the yoke.

it demanded our instant release, or his Raj would be
forfeited; and declared that if a hair of our heads were
touched, his life should be the penalty.

The Rajah was also incessantly urging the Dewan to
hasten us onwards as free men to Dorjiling, but the latter
took all remonstrances with assumed coolness, exercised his
ponies, played at bow and arrow, intruded on us at meal-
times to be invited to partake, and loitered on the road,
changing garments and hats, which he pestered us to buy.
Nevertheless, he was evidently becoming daily more nervous
and agitated.

From the Rungmi valley we crossed on the 14th south-
·ward to that of Runniok, and descended to Dikkeeling, a
large village of Dhurma Bhoteeas (Bhotanese), which is
much the most populous, industrious, and at the same time
turbulent, in Sikkim.   It is 4,950 feet above the sea, and
occupies many broad cultivated spurs facing the south.
This district once belonged to Bhotan, and was ceded to the
Sikkim Rajah by the Paro Pilo,* in consideration of some
military services, rendered by the former in driving off the
Tibetans, who had usurped it for the authorities of Lhassa.
Since then the Sikkim and Bhotan people have repeatedly
fallen out, and Dikkeeling has become a refuge for runaway
Bhotanese, and kidnapping is constantly practised on this
frontier.

The Dewan halted us here for three days, for no assigned
cause.   On the 16th, letters arrived, including a most kind
and encouraging one from Mr. Lushington, who had taken
charge of Campbell's office at Dorjiling.   Immediately after
arriving, the messenger was seized with violent vomitings
and gripings: we could not help suspecting poison, espe-

---

* The temporal sovereign, in contra-distinction to the Dhurma Rajah, or
spiritual sovereign of Bhotan.

cially as we were now amongst adherents of the Dewan, and the Bhotanese are notorious for this crime. Only one means suggested itself for proving this, and with Campbell's permission I sent my compliments to the Dewan, with a request for one of his hunting dogs to eat the vomit. It was sent at once, and performed its duty without any ill effects. I must confess to having felt a malicious pleasure in the opportunity thus afforded of showing our jailor how little we trusted him ; feeling indignant at the idea that he should suppose he was making any way in our good opinion by his familiarities, which we were not in circumstances to resist.

The crafty fellow, however, outwitted me by inviting us to dine with him the same day, and putting our stomachs and noses to a severe test. Our dinner was served in Chinese fashion, but most of the luxuries, such as *béche-de-mer*, were very old and bad. We ate, sometimes with chop-sticks, and at others with Tibetan spoons, knives, and two-pronged forks. After the usual amount of messes served in oil and salt water, sweets were brought, and a strong spirit. Thoba-sing, our filthy, cross-eyed spy, was waiter, and brought in every little dish with both hands, and raised it to his greasy forehead, making a sort of half bow previous to depositing it before us. Sometimes he undertook to praise its contents, always adding, that in Tibet none but very great men indeed partook of such sumptuous fare. Thus he tried to please both us and the Dewan, who conducted himself with pompous hospitality, showing off what he considered his elegant manners and graces. Our blood boiled within us at being so patronised by the squinting ruffian, whose insolence and ill-will had sorely aggravated the discomforts of our imprisonment.

Not content with giving us what he considered a magnificent dinner (and it had cost him some trouble), the

Dewan produced a little bag from a double-locked escritoire, and took out three dinner-pills, which he had received as a great favour from the Rimbochay Lama, and which were a sovereign remedy for indigestion and all other ailments; he handed one to each of us, reserving the third for himself. Campbell refused his; but there appeared no help for me, after my groundless suspicion of poison, and so I swallowed the pill with the best grace I could. But in truth, it was not poison I dreaded in its contents, so much as being compounded of some very questionable materials, such as the Rimbochay Lama blesses and dispenses far and wide. To swallow such is a sanctifying work, according to Boodhist superstition, and I believe there was nothing in the world, save his ponies, to which the Dewan attached a greater value.

To wind up the feast, we had pipes of excellent mild yellow Chinese tobacco called "Tseang," made from *Nicotiana rustica*, which is cultivated in East Tibet, and in West China according to MM. Huc and Gabet. It resembles in flavour the finest Syrian tobacco, and is most agreeable when the smoke is passed through the nose. The common tobacco of India (*Nicotiana Tabacum*) is much imported into Tibet, where it is called "Tamma," (probably a corruption of the Persian "Toombac,") and is said to fetch the enormous price of 30s. per lb. at Lhassa, which is sixty times its value in India. Rice at Lhassa, when cheap, sells at 2s. for 5 lbs.; it is, as I have elsewhere said, all bought up for rations for the Chinese soldiery.

The Bhotanese are more industrious than the Lepchas, and better husbandmen; besides having superior crops of all ordinary grains, they grow cotton, hemp, and flax. The cotton is cleansed here as elsewhere, with a simple gin. The Lepchas use no spinning wheel, but a spindle and distaff; their loom, which is Tibetan, is a very complicated one framed of bamboo; it is worked by hand, without beam, treddle, or shuttle.

On the 18th we were marched, three miles only, to Singdong (alt. 2,116 feet), and on the following day five miles farther, to Katong Ghat (alt. 750 feet), on the Teesta river, which we crossed with rafts, and camped on the opposite bank, a few miles above the junction of this river with the Great Rungeet. The water, which is sea-green in colour, had a temperature of 53° 5 at 4 P.M., and 51° 7 the following morning; its current was very powerful. The rocks, since leaving Tumloong, had been generally micaceous, striking north-west, and dipping north-east. The climate was hot, and the vegetation on the banks tropical; on the hills around, lemon-bushes ("Kucheala," Lepcha) were abundant, growing apparently wild.

The Dewan was now getting into a very nervous and depressed state; he was determined to keep up appearances before his followers, but was himself almost servile to us; he caused his men to make a parade of their arms, as if to intimidate us, and in descending narrow gullies we had several times the disagreeable surprise of finding some of his men at a sudden turn, with drawn bows and arrows pointed towards us. Others gesticulated with their long knives, and made fell swoops at soft plantain-stems; but these artifices were all as shallow as they were contemptible, and a smile at such demonstrations was generally answered with another from the actors.

From Katong we ascended the steep east flank of Tendong or Mount Ararat, through forests of Sal and long leaved pine, to Namten (alt. 4,483 feet), where we again halted two days. The Dingpun Tinli lived near, and

waited on us with a present, which, with all others
that had been brought, Campbell received officially, and
transferred to the authorities at Dorjiling.

The Dewan was thoroughly alarmed at the news here
brought in, that the Rajah's present of yaks, ponies, &c.,
which had been sent forward, had been refused at Dorjiling;
and equally so at the clamorous messages which reached
him from all quarters, demanding our liberation; and at
the desertion of some of his followers, on hearing that large
bodies of troops were assembling at Dorjiling. Repudiated
by his Rajah and countrymen, and paralysed between his
dignity and his ponies, which he now perceived would not
be welcomed at the station, and which were daily losing
flesh, looks, and value in these hot valleys, where there is no
grass pasture, he knew not what olive-branch to hold out
to our government, except ourselves, whom he therefore
clung to as hostages.

On the 22nd of December he marched us eight miles
further, to Cheadam, on a bold spur 4,653 feet high,
overlooking the Great Rungeet, and facing Dorjiling, from
which it was only twenty miles distant. The white
bungalows of our friends gladdened our eyes, while the
new barracks erecting for the daily arriving troops struck
terror into the Dewan's heart. The six Sepoys* who had
marched valiantly beside us for twenty days, carrying the
muskets given to the Rajah the year before by the
Governor-General, now lowered their arms, and vowed
that if a red coat crossed the Great Rungeet, they would
throw down their guns and run away. News arrived

---

* These Sepoys, besides the loose red jacket and striped Lepcha kirtle, wore a
very curious national black hat of felt, with broad flaps turned up all round : this
is represented in the right-hand figure. A somewhat similar hat is worn by some
classes of Nepal soldiery.

that the Bhotan inhabitants of Dorjiling headed by my
bold Sirdar Nimbo, had arranged a night attack for our
release; an enterprise to which they were quite equal, and
in which they have had plenty of practice in their own
misgoverned country.   Watch-fires gleamed amongst the
bushes, we were thrust into a doubly-guarded house, and

LEPCHA SEPOYS.   TIBETAN SEPOYS IN THE BACK-GROUND.

(See p. 160.)

bows and arrows were ostentatiously levelled so as to rake
the doorway, should we attempt to escape.   Some of the
ponies were sent back to Dikkeeling, though the Dewan
still clung to his merchandise and the feeble hope of traffic.
The confusion increased daily, but though Tchebu Lama
looked brisk and confident, we were extremely anxious;

scouts were hourly arriving from the road to the Great
Rungeet, and if our troops had advanced, the Dewan might
have made away with us from pure fear.

In the forenoon he paid us a long visit, and brought
some flutes, of which he gave me two very common ones
of apricot wood from Lhassa, producing at the same time a
beautiful one, which I believe he intended for Campbell,
but his avarice got the better, and he commuted his gift
into the offer of a tune, and pitching it in a high key, he
went through a Tibetan air that almost deafened us by its
screech.   He tried bravely to maintain his equanimity,
but as we preserved a frigid civility and only spoke when
addressed, the tears would start from his eyes in the
pauses of conversation.   In the evening he came again; he
was excessively agitated and covered with perspiration,
and thrust himself unceremoniously between us on the
bench we occupied.   As his familiarity increased, he
put his arm round my. neck, and as he was armed with
a small dagger, I felt rather uneasy about his intentions,
but he ended by forcing on my acceptance a coin,
value threepence, for he was in fact beside himself with
terror.

Next morning Campbell received a hint that this was
a good opportunity for a vigorous remonstrance.   The
Dewan came with Tchebu Lama, his own younger brother
(who was his pony driver), and the Lassoo Kajee. The latter
had for two months placed himself in an attitude of hostility
opposite Dorjiling, with a ragged company of followers,
but he now sought peace and friendship as much as the
Dewan; the latter told us he was waiting for a reply to a
letter addressed to Mr. Lushington, after which he would
set us free.   Campbell said: "As you appear to have
made up your mind, why not dismiss us at once?"   He

answered that we should go the next day at all events. Here I came in, and on hearing from Campbell what had passed, I added, that he had better for his own sake let us go at once; that the next day was our great and only annual Poojah (religious festival) of Christmas, when we all met; whereas he and his countrymen had dozens in the year. As for me, he knew I had no wife, nor children, nor any relation, within thousands of miles, and it mattered little where I was, he was only bringing ruin on himself by his conduct to me as the Governor-General's friend; but as regarded Campbell, the case was different; his home was at Dorjiling, which was swarming with English soldiers, all in a state of exasperation, and if he did not let us depart before Christmas, he would find Dorjiling too hot to hold him, let him offer what reparation he might for the injuries he had done us. I added: "We are all ready to go—dismiss us." The Dewan again turned to Campbell, who said, "I am quite ready; order us ponies at once, and send our luggage after us." He then ordered the ponies, and three men, including Meepo, to attend us; whereupon we walked out, mounted, and made off with all speed.

We arrived at the cane bridge over the Great Rungeet at 4 P.M., and to our chagrin found it in the possession of a posse of ragged Bhoteeas, though there were thirty armed Sepoys of our own at the guard-house above. At Meepo's order they cut the network of fine canes by which they had rendered the bridge impassable, and we crossed. The Sepoys at the guard-house turned out with their clashing arms and bright accoutrements, and saluted to the sound of bugles; scaring our three companions, who ran back as fast as they could go. We rode up that night to Dorjiling, and I arrived at 8 P.M. at Hodgson's house,

where I was taken for a ghost, and received with shouts of
welcome by my kind friend and his guest Dr. Thomson,
who had been awaiting my arrival for upwards of a
month.

Thus terminated our Sikkim captivity, and my last
Himalayan exploring journey, which in a botanical and
geographical point of view had answered my purposes
beyond my most sanguine expectations, though my collections had been in a great measure destroyed by so many
untoward events.   It enabled me to survey the whole
country, and to execute a map of it, and Campbell had
further gained that knowledge of its resources which the
British government should all along have possessed, as the
protector of the Rajah and his territories.

It remains to say a few words of the events that succeeded our release, in so far as they relate to my connection
with them.   The Dewan moved from Cheadam to Namtchi,
immediately opposite Dorjiling, where he remained
throughout the winter.   The supreme government of
Bengal demanded of the Rajah that he should deliver up
the most notorious offenders, and come himself to Dorjiling, on pain of an army marching to Tumloong to
enforce the demand; a step which would have been easy,
as there were neither troops, arms, ammunition, nor other
means of resistance, even had there been the inclination
to stop us, which was not the case.   The Rajah would in
all probability have delivered himself up at Tumloong,
throwing himself on our mercy, and the army would
have sought the culprits in vain, both the spirit and the
power to capture them being wanting on the part of the
people and their ruler.

The Rajah expressed his willingness, but pleaded his
inability to fulfil the demand, whereupon the threat was

repeated, and additional reinforcements were moved on to Dorjiling. The general officer in command at Dinapore was ordered to Dorjiling to conduct operations : his skill and bravery had been proved during the progress of the Nepal war so long ago as 1815. From the appearance of the country about Dorjiling, he was led to consider Sikkim to be impracticable for a British army This was partly owing to the forest-clad mountains, and partly to the fear of Tibetan troops coming to the Rajah's aid, and the Nepalese * taking the opportunity to attack us. With the latter we were in profound peace, and we had a resident at their court; and I have elsewhere shown the impossibility of a Tibet invasion, even if the Chinese or Lhassan authorities were inclined to interfere in the affairs of Sikkim, which they long ago formally declined doing in the case of aggressions of the Nepalese and Bhotanese, the Sikkim Rajah being under British protection.†

* Jung Bahadoor was at this time planning his visit to England, and to his honour I must say, that on hearing of our imprisonment he offered to the govern-ment at Calcutta to release us with a handful of men. This he would no doubt have easily effected, but his offer was wisely declined, for the Nepalese (as I have elsewhere stated) want Sikkim and Bhotan too, and we had undertaken the protection of the former country, mainly to keep the Nepalese out of it.

† The general officer considered that our troops would have been cut to pieces if they entered the country ; and the late General Sir Charles Napier has since given evidence to the same effect. Having been officially asked at the time whether I would guide a party into the country, and having drawn up (at the request of the general officer) plans for the purpose, and having given it as my opinion that it would not only have been feasible but easy to have marched a force in peace and safety to Tumloong, I feel it incumbent on me here to remark, that I think General Napier, who never was in Sikkim, and wrote from many hundred miles' distance, must have misapprehended the state of the case. Whether an invasion of Sikkim was either advisable or called for, was a matter in which I had no concern : nor do I offer an opinion as to the impregnability of the country if it were defended by natives otherwise a match for a British force, and having the advantage of position. I was not consulted with reference to any difference of opinion between the civil and military powers, such as seems to have called for the expression of Sir Charles Napier's opinion on this matter, and which appears to be considerably overrated in his evidence.

The general officer honoured me with his friendship at Dorjiling, and to

There were not wanting offers of leading a company of soldiers to Tumloong, rather than that the threat should have twice been made, and then withdrawn; but they were not accepted. A large body of troops was however marched from Dorjiling, and encamped on the north bank of the Great Rungeet for some weeks: but after that period they were recalled, without any further demonstration; the Dewan remaining encamped the while on the Namtchi hill, not three hours' march above them. The simple Lepchas daily brought our soldiers milk, fowls, and eggs, and would have continued to do so had they proceeded to Tumloong, for I believe both Rajah and people would have rejoiced at our occupation of the country.

After the withdrawal of the troops, the threat was modified into a seizure of the Terai lands, which the Rajah had originally received as a free gift from the British, and which were the only lucrative or fertile estates he possessed. This was effected by four policemen taking possession of the treasury (which contained exactly twelve shillings, I believe), and announcing to the villagers the confiscation of the territory to the British government, in which they gladly acquiesced. At the same time there was annexed to it the whole southern part of Sikkim, between the Great Rungeet and the plains of India, and from Nepal on the west to the Bhotan frontier and the Teesta river on the east; thus confining the Rajah to his

Mr. Lushington, I am, as I have elsewhere stated, under great obligations for his personal consideration and kindness, and vigorous measures during my detention. On my release and return to Dorjiling, any interference on my part would have been meddling with what was not my concern. I never saw, nor wished to see, a public document connected with the affair, and have only given as many of the leading features of the case as I can vouch for, and as were accessible to any other bystander.

mountains, and cutting off all access to the plains, except through the British territories. To the inhabitants (about 5000 souls) this was a matter of congratulation, for it only involved the payment of a small fixed tax in money to the treasury at Dorjiling, instead of a fluctuating one in kind, with service to the Rajah, besides exempting them from further annoyance by the Dewan. At the present time the revenues of the tract thus acquired have doubled, and will very soon be quadrupled: every expense of our detention and of the moving of troops, &c., has been already repaid by it, and for the future all will be clear profit; and I am given to understand that this last year it has realized upwards of 30,000 rupees (£3000).

Dr. Campbell resumed his duties immediately afterwards, and the newly-acquired districts were placed under his jurisdiction. The Rajah still begs hard for the renewal of old friendship, and the restoration of his Terai land, or the annual grant of £300 a year which he formerly received. He has forbidden the culprits his court, but can do no more. The Dewan, disgraced and turned out of office, is reduced to poverty, and is deterred from entering Tibet by the threat of being dragged to Lhassa with a rope round his neck. Considering, however, his energy, a rare quality in these countries, I should not be surprised at his yet cutting a figure in Bhotan, if not in Sikkim itself: especially if, at the Rajah's death, the British government should refuse to take the country under its protection. The Singtam Soubah and the other culprits live disgraced at their homes. Tchebu Lama has received a handsome reward, and a grant of land at Dorjiling, where he resides, and whence he sends me his salaams by every opportunity.

CHAPTER XXVII.

I was chiefly occupied during January and February of
1850, in arranging and transmitting my collections to
Calcutta, and completing my manuscripts, maps, and surveys.   My friend Dr. Thomson having joined me here, for
the purpose of our spending a year in travelling and botanising together, it became necessary to decide on the best
field for our pursuits.   Bhotan offered the most novelty,
but it was inaccessible to Europeans; and we therefore
turned our thoughts to Nepal, and failing that, to the
Khasia mountains.

The better to expedite our arrangements, I made a trip
to Calcutta in March, where I expected to meet both Lord
Dalhousie, on his return from the Straits of Malacca, and
Jung Bahadoor (the Nepalese minister), who was then *en
route* as envoy to England.   I staid at Government House,

where every assistance was afforded me towards obtaining
the Nepal Rajah's permission to proceed through the
Himalaya from Dorjiling to Katmandu. Jung Bahadoor
received me with much courtesy, and expressed his great
desire to serve me; but begged me to wait until his return
from England, as he could not be answerable for my personal safety when travelling during his absence; and he

DR. FALCONER'S RESIDENCE, CALCUTTA BOTANIC GARDENS, FROM SIR L. PEEL'S GROUNDS.

referred to the permission he had formerly given me (and
such was never before accorded to any European) in earnest
of his disposition, which was unaltered. We therefore
determined upon spending the season of 1850 in the
Khasia mountains in eastern Bengal, at the head of the
great delta of the Ganges and Burrampooter.

I devoted a few days to the Calcutta Botanic Gardens, where I found my kind friend Dr. Falconer established, and very busy. The destruction of most of the palms, and of all the noble tropical features of the gardens, during Dr. Griffith's incumbency, had necessitated the replanting of the greater part of the grounds, the obliteration of old walks, and the construction of new: it was also necessary to fill up tanks whose waters, by injudicious cuttings, were destroying some of the most valuable parts of the land, to drain many acres, and to raise embankments to prevent the encroachments of the Hoogly: the latter being a work attended with great expense, now cripples the resources of the garden library, and other valuable adjuncts; for the trees which were planted for the purpose having been felled and sold, it became necessary to buy timber at an exorbitant price.

The avenue of Cycas trees (*Cycas circinalis*), once the admiration of all visitors, and which for beauty and singularity was unmatched in any tropical garden, had been swept away by the same unsparing hand which had destroyed the teak, mahogany, clove, nutmeg, and cinnamon groves. In 1847, when I first visited the establishment, nothing was to be seen of its former beauty and grandeur, but a few noble trees or graceful palms rearing their heads over a low ragged jungle, or spreading their broad leaves or naked limbs over the forlorn hope of a botanical garden, that consisted of open clay beds, disposed in concentric circles, and baking into brick under the fervid heat of a Bengal sun.

The rapidity of growth is so great in this climate, that within eight months from the commencement of the improvements, a great change had already taken place. The grounds bore a park-like appearance; broad shady walks had replaced the narrow winding paths that ran in distorted lines over the ground, and a large Palmetum, or collection of tall and graceful palms of various kinds, occupied several acres at one side of the garden; whilst a still larger portion of ground was being appropriated to a picturesque assemblage of certain closely allied families of plants, whose association promised to form a novel and attractive object of study to the botanist, painter, and landscape gardener. This, which the learned Director called in scientific language a Thamno-Endogenarium, consists of groups of all kinds of bamboos, tufted growing palms, rattan canes (*Calami*), *Dracænæ*, plantains, screw-pines, (*Pandani*), and such genera of tropical monocotyledonous plants. All are evergreens of most vivid hue, some of which, having slender trailing stems, form magnificent masses; others twine round one another, and present impenetrable hillocks of green foliage; whilst still others shoot out broad long wavy leaves from tufted roots; and a fourth class is supported by aerial roots, diverging on all sides and from all heights on the stems, every branch of which is crowned with an enormous plume of grass-like leaves.*

The great *Amherstia* tree had been nearly killed by injudicious treatment, and the baking of the soil above its roots. This defect was remedied by sinking bamboo pipes four feet and a half in the earth, and watering through them—a plan first recommended by Major M'Farlane of Tavoy. Some fine *Orchideæ* were in flower in the gardens, but few of them fruit;

* Since I left India, these improvements have been still further carried out, and now (in the spring of 1853) I read of five splendid *Victoria* plants flowering at once, with *Euryale ferox*, white, blue, and red water-lilies, and white, yellow and scarlet lotus, rendering the tanks gorgeous, sunk as their waters are in frames of green grass, ornamented with clumps of *Nipa fruticans* and *Phœnix paludosa*.

and those *Dendrobiums* which bear axillary viviparous buds never do. Some of the orchids appear to be spread by birds amongst the trees; but the different species of *Vanda* are increasing so fast, that there seems no doubt that this tribe of air-plants grows freely from seed in a wild state, though we generally fail to rear them in England.

The great Banyan' tree (*Ficus Indica*) is still the pride and ornament of the garden. Dr. Falconer has ascertained satisfactorily that it is only seventy-five years old: annual rings, size, &c., afford no evidence in such a case, but people were alive a few years ago who remembered well its site being occupied in 1782 by a Kujoor (Date-palm), out of whose crown the Banyan sprouted, and beneath which a Fakir sat. It is a remarkable fact that the banyan hardly ever vegetates on the ground; but its figs are eaten by birds, and the seeds deposited in the crowns of palms, where they grow, sending down roots that embrace and eventually kill the palm, which decays away. This tree is now eighty feet high, and throws an area 300 feet * in diameter into a dark, cool shade. The gigantic limbs spread out about ten feet above the ground, and from neglect during Dr. Wallich's absence, there were on Dr. Falconer's arrival no more than eighty-nine descending roots or props; there are now several hundreds, and the growth of this grand mass of vegetation is proportionably stimulated and increased. The props are induced to sprout by wet clay and moss tied to the branches, beneath which

---

* Had this tree been growing in 1849 over the great palm-stove at Kew, only thirty feet of each end of that vast structure would have been uncovered: its increase was proceeding so rapidly, that by this time it could probably cover the whole. Larger banyans are common in Bengal; but few are so symmetrical in shape and height. As the tree gets old, it breaks up into separate masses, the original trunk decaying, and the props becoming separate trunks of the different portions.

a little pot of water is hung, and after they have made some progress, they are inclosed in bamboo tubes, and so coaxed down to the ground. They are mere slender whip-cords before reaching the earth, where they root, remaining very lax for several months ; but gradually, as they grow and swell to the size of cables, they tighten, and eventually become very tense. This is a curious pheno- menon, and so rapid, that it appears to be due to the rooting part mechanically dragging down the aerial. The branch meanwhile continues to grow outwards, and being supplied by its new support, thickens beyond it, whence the props always slant outwards from the ground towards the circumference of the tree.

Cycas trees abound in the gardens, and, though generally having only one, or rarely two crowns, they have sometimes sixteen, and their stems are everywhere covered with leafy buds, which are developed on any check being given to the growth of the plant, as by the operation of trans- plantation, which will cause as many as 300 buds to appear in the course of a few years, on a trunk eight feet high.

During my stay at the gardens, Dr. Falconer received a box of living plants packed in moss, and transported in a frozen state by one of the ice ships from North America : * they left in November, and arriving in March, I was present at the opening of the boxes, and saw 391 plants (the whole contents) taken out in the most perfect state. They were chiefly fruit-trees, apples, pears, peaches, currants, and gooseberries, with beautiful plants of the Venus' fly-trap (*Dionæa muscipula*). More perfect success never attended an experiment : the plants were in vigorous

---

* The ice from these ships is sold in the Calcutta market for a penny a pound, to great profit; it has already proved an invaluable remedy in cases of inflammation and fever, and has diminished mortality to a very appreciable extent.

bud, and the day after being released from their icy bonds, the leaves sprouted and unfolded, and they were packed in Ward's cases for immediate transport to the Himalaya mountains.

My visit to Calcutta enabled me to compare my instruments with the standards at the Observatory, in which I was assisted by my friend, Capt. Thuillier, to whose kind offices on this and many other occasions I am greatly indebted.

I returned to Dorjiling on the 17th of April, and Dr. Thomson and I commenced our arrangements for proceeding to the Khasia mountains. We started on the 1st of May, and I bade adieu to Dorjiling with no light heart; for I was leaving the kindest and most disinterested friends I had ever made in a foreign land, and a country whose mountains, forests, productions, and people had all become endeared to me by many ties and associations. The prospects of Dorjiling itself are neither doubtful nor insignificant. Whether or not Sikkim will fall again under the protection of Britain, the station must prosper, and that very speedily. I had seen both its native population and its European houses doubled in two years; its salubrious climate, its scenery, and accessibility, ensure it so rapid a further increase that it will become the most populous hill-station in India. Strong prejudices against a damp climate, and the complaints of loungers and idlers who only seek pleasure, together with a groundless fear of the natives, have hitherto retarded its progress; but its natural advantages will outweigh these and all other obstacles.

I am aware that my opinion of the ultimate success of Dorjiling is not shared by the general public of India, and must be pardoned for considering their views in this matter short-sighted. With regard to the disagreeables of its climate, I can sufficiently appreciate them, and shall be considered by the residents to have over-estimated the amount and constancy of mist, rain, and humidity, from the two seasons I spent there being exceptional in these respects. Whilst on the one hand I am willing to admit the probability of this,* I may be allowed on the other to say that I have never visited any spot under the sun, where I was not told that the season was exceptional, and generally for the worse; added to which there is no better and equally salubrious climate east of Nepal, accessible from Calcutta.

All climates are comparative, and fixed residents naturally praise their own. I have visited many latitudes, and can truly say that I have found no two climates resembling each other, and that all alike are complained of. That of Dorjiling is above the average in point of comfort, and for perfect salubrity rivals any; while in variety, interest, and grandeur, the scenery is unequalled.

From Sikkim to the Khasia mountains our course was by boat down the Mahanuddy to the upper Gangetic delta, whose many branches we followed eastwards to the Megna; whence we ascended the Soormah to the Silhet district. We arrived at Kishengunj, on the Mahanuddy, on the 3rd of May, and were delayed two days for our boat, which should have been waiting here to take us to Berhampore on the Ganges: we were, however, hospitably received by Mr. Perry's family.

The approach of the rains was indicated by violent easterly storms of thunder, lightning, and rain; the thermometer ranging from 70° to 85°. The country around Kishengunj

---

* I am informed that hardly a shower of rain has fallen this season, between November 1852, and April 1853; and a very little snow in February only.

is flat and very barren ; it is composed of a deep sandy soil,
covered with a short turf, now swarming with cockchafers.
Water is found ten or twelve feet below the surface, and
may be supplied by underground streams from the
Himalaya, distant forty-five miles. The river, which at this
season is low, may be navigated up to Titalya during the
rains; its bed averages 60 yards in width, and is extremely
tortuous ; the current is slight, and, though shallow, the
water is opaque. We slowly descended to Maldah, where
we arrived on the 11th : the temperature both of the water
and of the air increased rapidly to upwards of 90°; the
former was always a few degrees cooler than the air by day,
and warmer by night. The atmosphere became drier as
we receded from the mountains.

The boatmen always brought up by the shore at night;
and our progress was so slow, that we could keep up
with the boat when walking along the bank. So long as
the soil and river-bed continued sandy, few bushes or
herbs were to be found, and it was difficult to collect a
hundred kinds of plants in a day : gradually, however,
clumps of trees appeared, with jujube bushes, *Trophis*,
Acacia, and *Buddleia*, a few fan-palms, bamboos, and Jack-
trees   A shell (*Anodon*) was the only one seen in the river,
which harboured few water-plants or birds, and neither
alligators nor porpoises ascend so high.

On the 7th of May, about eighty miles in a straight line
from the foot of the Himalaya, we found the stratified sandy
banks, which had gradually risen to a height of thirteen feet,
replaced by the hard alluvial clay of the Gangetic valley,
which underlies the sand : the stream contracted, and the
features of its banks were materially improved by a jungle
of tamarisk, wormwood (*Artemisia*), and white rose-bushes
(*Rosa involucrata*), whilst mango trees became common,

with tamarinds, banyan, and figs. Date and *Caryota* palms, and rattan canes, grew in the woods, and parasitic Orchids on the trees, which were covered with a climbing fern (*Acrosticum scandens*), so that we easily doubled our flora of the river banks before arriving at Maldah.

This once populous town is, like Berhampore, now quite decayed, since the decline of its silk and indigo trades : the staple product, called "Maldy," a mixture of silk and cotton, very durable, and which washes well, now forms its only trade, and is exported through Sikkim to the north-west provinces and Tibet. It is still famous for the size and excellence of its mangos, which ripen late in May; but this year the crop had been destroyed by the damp heats of spring, the usual north-west dry winds not having prevailed.

The ruins of the once famous city of Gour, a few miles distant, are now covered with jungle, and the buildings are fast disappearing, owing to the bricks being carried away to be used elsewhere

Below Maldah the river gets broader, and willow becomes common. We found specimens of a *Planorbis* in the mud of the stream, and saw apparently a boring shell in the alluvium, but could not land to examine it. Chalky masses of alligators' droppings, like coprolites, are very common, buried in the banks, which become twenty feet high at the junction with the Ganges, where we arrived on the 14th. The waters of this great river were nearly two degrees cooler than those of the Mahanuddy.

Rampore-Bauleah is a large station on the north bank of the Ganges, whose stream is at this season fully a mile wide, with a very slow current ; its banks are thirty feet above the water. We were most kindly received by Mr. Bell, the collector of the district, to whom we were greatly indebted for furthering us on our voyage : boats being very difficult to procure, we were, however, detained here from the 16th to the 19th. I was fortunate in being able to compare my barometers with a first-rate standard instrument, and in finding no appreciable alteration since leaving Calcutta in the previous April. The elevation of the station is 130 feet above the sea, that of Kishengunj I made 131 ; so that the Gangetic valley is nearly a dead level for fully a hundred miles north, beyond which it rises ; Titalya, 150 miles to the north, being 360 feet, and Siligoree, at the margin of the Terai, rather higher. The river again falls more considerably than the land ; the Mahanuddy, at Kishengunj, being about twenty feet below the level of the plains, or 110 above the sea ; whereas the Ganges, at Rampore, is probably not more than eighty feet, even when the water is highest.

The climate of Rampore is marked by greater extremes than that of Calcutta : during our stay the temperature rose above 106°, and fell to 78° at night : the mean was $2\frac{1}{2}$° higher than at Calcutta, which is 126 miles further south. Being at the head of the Gangetic delta, which points from the Sunderbunds obliquely to the north-west, it is much damper than any locality further west, as is evidenced by two kinds of *Calamus* palm abounding, which do not ascend the Ganges beyond Monghyr. Advancing eastwards, the dry north west wind of the Gangetic valley, which blows here in occasional gusts, is hardly felt ; and easterly winds, rising after the sun (or, in other words, following the heating of the open dry country), blow down the great valley of the Burrampooter, or south-easterly ones come up from the Bay of Bengal. The western head of the Gangetic delta is thus placed in what are called " the variables" in naval phraseology ; but only so far as its superficial winds are concerned, for its great atmospheric current always blows from the Bay of Bengal, and flows over all northern India, to the lofty regions of Central Asia.

At Rampore I found the temperature of the ground, at three feet depth, varied from 87° 8 to 89° 8, being considerably lower than that of the air (94° 2), whilst that of a fine ripening shaddock, into which I plunged a thermometer bulb, varied little from 81°, whether the sun shone on it or not. From this place we made very slow progress south-eastwards, with a gentle current, but against constant easterly winds, and often violent gales and thunder-storms, which obliged us to bring up under shelter of banks and islands of sand. Sometimes we sailed along the broad river, whose opposite shores were rarely both visible at once, and at others tracked the boat through narrow creeks that unite the many Himalayan streams, and form a network soon after leaving their mountain valleys.

A few miles beyond Pubna we passed from a narrow canal at once into the main stream of the Burrampooter at Jaffergunj: our maps had led us to expect that it flowed fully seventy miles to the eastward in this latitude ; and we were surprised to hear that within the last twenty years the main body of that river had shifted its course thus far to the westward. This alteration was not effected by the gradual working westwards of the main stream, but by the old eastern channel so rapidly silting up as to be now unnavigable ; while the Jummul, which receives the Teesta, and which is laterally connected by branches with the Burrampooter, became consequently wider and deeper, and eventually the principal stream.

Nothing can be more dreary and uninteresting than the scenery of this part of the delta. The water is clay-coloured and turbid, always cooler than the air, which

again was 4° or 5° below that of Calcutta, with a
damper atmosphere. The banks are of stratified sand
and mud, hardly raised above the mean level of the country,
and consequently unlike those bordering most annually
flooded rivers; for here the material is so unstable,
that the current yearly changes its course. A wiry grass
sometimes feebly binds the loose soil, on which there are
neither houses nor cultivation.

Ascending the Jummul (now the main channel of
the Burrampooter) for a few miles, we turned off
into a narrower channel, sixty miles long, which passes
by Dacca, where we arrived on the 28th, and where
we were again detained for boats, the demand for which
is rapidly increasing with the extended cultivation of the
Sunderbunds and Delta. We stayed with Mr. Atherton,
and botanised in the neighbourhood of the town, which
was once very extensive, and is still large, though
not flourishing. The population is mostly Mahometan;
the site, though beautiful and varied, is unhealthy for
Europeans. Ruins of great Moorish brick buildings
still remain, and a Greek style of ornamenting the houses
prevails to a remarkable degree.

The manufacture of rings for the arms and ancles, from
conch-shells imported from the Malayan Archipelago, is
still almost confined to Dacca : the shells are sawn across
for this purpose by semicircular saws, the hands and toes
being both actively employed in the operation. The introduction of circular saws has been attempted by some
European gentlemen, but steadily resisted by the natives,
despite their obvious advantages. The Dacca muslin
manufacture, which once employed thousands of hands, is
quite at an end, so that it was with great difficulty that the
specimens of these fabrics sent to the Great Exhibition

of 1851, were procured.   The kind of cotton (which is
very short in the staple) employed, is now hardly grown,
and scarcely a loom exists which is fit for the finest fabrics.
The jewellers still excel in gold and silver filagree.

Pine-apples, plantains, mangos, and oranges, abound in
the Dacca market, betokening a better climate for tropical
fruits than that of Western Bengal; and we also saw the
fruit of *Euryale ferox*,* which is round, soft, pulpy, and
the size of a small orange; it contains from eight to fifteen
round black seeds as large as peas, which are full of flour,
and are eaten roasted in India and China, in which latter
country the plant is said to have been in cultivation for·
upwards of 3000 years.

The native vegetation is very similar to that of the
Hoogly, except that the white rose is frequent here.   The
fact of a plant of this genus being as common on the plains
of Bengal as a dog-rose is in England, and associated with
cocoa-nuts, palms, mangos, plantains, and banyans, has
never yet attracted the attention of botanists, though the
species was described by Roxburgh.   As a geographical
fact it is of great importance, for the rose is usually considered a northern genus, and no kind but this inhabits a
damp hot tropical climate.   Even in mountainous countries
situated near the equator, as in the Himalaya and Andes,
wild roses are very rare, and only found at great elevations, whilst they are unknown in the southern hemisphere.
It is curious that this rose, which is also a native of Birma
and the Indian Peninsula, does not in this latitude grow

---

* An Indian water-lily with a small red flower, covered everywhere with
prickles, and so closely allied to *Victoria regia* as to be scarcely generically distinguishable from it.   It grows in the eastern Sunderbunds, and also in Kashmir.
The discoverer of Victoria called the latter "*Euryale Amazonica.*"   These interesting plants are growing side by side in the new Victoria house at Kew.   The
Chinese species has been erroneously considered different from the Indian one.

west of the meridian of 87°; it is confined to the upper
Gangetic delta, and inhabits a climate in which it would
least of all be looked for.

I made the elevation of Dacca by barometer only
seventy-two feet above the sea; and the banks of
the Dallisary being high, the level of its waters at this
season is scarcely above that of the Bay of Bengal.
The mean temperature of the air was $86\frac{3}{4}$ during our
stay, or half a degree lower than Calcutta at the same
period.

We pursued our voyage on the 30th of May, to the old
bed of the Burrampooter, an immense shallow sheet of
water, of which the eastern bank is for eighty miles occupied
by the delta of the Soormah. This river rises on the
Munnipore frontier, and flows through Cachar, Silhet, and
the Jheels of east Bengal, receiving the waters of the
Cachar, Jyntea, Khasia, and Garrow mountains (which
bound the Assam valley to the south), and of the Tipperah
hills, which stretch parallel to them, and divide the Soormah
valley from the Bay of Bengal. The immense area thus
drained by the Soormah is hardly raised above the level of
the sea, and covers about 10,000 square miles. The anastomosing rivers that traverse it, flow very gently, and do not
materially alter their course; hence their banks gradually
rise above the mean level of the surrounding country, and
on them the small villages are built, surrounded by extensive rice-fields that need no artificial irrigation. At this
season the general surface of the Jheels is marshy; but
during the rains, which are excessive on the neighbouring
mountains, they resemble an inland sea, the water rising
gradually to within a few inches of the floor of the huts;
as, however, it subsides as slowly in autumn, it commits
no devastation. The communication is at all seasons by

boats, in the management of which the natives (chiefly Mahometans) are expert.

The want of trees and shrubs is the most remarkable feature of the Jheels; in which respect they differ from the Sunderbunds, though the other physical features of each are similar, the level being exactly the same : for this difference there is no apparent cause, beyond the influence of the tide and sea atmosphere. Long grasses of tropical genera (*Saccharum, Donax, Andropogon*, and *Rottbœllia*) ten feet high, form the bulk of the vegetation, with occasional low bushes along the firmer banks of the natural canals that everywhere intersect the country ; amongst these the rattan cane (*Calamus*), rose, a laurel, *Stravadium*, and fig, are the most common ; while beautiful convolvuli throw their flowering shoots across the water.

The soil, which is sandy along the Burrampooter, is more muddy and clayey in the centre of the Jheels, with immense spongy accumulations of vegetable matter in the marshes, through which we poked the boat-staves without finding bottom : they were for the most part formed of decomposed grass roots, with occasionally leaves, but no quantity of moss or woody plants. Along the courses of the greater streams drift timber and various organic fragments are no doubt imbedded, but as there is no current over the greater part of the flooded surface, there can be little or no accumulation, except perhaps of old canoes, or of such vegetables as grow on the spot. The waters are dark-coloured, but clear and lucid, even at their height.

We proceeded up the Burrampooter, crossing it obliquely ; its banks were on the average five miles apart, and formed of sand, without clay, and very little silt or mud : the water was clear and brown, like that of the Jheels, and very different from that of the Jummul. We thence turned eastwards into the delta of the Soormah, which we traversed in a north-easterly direction to the stream itself. We often passed through very narrow channels, where the grasses towered over the boats : the boatmen steered in and out of them as they pleased, and we were utterly at a loss to know how they guided themselves, as they had neither compass nor map, and there were few villages or landmarks; and on climbing the mast we saw multitudes of other masts and sails peering over the grassy marshes, doing just the same as we did. All that go up have the south-west wind in their favour, and this helps them to their course, but beyond this they have no other guide but that instinct which habit begets. Often we had to retreat from channels that promised to prove short cuts, but which turned out to be blind alleys. Sometimes we sailed up broader streams of ches-nut-brown water, accompanied by fleets of boats repairing to the populous districts at the foot of the Khasia, for rice, timber, lime, coal, bamboos, and long reeds for thatching, all of which employ an inland navy throughout the year in their transport to Calcutta.

Leeches and mosquitos were very troublesome, the latter appearing in clouds at night; during the day they were rarer, but the species was the same. A large cray-fish was common, but there were few birds and no animals to be seen.

Fifty-four barometric observations, taken at the level of the water on the voyage between Dacca and the Soormah, and compared with Calcutta, showed a gradual rise of the mercury in proceeding eastwards ; for though the pressure at Calcutta was ·055 of an inch higher than at Dacca, it was ·034 lower than on the Soormah : the mean difference between all these observations and the cotemporaneous ones at Calcutta was + ·003 in favour of Calcutta, and the temperature half a degree lower; the dew-point and humidity were nearly the same at both places. This being the driest season of the year, it is very probable that the mean level of the water at this part of the delta is not higher than that of the Bay of Bengal; but as we advanced northwards towards the Khasia, and entered the Soormah itself, the atmospheric pressure increased further, thus appearing to give the bed of that stream a depression of thirty-five feet below the Bay of Bengal, into which it flows! This was no doubt the result of unequal atmospheric pressure at the two localities, caused by the disturbance of the column of atmosphere by the Khasia mountains; for in December of the same year, thirty-eight observations on the surface of the Soormah made its bed forty-six feet *above* the Bay of Bengal, whilst from twenty-three observations on the Megna, the pressure only differed + 0·020 of an inch from that of the barometer at Calcutta, which is eighteen feet above the sea-level.

These barometric levellings, though far from satisfactory as compared with trigonometric, are extremely interesting in the absence of the latter. In a scientific point of view nothing has been done towards determining the levels of the land and waters of the great Gangetic delta, since Rennell's time, yet no geodetical operation promises more valuable results in geography and physical geology than running three lines of level across its area; from Chittagong to Calcutta, from Silhet to Rampore, and from Calcutta to Silhet. The foot of the Sikkim Himalaya has, I believe, been connected with Calcutta by the great trigonometrical survey, but I am given to understand that the results are not published.

My own barometric levellings would make the bed of the

Mahanuddy and Ganges at the western extremity of the delta, considerably higher than I should have expected, considering how gentle the current is, and that the season was that of low water. If my observations are correct, they probably indicate a diminished pressure, which is not easily accounted for, the lower portion of the atmospheric column at Rampore being considerably drier and therefore heavier than at Calcutta. At the eastern extremity again, towards Silhet, the atmosphere is much damper than at Calcutta, and the barometer should therefore have stood lower, indicating a higher level of the waters than is the case.

To the geologist the Jheels and Sunderbunds are a most instructive region, as whatever may be the mean elevation of their waters, a permanent depression of ten to fifteen feet would submerge an immense tract, which the Ganges, Burrampooter, and Soormah would soon cover with beds of silt and sand. There would be extremely few shells in the beds thus formed, the southern and northern divisions of which would present two very different floras and faunas, and would in all probability be referred by future geologists to widely different epochs. To the north, beds of peat would be formed by grasses, and in other parts, temperate and tropical forms of plants and animals would be preserved in such equally balanced proportions as to confound the palæontologist; with the bones of the long-snouted alligator, Gangetic porpoise, Indian cow, buffalo, rhinoceros, elephant, tiger, deer, boar, and a host of other animals, he would meet with acorns of several species of oak, pine-cones and magnolia fruits, rose seeds, and *Cycas* nuts, with palm nuts, screw-pines, and other tropical productions. On the other hand, the Sunderbunds portion, though containing also the bones of the tiger, deer, and buffalo, would have none of the Indian cow, rhinoceros, or elephant; there would be different species of porpoise, alligator, and deer, and none of the above mentioned plants (*Cycas*, oak, pine, magnolia and rose), which would be replaced by numerous others, all distinct from those of the

VIEW IN THE JHEELS.

Jheels, and many of them indicative of the influence of salt water, whose proximity (from the rarity of sea-shells) might not otherwise be suspected.

On the 1st of June we entered the Soormah, a full and muddy stream flowing west, a quarter of a mile broad, with banks of mud and clay twelve or fifteen feet high, separating it from marshes, and covered with betel-nut and cocoa-nut palms, figs, and banyans. Many small

villages were scattered along the banks, each with a swarm
of boats, and rude kilns for burning the lime brought from
the Khasia mountains, which is done with grass and bushes.
We ascended to Chattuc, against a gentle current, arriving
on the 9th.

From this place the Khasia mountains are seen as a long
table-topped range running east and west, about 4000 to
5000 feet high, with steep faces towards the Jheels, out of
which they appear to rise abruptly. Though twelve miles
distant, large waterfalls are very clearly seen precipitating
themselves over the cliffs into a bright green mass of
foliage, that seems to creep half way up their flanks. The
nearly horizontal arrangement of the strata is as conspicuous here, as in the sandstone of the Kymore hills in
the Soane valley, which these mountains a good deal
resemble; but they are much higher, and the climate is
widely different. Large valleys enter the hills, and are
divided by hog-backed spurs, and it is far within these
valleys that the waterfalls and precipices occur; but the
nearer and further cliffs being thrown by perspective
into one range, they seem to rise out of the Jheels so
abruptly as to remind one of some precipitous island in
the ocean.

Chattuc is mainly indebted for its existence to the late
Mr. Inglis, who resided there for upwards of sixty years,
and opened a most important trade between the Khasia
and Calcutta in oranges, potatos, coal, lime, and timber.
We were kindly received by his son, whose bungalow
occupies a knoll, of which there are several, which
attracted our attention as being the only elevations fifty
feet high which we had ascended since leaving the foot of
the Sikkim Himalaya. They rise as islets (commonly
called Teela, Beng.) out of the Jheels, within twelve to

twenty miles of the Khasia; they are chiefly formed of stratified gravel and sand, and are always occupied by villages and large trees. They seldom exceed sixty feet in height, and increase in number and size as the hills are approached; they are probably the remains of a deposit that was once spread uniformly along the foot of the mountains, and they in all respects resemble those I have described as rising abruptly from the plains near Titalya (see vol. i. p. 382).

The climate of Chattuc is excessively damp and hot throughout the year, but though sunk amid interminable swamps, the place is perfectly healthy! Such indeed is the character of the climate throughout the Jheels, where fevers and agues are rare; and though no situations can appear more malarious to the common observer than Silhet and Cachar, they are in fact eminently salubrious. These facts admit of no explanation in the present state of our knowledge of endemic diseases. Much may be attributed to the great amount and purity of the water, the equability of the climate, the absence of forests and of sudden changes from wet to dry; but such facts afford no satisfactory explanation. The water, as I have above said, is of a rich chesnut-brown in the narrow creeks of the Jheels, and is golden yellow by transmitted light, owing no doubt, as in bog water and that of dunghills, to a vegetable extractive and probably the presence of carburetted hydrogen. Humboldt mentions this dark-coloured water as prevailing in some of the swamps of the Cassiquares, at the junction of the Orinoco and Amazon, and gives much curious information on its accompanying features of animal and vegetable life.

The rains generally commence in May: they were unusually late this year, though the almost daily gales and

thunder-storms we experienced, foretold their speedy arrival.
From May till October they are unremitting, and the
country is under water, the Soormah rising about fifty feet.
North-easterly winds prevail, but they are a local current
reflected from the Khasia, against which the southerly
perennial trade-wind impinges.   Westerly winds are very
rare, but the dry north-west blasts of India have been
known to traverse the delta and reach this meridian, in one
or two short hot dry puffs during March and April. Hoar-
frost is unknown.*

China roses and tropical plants (*Bignoniæ*, *Asclepia-
deæ*, and *Convolvuli*) rendered Mr. Inglis' bungalow gay,
but little else will grow in the gardens.   Pine-apples are
the best fruit, and oranges from the foot of the Khasia:
plantains ripen imperfectly, and the mango is always acid,
attacked by grubs, and having a flavour of turpentine.
The violent hailstorms of the vernal equinox cut both
spring and cold season flowers and vegetables, and the
rains destroy all summer products.   The soil is a wet clay,
in which some European vegetables thrive well if planted
in October or November.   We were shown marrowfat peas
that had been grown for thirty years without degenerating
in size, but their flavour was poor.

Small long canoes, paddled rapidly by two men, were
procured here, whereby to ascend the narrow rivers that
lead up to the foot of the mountains: they each carry one
passenger, who lies along the bottom, protected by a bamboo platted arched roof.   We started at night, and early
the next morning arrived at Pundua,† where there is a

---

* It however forms further south, at the very mouth of the Megna, and is the
effect of intense radiation when the thermometer in the shade falls to 45°.

† Pundua, though an insignificant village, surrounded by swamps, has enjoyed
an undue share of popularity as a botanical region. Before the geographical
features of the country north of Silhet were known, the plants brought from

dilapidated bungalow : the inhabitants are employed in the debarkation of lime, coal, and potatos.   Large fleets of boats crowded the narrow creeks, some of the vessels being of several tons burden.

Elephants were kindly sent here for us by Mr. H. Inglis, to take us to the foot of the mountains, about three miles distant, and relays of mules and ponies to ascend to Churra, where we were received with the greatest hospitality by that gentleman, who entertained us till the end of June, and procured us servants and collectors.   To his kind offices we were also indebted throughout our travels in the Khasia, for much information, and for facilities and necessaries of all kinds : things in which the traveller is more dependent on his fellow countrymen in India, than in any other part of the world.

We spent two days at Pundua, waiting for our great boats (which drew several feet of water), and collecting in the vicinity.   The old bungalow, without windows and with the roof falling in, was a most miserable shelter; and whichever way we turned from the door, a river or a swamp lay before us.   Birds, mosquitos, leeches, and large wasps swarmed, also rats and sandflies.   A more pestilential hole cannot be conceived; and yet people traverse this district, and sleep here at all seasons of the year with impunity.   We did so ourselves in the month of June, when the Sikkim and all other Terais are deadly : we returned in September, traversing the Jheels and nullahs at the very foot of the hills during a short break of fine weather in the middle of the rains; and we again

those hills by native collectors were sent to the Calcutta garden (and thence to Europe) as from Pundua.   Hence Silhet mountains and Pundua mountains, both very erroneous terms, are constantly met with in botanical works, and generally refer to plants growing in the Khasia mountains.

slept here in November, * always exposed in the heat of
the day to wet and fatigue, and never having even a
soupçon of fever, ague, or rheumatism. This immunity
does not, however, extend to the very foot of the hills, as
it is considered imprudent to sleep at this season in the
bungalow of Terrya, only three miles off.

The elevation of Pundua bungalow is about forty feet
above the sea, and that of the waters surrounding it, from
ten to thirty, according to the season. In June the mean
of the barometer readings at the bungalow was absolutely
identical with that of the Calcutta barometer. In September it was 0·016 inch lower, and in November 0·066
lower. The mean annual temperature throughout the
Jheels is less than 2° below that of Calcutta.

Terrya bungalow lies at the very foot of the first rise of
the mountains; on the way we crossed many small streams
upon the elephants, and one large one by canoes: the water
in all was cool† and sparkling, running rapidly over boulders
and pebbles. Their banks of sandy clay were beautifully
fringed with a willow-like laurel, *Ehretia* bushes, bamboos,
palms, *Bauhinia*, *Bombax*, and *Erythrina*, over which
Calamus palm (rattan) and various flowering plants
climbed. The rock at Terrya is a nummulitic limestone,
worn into extensive caverns. This formation is said to
extend along the southern flank of the Khasia, Garrow,
and Jyntea mountains, and to be associated with sandstone
and coal: it is extensively quarried in many places, several

---

* At the north foot of the Khasia, in the heavily timbered dry Terai stretching
for sixty miles to the Burrampooter, it is almost inevitable death for a European
to sleep, any time between the end of April and of November. Many have crossed
that tract, but not one without taking fever: Mr. H. Inglis was the only survivor
of a party of five, and he was ill from the effects for upwards of two years, after
having been brought to death's door by the first attack, which came on within
three weeks of his arrival at Churra, and by several relapses.

† Temperature in September 77° to 80°; and in November 75·7°.

thousand tons being annually shipped for Calcutta and
Dacca.   It is succeeded by a horizontally stratified sandstone, which is continued up to 4000 feet, where it is over-
lain by coal-beds and then by limestone again.

The sub-tropical scenery of the lower and outer Sikkim
Himalaya, though on a much more gigantic scale, is
not comparable in beauty and luxuriance with the really
tropical vegetation induced by the hot, damp, and insular
climate of these perennially humid mountains.   At the
Himalaya forests of gigantic trees, many of them deciduous,
appear from a distance as masses of dark gray foliage,
clothing mountains 10,000 feet high : here the individual
trees are smaller, more varied in kind, of a brilliant green,
and contrast with gray limestone and red sandstone rocks
and silvery cataracts.   Palms are more numerous here ; *
the cultivated *Areca* (betel-nut) especially, raising its graceful
stem and feathery crown, " like an arrow shot down from
heaven," in luxuriance and beauty above the verdant slopes.
This difference is at once expressed to the Indian botanist
by defining the Khasia flora as of Malayan character ;
by which is meant the prevalence of brilliant glossy-leaved
evergreen tribes of trees (as *Euphorbiaceæ* and *Urticeæ*),
especially figs, which abound in the hot gulleys, where the
property of their roots, which inosculate and form natural
grafts, is taken advantage of in bridging streams, and in
constructing what are called living bridges, of the most
picturesque forms.   *Combretaceæ*, oaks, oranges, *Garcinia*
(gamboge), *Diospyros*, figs, Jacks, plantains, and *Pandanus*,
are more frequent here, together with pinnated leaved
Leguminosæ, *Meliaceæ*, vines and peppers, and above all

* There are upwards of twenty kinds of Palm in this district, including
Chamærops, three species of *Areca*, two of *Wallichia*, *Arenga*, *Caryota*, three of
Phœnix, *Plectocomia*, *Licuala*, and many species of *Calamus*.   Besides these there
are several kinds of *Pandanus*, and the *Cycas pectinata*.

palms, both climbing ones with pinnated shining leaves (as *Calamus* and *Plectocomia*), and erect ones with similar leaves (as cultivated cocoa-nut, *Areca* and *Arenga*), and the broader-leaved wild betel-nut, and beautiful *Caryota* or wine-palm, whose immense decompound leaves are twelve feet long. Laurels and wild nutmegs, with *Henslowia, Itea,* &c., were frequent in the forest, with the usual prevalence of para-sites, mistleto, epiphytical *Orchideæ, Æschynanthus*, ferns, mosses, and *Lycopodia ;* and on the ground were *Rubiaceæ, Scitamineæ,* ferns, *Acanthaceæ,* beautiful balsams, and her-baceous and shrubby nettles. Bamboos* of many kinds are very abundant, and these hills further differ remarkably from those of Sikkim in the great number of species of grasses.

The ascent was at first gradual, along the sides of a sand-stone spur. At 2000 feet the slope suddenly became steep and rocky, at 3000 feet tree vegetation disappeared, and we opened a magnificent prospect of the upper scarped flank of the valley of Moosmai, which we were ascending, with four or five beautiful cascades rolling over the table top of the hills, broken into silvery foam as they leapt from ledge to ledge of the horizontally stratified precipice, and throwing a veil of silver gauze over the gulf of emerald green vegetation, 2000 feet below. The views of the many

---

* The natives enumerate about fourteen different kinds of bamboo, of which we found five in flower, belonging to three very distinct genera. Uspár, Uspét, Uspít, Uskén, Uskóng, Uktáng, Ustó, Silee, Namlang, Tirra, and Battooba are some of the names of Bamboos vouched for by Mr. Inglis as correctly spelt. Of other Khasia names of plants, Wild Plantains are called Kairem, and the culti-vated Kakesh ; the latter are considered so nourishing that they are given to new-born infants. Senteo is a flower in Khas, So a fruit, Ading a tree, and Te a leaf. *Pandanus* is Kashelan. *Plectocomia*, Usmole. *Licuala*, Kuslow. *Caryota*, Kalai-katang. *Wallichia*, Kalai-nili. *Areca*, Waisola. Various *Calami* are Rhimét, Uriphin, Ureek hilla, Tindrio, &c. This list will serve as a specimen: I might increase it materially, but as I have elsewhere observed, the value attached to the supposed definite application of native names to natural objects is greatly over-rated, and too much reliance on them has introduced a prodigious amount of confusion into scientific works and philological inquiries.

"LIVING BRIDGE" FORMED OF THE AËRIAL ROOTS OF THE INDIA-RUBBER AND OTHER KINDS OF FIGS.

cataracts of the first class that are thus precipitated over the bare table-land on which Churra stands, into the valleys on either side, surpass anything of the kind that I have elsewhere seen, though in many respects vividly recalling the scenery around Rio de Janeiro: nor do I know any spot in the world more calculated to fascinate the naturalist who, while appreciating the elements of which a landscape is composed, is also keenly alive to the beauty and grandeur of tropical scenery.

At the point where this view opens, a bleak stony region commences, bearing numberless plants of a temperate flora and of European genera, at a comparatively low elevation; features which continue to the top of the flat on which the station is built, 4000 feet above the sea.

DEWAN'S EAR-RING.

CHAPTER XXIX.

CHURRA POONJI is said to be so called from the number
of streams in the neighbourhood, and poonji, "a village"
(Khas.) : it was selected for a European station, partly from
the elevation and consequent healthiness of the spot, and
partly from its being on the high road from Silhet to
Gowahatty, on the Burrampooter, the capital of Assam,
which is otherwise only accessible by ascending that river,
against both its current and the perennial east wind.
A rapid postal communication is hereby secured : but the
extreme unhealthiness of the northern foot of the
mountains effectually precludes all other intercourse for
nine months in the year.

On the first opening up of the country, the Europeans
were brought into sanguinary collision with the Khasias,
who fought bravely with bows and arrows, displaying a
most blood-thirsty and cruel disposition.   This is indeed

natural to them ; and murders continued very frequent as
preludes to the most trifling robberies, until the extreme
penalty of our law was put in force. Even now, some of
the tributary Rajahs are far from quiet under our rule, and
various parts of the country are not safe to travel in. The
Garrows, who occupy the western extremity of this range,
at the bend of the Burrampooter, are still in a savage
state. Human sacrifices and polyandry are said to be
frequent amongst them, and their orgies are detestable.
Happily we are hardly ever brought into collision with them,
except by their occasional depredations on the Assam and
Khasia frontier : their country is very unhealthy, and is said
to contain abundance of coal, iron, and lime.

We seldom employed fewer than twelve or fourteen of
the natives as collectors, and when travelling, from thirty
to forty as coolies, &c. They are averse to rising early,
and are intolerably filthy in their persons, though not so
in their cottages, which are very poor, with broad grass
roofs reaching nearly to the ground, and usually encircled
by bamboo fences; the latter custom is not common in
savage communities, and perhaps indicates a dread of
treachery. The beams are of hewn wood (they do not use
saws), often neatly carved, and the doors turn on good
wooden pivots. They have no windows, and the fire is
made on the floor : the utensils, &c. are placed on hanging
shelves and in baskets.

The Khasia people are of the Indo-Chinese race ; they
are short, very stout, and muscular, with enormous calves
and knees, rather narrow eyes and little beard, broad, high
cheekbones, flat noses, and open nostrils. I believe that a
few are tattooed. The hair is gathered into a top-knot,
and sometimes shaved off the forehead and temples. A
loose cotton shirt, often striped blue and red, without

sleeves and bordered with long thread fringes, is their
principal garment; it is gathered into a girdle of silver
chains by people of rank. A cotton robe is sometimes
added, with a large cotton turban or small skull-cap. The
women wear a long cloth tied in a knot across the breast.
During festivals both men and women load themselves
with silk robes, fans, peacock's feathers, and gold and
silver ornaments of great value, procured from Assam,
many of which are said to be extremely curious, but I
regret to say that I never saw any of them. On these
occasions spirits are drunk, and dancing kept up all night:
the dance is described as a slow ungraceful motion, the
women being tightly swathed in cloths.

All their materials are brought from Assam; the only
articles in constant use, of their own manufacture, being a
rude sword or knife with a wooden handle and a long,
narrow, straight blade of iron, and the baskets with head-
straps, like those used by the Lepchas, but much neater;
also a netted bag of pine-apple fibre (said to come from
Silhet) which holds a clasp-knife, comb, flint, steel, and
betel-nut box. They are much addicted to chewing pawn
(betel-nut, pepper leaves, and lime) all day long, and their
red saliva looks like blood on the paths. Besides the
sword I have described, they carry bows and arrows, and
rarely a lance, and a bamboo wicker-work shield.

We found the Khasias to be sulky intractable fellows,
contrasting unpleasantly with the Lepchas; wanting in quickness, frankness, and desire to please, and obtrusively
independent in manner; nevertheless we had a head man
who was very much the reverse of this, and whom we had
never any cause to blame. Their language is, I believe,
Indo-Chinese and monosyllabic: it is disagreeably nasal
and guttural, and there are several dialects and accents in

contiguous villages. All inflections are made by prefixing syllables, and when using the Hindoo language, the future is invariably substituted for the past tense. They count up to a hundred, and estimate distances by the number of mouthfuls of pawn they eat on the road.

Education has been attempted by missionaries with partial success, and the natives are said to have shown themselves apt scholars. Marriage is a very loose tie amongst them, and hardly any ceremony attends it. We were informed that the husband does not take his wife home, but enters her father's household, and is entertained there. Divorce and an exchange of wives is common, and attended with no disgrace: thus the son often forgets his father's name and person before he grows up, but becomes strongly attached to his mother. The sister's son inherits both property and rank, and the proprietors' or Rajahs' offspring are consequently often reared in poverty and neglect. The usual toy of the children is the bow and arrow, with which they are seldom expert; they are said also to spin pegtops like the English, climb a greased pole, and run round with a beam turning horizontally on an upright, to which it is attached by a pivot.

The Khasias eat fowls, and all meat, especially pork, potatos and vegetables, dried and half putrid fish in abundance, but they have an aversion to milk, which is very remarkable, as a great proportion of their country is admirably adapted for pasturage. In this respect, however, they assimilate to the Chinese, and many Indo-Chinese nations who are indifferent to milk, as are the Sikkim people. The Bengalees, Hindoos, and Tibetans, on the other hand, consume immense quantities of milk. They have no sheep, and few goats or cattle, the latter of which are kept for slaughter; they have, however, plenty

T 2

of pigs and fowls. Eggs are most abundant, but used
for omens only, and it is a common, but disgusting
occurrence, to see large groups employed for hours in
breaking them upon stones, shouting and quarrelling,
surrounded by the mixture of yellow yolks and their red
pawn saliva.

The funeral ceremonies are the only ones of any importance, and are often conducted with barbaric pomp and
expense; and rude stones of gigantic proportions are erected
as monuments, singly or in rows, circles, or supporting one
another, like those of Stonehenge, which they rival in
dimensions and appearance. The body is burned, though
seldom during the rains, from the difficulty of obtaining
a fire; it is therefore preserved in honey (which is abundant and good) till the dry season: a practice I have read of
as prevailing among some tribes in the Malay peninsula.
Spirits are drunk on these occasions; but the hill Khasia is
not addicted to drunkenness, though some of the natives of
the low valleys are very much so. These ascend the rocky
faces of the mountains by ladders, to the Churra markets,
and return loaded at night, apparently all but too drunk
to stand; yet they never miss their footing in places
which are most dangerous to persons unaccustomed to
such situations.

The Khasias are superstitious, but have no religion; like
the Lepchas, they believe in a supreme being, and in deities
of the grove, cave, and stream. Altercations are often
decided by holding the disputants' heads under water,
when the longest winded carries his point. Fining is a
common punishment, and death for grave offences. The
changes of the moon are accounted for by the theory
that this orb, who is a man, monthly falls in love
with his wife's mother, who throws ashes in his face.

Pl. X.

J.D.H. del.

John Murray, Albemarle Street. 1854.

W.L. Walton, lith.

Churra Station and the Jheels from the Khasia Mountains.

The sun is female ; and Mr. Yule * (who is my authority) says that the Pleiades are called " the Hen-man " (as in Italy "the chickens"); also that they have names for the twelve months; they do not divide their time by weeks, but hold a market every four days.   These people are industrious, and good cultivators of rice, millet, and legumes of many kinds.   Potatoes were introduced amongst them about twenty years ago by Mr. Inglis, and they have increased so rapidly that the Calcutta market is now supplied by their produce.   They keep bees in rude hives of logs of wood.

The flat table-land on which Churra Poonji is placed, is three miles long and two broad, dipping abruptly in front and on both sides, and rising behind towards the main range, of which it is a spur.   The surface of this area is everywhere intersected by shallow, rocky water-courses, which are the natural drains for the deluge that annually visits it.   The western part is undulated and hilly, the southern rises in rocky ridges of limestone and coal, and the eastern is very flat and stony, broken only by low isolated conical mounds.

The scenery varies extremely at different parts of the surface.   Towards the flat portion, where the English reside, the aspect is as bleak and inhospitable as can be imagined : a thin stratum of marshy or sandy soil covers a tabular mass of cold red sandstone ; and there is not a tree, and scarcely a shrub to be seen, except occasional clumps of *Pandanus*.   The low white bungalows are few in number, and very scattered, some of them being a mile asunder, enclosed with stone walls and shrubs; and a small white

* I am indebted to Mr. Inglis for most of this information relating to the Khasias, which I have since found, with much more that is curious and interesting, in a paper by Lieut. Yule in Bengal Asiat. Soc. Journal.

church, disused on account of the damp, stands lonely in
the centre of all.

The views from the margins of this plateau are magnificent : 4000 feet below are bay-like valleys, carpetted as
with green velvet, from which rise tall palms, tree-ferns
with spreading crowns, and rattans shooting their pointed
heads, surrounded with feathery foliage, as with ostrich
plumes, far above the great trees. Beyond are the Jheels,
looking like a broad shallow sea with the tide half out,
bounded in the blue distance by the low hills of Tipperah.
To the right and left are the scarped red rocks and roaring
waterfalls, shooting far over the cliffs, and then arching
their necks as they expand in feathery foam, over which
rainbows float, forming and dissolving as the wind sways
the curtains of spray from side to side.

To the south of Churra the lime and coal measures rise
abruptly in flat-topped craggy hills, covered with brushwood and small trees. Similar hills are seen far westward
across the intervening valleys in the Garrow country,
rising in a series of steep isolated ranges, 300 to 400 feet
above the general level of the country, and always skirting
the south face of the mountains. Considerable caverns
penetrate the limestone, the broken surface of which rock
presents many picturesque and beautiful spots, like the
same rocks in England.

Westward the plateau becomes very hilly, bare, and
grassy, with the streams broad and full, but superficial
and rocky, precipitating themselves in low cascades over
tabular masses of sand-stone. At Mamloo their beds are
deeper, and full of brushwood, and a splendid valley and
amphitheatre of red cliffs and cascades, rivalling those of
Moosmai (p. 261), bursts suddenly into view. Mamloo is
a large village, on the top of a spur, to the westward : it

is buried in a small forest, particularly rich in plants, and
is defended by a stone wall behind : the only road is tun-
nelled through the sandstone rock, under the wall ; and
the spur on either side dips precipitously, so that the place
is almost impregnable if properly defended.   A sanguinary

MAMLOO CASCADES.

conflict took place here between the British and the
Khasias, which terminated in the latter being driven over
the precipices, beneath which many of them were shot.
The fan-palm, *Chamærops Khasiana* (" Pakha," Khas.),
grows on the cliffs near Mamloo : it may be seen on
looking over the edge of the plateau, its long curved trunk
rising out of the naked rocks, but its site is generally

inaccessible;* while near it grows the *Saxifragis ciliaris*
of our English gardens, a common plant in the north-
west Himalaya, but extremely scarce in Sikkim and the
Khasia mountains.

The descent of the Mamloo spur is by steps, alternating
with pebbly flats, for 1500 feet, to a saddle which connects
the Churra hills with those of Lisouplang to the westward.
The rise is along a very steep narrow ridge to a broad long
grassy hill, 3,500 feet high, whence an extremely steep
descent leads to the valley of the Boga-panee, and the great
mart of Chela, which is at the embouchure of that river.
The transverse valley thus formed by the Mamloo spur, is
full of orange groves, whose brilliant green is particularly
conspicuous from above. At the saddle below Mamloo
are some jasper rocks, which are the sandstone altered by
basalt. Fossil shells are recorded to have been found by
Dr. M'Lelland† on some of the flats, which he considers to
be raised beaches: but we sought in vain for any evidence
of this theory beyond the pebbles, whose rounding we
attributed to the action of superficial streams.

It is extremely difficult to give within the limits of this
narrative any idea of the Khasia flora, which is, in extent
and number of fine plants, the richest in India, and probably in all Asia. We collected upwards of 2000 flowering
plants within ten miles of the station of Churra, besides 150
ferns, and a profusion of mosses, lichens, and fungi. This
extraordinary exuberance of species is not so much attributable to the elevation, for the whole Sikkim Himalaya

---

* This species is very closely allied to, if not identical with *P. Martiana* of Nepal,
which ascends to 8000 feet in the western Himalaya, where it is annually covered
with snow: it is not found in Sikkim, but an allied species occurs in Affghanistan,
called *P. Ritcheana:* the dwarf palm of southern Europe is a fourth species.

† See a paper on the geology of the Khasia mountains by Dr. M'Lelland in
the "Bengal Asiatic Society's Journal."

(three times more elevated) does not contain 500 more
flowering plants, and far fewer ferns, &c. ; but to the variety
of exposures ; namely, 1. the Jheels , 2. the tropical jungles,
both in deep, hot, and wet valleys, and on drier slopes ;
3. the rocks; 4. the bleak table-lands and stony soils;
5. the moor-like uplands, naked and exposed, where many
species and genera appear at 5000 to 6000 feet, which are
not found on the outer ranges of Sikkim under 10,000.*
In fact, strange as it may appear, owing to this last cause,
the temperate flora descends fully 4000 feet lower in the
latitude of Khasia (25° N.) than in that of Sikkim (27° N.),
though the former is two degrees nearer the equator.

The *Pandanus* alone forms a conspicuous feature in the
immediate vicinity of Churra; while the small woods about
Mamloo, Moosmai, and the coal-pits, are composed of *Symplocos*, laurels, brambles, and jasmines, mixed with small
oaks and *Photinia*, and many tropical genera of trees and
shrubs.

Orchideæ are, perhaps, the largest natural order in the
Khasia, where fully 250 kinds grow, chiefly on trees
and rocks, but many are terrestrial, inhabiting damp
woods and grassy slopes. I doubt whether in any other
part of the globe the species of orchids outnumber those of
any other natural order, or form so large a proportion of
the flora. Balsams are next in relative abundance (about
twenty-five), both tropical and temperate kinds, of great
beauty and variety in colour, form, and size of blossom.
Palms amount to fourteen, of which the *Chamærops* and
Arenga are the only genera not found in Sikkim. Of
bamboos there are also fifteen, and of other grasses 150,
which is an immense proportion, considering that the

* As *Thalictrum*, *Anemone*, primrose, cowslip, *Tofieldia*, Yew, Pine, Saxifrage,
Delphinium, *Pedicularis*.

Indian flora (including those of Ceylon, Kashmir, and all the Himalaya), hardly contains 400. *Scitamineæ* also are abundant, and extremely beautiful; we collected thirty-seven kinds.

No rhododendron grows at Churra, but several species occur a little further north : there is but one pine (*P. Khasiana*) besides the yew, (and two *Podocarpi*), and that is only found in the drier interior regions. Singular to say, it is a species not seen in the Himalaya or elsewhere, but very nearly allied to *Pinus longifolia*,* though more closely resembling the Scotch fir than that tree does.

The natural orders whose rarity is most noticeable, are *Cruciferæ*, represented by only three kinds, and *Caryophylleæ*. Of *Ranunculaceæ*, there are six or seven species of *Clematis*, two of *Anemone*, one *Delphinium*, three of *Thalictrum*, and two *Ranunculi*. *Compositæ* and *Leguminosæ* are far more numerous than in Sikkim.

The climate of Khasia is remarkable for the excessive rain-fall. Attention was first drawn to this by Mr. Yule, who stated, that in the month of August, 1841, 264 inches fell, or twenty-two feet; and that during five successive days, thirty inches fell in every twenty-four hours! Dr. Thomson and I also recorded thirty inches in one day and night, and during the seven months of our stay, upwards of 500 inches fell, so that the total annual fall perhaps greatly exceeded 600 inches, or fifty feet, which

* Cone-bearing pines with long leaves, like the common Scotch fir, are found in Asia, and as far south as the Equator (in Borneo) and also inhabit Arracan, the Malay Peninsula, Sumatra, and South China. It is a very remarkable fact that no Gymnospermous tree inhabits the Peninsula of India; not even the genus *Podocarpus*, which includes most of the tropical Gymnosperms, and is technically coniferous, and has glandular woody fibre; though like the yew it bears berries. Two species of this genus are found in the Khasia, and one advances as far west as Nepal. The absence of oaks and of the above genera (*Podocarpus* and *Pinus*) is one of the most characteristic differences between the botany of the east and west shores of the Bay of Bengal.

has been registered in succeeding years! From April, 1849, to April, 1850, 502 inches (forty-two feet) fell. This unparalleled amount is attributable to the abruptness of the mountains which face the Bay of Bengal, from which they are separated by 200 miles of Jheels and Sunderbunds.

This fall is very local: at Silhet, not thirty miles further south, it is under 100 inches; at Gowahatty, north of the Khasia in Assam, it is about 80; and even on the hills, twenty miles inland from Churra itself, the fall is reduced to 200. At the Churra station, the distribution of the rain is very local; my gauges, though registering the same amount when placed beside a good one in the station; when removed half a mile, received a widely different quantity, though the different gauges gave nearly the same mean amount at the end of each whole month.

The direct effect of this deluge is to raise the little streams about Churra fourteen feet in as many hours, and to inundate the whole flat; from which, however, the natural drainage is so complete, as to render a tract, which in such a climate and latitude should be clothed with exuberant forest, so sterile, that no tree finds support, and there is no soil for cultivation of any kind whatsoever, not even of rice. Owing, however, to the hardness of the horizontally stratified sandstone, the streams have not cut deep channels, nor have the cataracts worked far back into the cliffs. The limestone alone seems to suffer, and the turbid streams from it prove how rapidly it is becoming denuded. The great mounds of angular gravel on the Churra flat, are perhaps the remains of an extensive deposit, fifty feet thick, elsewhere washed away by these rains; and I have remarked traces of the same over many slopes of the hills around.

The mean temperature of Churra (elev. 4000 feet) is about 66°, or 16° below that of Calcutta; which, allowing for $2\frac{1}{2}$° of northing, gives 1° of temperature to every 290 to 300 feet of ascent. In summer the thermometer often rises to 88° and 90°; and in the winter, owing to the intense radiation, hoar-frost is frequent. Such a climate is no less inimical to the cultivation of plants, than is the wretched soil: of this we saw marked instances in the gardens of two of the resident officers, Lieutenants Raban and Cave, to whom we were indebted for the greatest kindness and hospitality. These gentlemen are indefatigable horticulturists, and took a zealous interest in our pursuits, accompanying us in our excursions, enriching our collections in many ways, and keeping an eye to them and to our plant-driers during our absence from the station. In their gardens the soil had to be brought from a considerable distance, and dressed copiously with vegetable matter. Bamboo clumps were planted for shelter within walls, and native shrubs, rhododendrons, &c., introduced. Many *Orchideæ* throve well on the branches of the stunted trees which they had planted, and some superb kinds of *Hedychium* in the ground; but a very few English garden plants throve in the flower-beds. Even in pots and frames, geraniums, &c., would rot, from the rarity of sunshine, which is as prejudicial as the damp and exposure. Still many wild shrubs of great interest and beauty flourished, and some European ones succeeded with skill and management; as geraniums, *Salvia*, *Petunia*, nasturtium, chrysanthemum, *Kennedya rubicunda*, *Maurandya*, and Fuchsia. The daisy seed sent from England as double, came up very poor and single. Dahlias do not thrive, nor double balsams. Now they have erected small but airy green-houses, and sunlight is the only desideratum.

At the end of June, we started for the northern or Assam face of the mountains. The road runs between the extensive and populous native village, or poonji, on the left, and a deep valley on the right, and commands a beautiful view of more waterfalls. Beyond this it ascends steeply, and the sandstone on the road itself is curiously divided into parallelograms, like hollow bricks,* enclosing irregularly shaped nodules, while in other places it looks as if it had been run or fused: spherical concretions of sand, coloured concentrically by infiltration, are common in it, which have been regarded as seeds, shells, &c.; it also contained spheres of iron pyrites. The general appearance of much of this rock is as if it had been bored by *Teredines* (ship worms), but I never detected any trace of fossils. It is often beautifully ripple-marked, and in some places much honeycombed, and full of shales and narrow seams of coal, resting on a white under-clay full of root-fibres, like those of *Stigmaria*.

At about 5000 feet the country is very open and bare, the ridges being so uniform and flat-topped, that the broad valleys they divide are hidden till their precipitous edges are reached; and the eye wanders far east and west over a desolate level grassy country, unbroken, save by the curious flat-topped hills I have described as belonging to the limestone formation, which lie to the south-west. These features continue for eight miles, when a sudden descent of 600 or 700 feet, leads into the valley of the Kala-panee (Black water) river, where there is a very dark and damp bungalow, which proved a very great accommodation to us.†

* I have seen similar bricks in the sandstones of the coal-districts of Yorkshire; they are very puzzling, and are probably due to some very obscure crystalline action analogous to jointing and cleavage.

† It may be of use to the future botanist in this country to mention a small

Lailang-kot is another village full of iron forges, from a height near which a splendid view is obtained over the Churra flat. A few old and very stunted shrubs of laurel and *Symplocos* grow on its bleak surface, and these are often sunk from one to three feet in a well in the horizontally stratified sandstone. I could only account for this by supposing it to arise from the drip from the trees, and if so, it is a wonderful instance of the wearing effects of water, and of the great age which small bushes sometimes attain.

The vegetation is more alpine at Kala-panee (elevation, 5,300 feet); *Benthamia, Kadsura, Stauntonia, Illicium, Actinidia, Helwingia, Corylopsis,* and berberry—all Japan and Chinese, and most of them Dorjiling genera—appear here, with the English yew, two rhododendrons, and *Bucklandia.* There are no large trees, but a bright green jungle of small ones and bushes, many of which are very rare and curious. *Luculia Pinceana* makes a gorgeous show here in October.

The sandstone to the east of Kala-panee is capped by some beds, forty feet thick, of conglomerate worn into cliffs; these are the remains of a very extensive horizontally stratified formation, now all but entirely denuded. In the valley itself, the sandstone alternates with alum shales, which rest on a bed of quartz conglomerate, and the latter on black greenstone. In the bed of the river, whose waters are beautifully clear, are hornstone rocks, dipping north-east, and striking north-west. Beyond the Kala-panee the road ascends about 600 feet, and is well quarried in hard greenstone; and passing through a narrow gap of

wood on the right of this road, near the village of Surureem, as an excellent botanical station: the trees are chiefly *Rhododendron arboreum,* figs, oaks, laurels, magnolias, and chestnuts, on whose limbs are a profusion of *Orchideæ,* and amongst which a Rattan palm occurs.

conglomerate rock,* enters a shallow, wild, and beautiful
valley, through which it runs for several miles.   The hills
on either side are of greenstone capped by tabular sandstone, immense masses of which have been precipitated on
the floor of the valley, producing a singularly wild and
picturesque scene.   In the gloom of the evening it is not
difficult for a fertile imagination to fancy castles and cities
cresting the heights above.†

There is some cultivation here of potatoes, and of *Rhysi-
cosia vestita* a beautiful purple-flowered leguminous plant,
with small tuberous roots.   Beyond this, a high ridge is
gained above the valley of the Boga panee, the largest river
in the Khasia; from this the Bhotan Himalaya may
be seen in clear weather, at the astonishing distance of
from 160 to 200 miles!   The vegetation here suddenly
assumes a different aspect, from the quantity of stunted
fir-trees clothing the north side of the valley, which rises
very steeply 1000 feet above the river: quite unaccount-
ably, however, not one grows on the south face.   A new
oak also appears abundantly; it has leaves like the English,
whose gnarled habit it also assumes.

The descent is very steep, and carried down a slope of
greenstone; ‡ the road then follows a clear affluent of the
Boga-panee, and afterwards winds along the margin of that
river, which is a rapid turbulent stream, very muddy, and

---

* Formed of rolled masses of greenstone and sandstone, united by a white and
yellow cement.

† *Hydrangea* grows here, with ivy, *Mussænda, Pyrus,* willow, *Viburnum, Parnassia, Anemone, Leycesteria formosa, Neillia, Rubus, Astilbe,* rose, *Panax,* apple,
Bucklandia, Daphne, pepper, *Scindapsus, Pieris,* holly, *Lilium giganteum* (" Kalang
tatti," Khas.), *Camellia, Elæocarpus, Buddleia, &c.*   Large bees' nests hang from
the rocks.

‡ This greenstone decomposes into a thick bed of red clay; it is much intersected by fissures or cleavage planes at all angles, whose surfaces are covered with
a shining polished superficial layer; like the fissures in the cleavage planes of
the gneiss granite of Kinchinjhow, whose adjacent surfaces are coated with a

hence contrasting remarkably with the Kala-panee. It derives its mud from the decomposition of granite, which is washed by the natives for iron, and in which rock it rises to the eastward. Thick beds of slate crop out by the roadside (strike north-east and dip north-west), and are continued along the bed of the river, passing into conglomerates, chert, purple slates, and crystalline sandstones, with pebbles, and angular masses of schist. Many of these rocks are much crumpled, others quite flat, and they are overlaid by soft, variegated gneiss, which is continued alternately with the slates to the top of the hills on the opposite side.

Small trees of hornbeam grow near the river, with *Rhus*, *Xanthoxylon*, *Vaccinium*, *Gualtheria*, and *Spiræa*, while many beautiful ferns, mosses, and orchids cover the rocks. An elegant iron suspension-bridge is thrown across the stream, from a rock matted with tufts of little parasitic *Orchideæ*. Crossing it, we came on many pine-trees; these had five-years' old cones on them, as well as those of all succeeding years; they bear male flowers in autumn, which impregnate the cones formed the previous year. Thus, the cones formed in the spring of 1850 are fertilised in the following autumn, and do not ripen their seeds till the second following autumn, that of 1852.

A very steep ascent leads to the bungalow of Moflong, on a broad, bleak hill-top, near the axis of the range (alt. 6,062 feet). Here there is a village, and some cultivation, surrounded by hedges of *Erythrina*, *Pieris*, *Viburnum*,

---

glassy waved layer of hornblende. This polishing of the surfaces is generally attributed to their having been in contact and rubbed together, an explanation which is wholly unsatisfactory to me; no such motion could take place in cleavage planes which often intersect, and were it to occur, it would not produce two polished surfaces of an interposed layer of a softer mineral. It is more probably due to metamorphic action.

Pyrus, *Colquhounia*, and *Corylopsis*, amongst which grew an autumn-flowering lark-spur, with most fœtid flowers.* The rocks are much contorted slates and gneiss (strike north-east and dip south-east). In a deep gulley to the northward, greenstone appears, with black basalt and jasper, the latter apparently altered gneiss : beyond this the rocks strike the opposite way, but are much disturbed.

We passed the end of June here, and experienced the same violent weather, thunder, lightning, gales, and rain, which prevailed during every midsummer I spent in India. A great deal of *Coix* (Job's tears) is cultivated about Moflong : it is of a dull greenish purple, and though planted in drills, and carefully hoed and weeded, is a very ragged crop. The shell of the cultivated sort is soft, and the kernel is sweet ; whereas the wild *Coix* is so hard that it cannot be broken by the teeth. Each plant branches two or three times from the base, and from seven to nine plants grow in each square yard of soil : the produce is small, not above thirty or forty fold.

From a hill behind Moflong bungalow, on which are some stone altars, a most superb view is obtained of the Bhotan Himalaya to the northward, their snowy peaks stretching in a broken series from north 17° east to north 35° west ; all are below the horizon of the spectator, though from 17,000 to 20,000 feet above his level. The finest view in the Khasia mountains, and perhaps a more extensive one than has ever before been described, is that from

---

* There is a wood a mile to the west of the bungalow, worth visiting by the botanist : besides yew, oak, *Sabia* and *Camellia*, it contains *Olea, Euonymus,* and *Sphærocarya,* a small tree that bears a green pear shaped sweet fruit, with a large stone : it is pleasant, but leaves a disagreeable taste in the mouth. On the grassy flats an *Astragalus* occurs, and *Roscoea purpurea, Tofieldia,* and various other fine plants are common.

Chillong hill, the culminant point of the range, about six
miles north-east from Moflong bungalow.   This hill, 6,660
feet above the sea, rises from an undulating grassy country,
covered with scattered trees and occasional clumps of
wood; the whole scenery about being park-like, and as
little like that of India at so low an elevation as it is possible to be.

I visited Chillong in October with Lieutenant Cave;
starting from Churra, and reaching the bungalow, two
miles from its top, the same night, with two relays of
ponies, which he had kindly provided.   We were unfortunate in not obtaining a brilliant view of the snowy
mountains, their tops being partially clouded; but the
coup d'œil was superb.   Northward, beyond the rolling
Khasia hills, lay the whole Assam valley, seventy miles
broad, with the Burrampooter winding through it, fifty
miles distant, reduced to a thread.   Beyond this, banks of
hazy vapour obscured all but the dark range of the Lower
Himalaya, crested by peaks of frosted silver, at the
immense distance of from 100 to 220 miles from Chillong.
All are below the horizon of the observer; yet so false is
perspective, that they seem high in the air.   The mountains occupy sixty degrees of the horizon, and stretch
over upwards of 250 miles, comprising the greatest extent
of snow visible from any point with which I am
acquainted.

Westward from Chillong the most distant Garrow hills
visible are about forty miles off; and eastward those of
Cachar, which are loftier, are about seventy miles.   To the
south the view is limited by the Tipperah hills, which,
where nearest, are 100 miles distant; while to the south-
west lies the sea-like Gangetic delta, whose horizon, lifted
by refraction, must be fully 120.   The extent of this view is

therefore upwards of 340 miles in one direction, and the
visible horizon of the observer encircles an area of fully thirty
thousand square miles, which is greater than that of Ireland!

Scarlet-flowered rhododendron bushes cover the north
side of Chillong,* whilst the south is grassy and quite bare;
and except some good *Orchideæ* on the trees, there is little
to reward the botanist. The rocks appeared to be sandstone at the summit, but micaceous gneiss all around.

Continuing northward from Moflong, the road, after five
miles, dips into a very broad and shallow flat-floored valley,
fully a mile across, which resembles a lake-bed: it is
bounded by low hills, and is called " Lanten-tannia," and
is bare of aught but long grass and herbs ; amongst these
are the large groundsel (*Senecio*), *Dipsacus*, *Ophelia*, and
Campanula.   On its south flank the micaceous slates strike
north-east, and dip north-west, and on the top repose beds,
a foot in thickness, of angular water-worn gravel, indicating an ancient water-level, 400 feet above the floor of
the valley.   Other smaller lake-beds, in the lateral valleys,
are equally evident.

A beautiful blue-flowered *Clitoria* creeps over the path,
with the ground-raspberry of Dorjiling. From the top a
sudden descent of 400 feet leads to another broad flat
valley, called " Syong " (elevation, 5,725 feet), in which is
a good bungalow, surrounded by hedges of *Prinsepia
utilis*, a common north-west Himalayan plant, only found
at 8000 feet in Sikkim. The valley is grassy, but otherwise bare. Beyond this the road passes over low rocky hills,
wooded on their north or sheltered flanks only, dividing
flat-floored valleys : a red sandy gneiss is the prevalent
rock, but boulders of syenite are scattered about. Extensive

* These skirt a wood of prickly bamboo, in which occur fig, laurel, *Aralia*,
Bœhmeria, *Smilax*, *Toddalia*, wild cinnamon, and three kinds of oak.

moors (elevation, 6000 feet) succeed, covered with stunted
pines, brake, and tufts of harsh grasses.*

Near the Dengship-oong (river), which flows in a narrow
valley, is a low dome of gneiss altered by syenite.   The prevalent dip is uniformly south-east, and the strike north-east;
and detached boulders of syenite become more frequent,
resting on a red gneiss, full of black garnets, till the descent
to the valley of Myrung, one of the most beautiful spots in
the Khasia, and a favourite resort, having an excellent
bungalow which commands a superb view of the Himalaya :
it is 5,650 feet above the sea, and is placed on the north
flank of a very shallow marshy valley, two miles broad,
and full of rice cultivation, as are the flat heads of all
the little valleys that lead into it.   There is a guard
here of light infantry, and a little garden, boasting a
gardener and some tea-plants, so that we had vegetables
during our four visits to the place, on two of which occasions we stayed some days.

From Kala-panee to Myrung, a distance of thirty-two
miles, the road does not vary 500 feet above or below
the mean level of 5,700 feet, and the physical features
are the same throughout, of broad flat-floored, steep-sided
valleys, divided by bleak, grassy, tolerably level-topped
hills.   Beyond Myrung the Khasia mountains slope to
the southward in rolling loosely-wooded hills, but the
spurs do not dip suddenly till beyond Nunklow, eight miles
further north.

On the south side of the Myrung valley is Nungbree
wood, a dense jungle, occupying, like all the other woods,

---

* These are principally *Andropogon* and *Brachypodium*, amongst which grow
yellow *Corydalis, Thalictrum, Anemone, Parnassia, Prunella,* strawberry, *Eupatorium, Hypericum,* willow, a *Polygonum* like *Bistorta, Osmunda regalis* and another
species, *Lycopodium alpinum,* a *Senecio* like *Jacobœa,* thistles, *Gnaphalium,* Gentians, *Iris, Paris, Sanguisorba* and *Agrimonia.*

the steep north exposure of the hill; many good plants grow in it, including some gigantic *Balanophoræ*, *Pyrola*, and *Monotropa*. The bungalow stands on soft, contorted, decomposing gneiss, which is still the prevalent rock, striking north-east. On the hills to the east of it, enormous hard blocks lie fully exposed, and are piled on one another, as if so disposed by glacial action; and it is difficult to account for them by denudation, though their surface scales, and similar blocks are scattered around Myrung exactly similar to the syenite blocks of Nunklow, and the granite ones of Nonkreem, to be described hereafter, and which are undoubtedly due to the process of weathering. A great mass of flesh-coloured crystalline granite rises in the centre of the valley, to the east of the road: it is fissured in various directions, and the surface scales concentrically; it is obscurely stratified in some parts, and appears to be half granite and half gneiss in mineralogical character.

We twice visited a very remarkable hill, called Kollong, which rises as a dome of granite 5,400 feet high, ten or twelve miles south-west of Myrung, and conspicuous from all directions. The path to it turns off from that to Nunklow, and strikes westerly along the shallow valley of Monai, in which is a village, and much rice and other cultivation. Near this there is a large square stockade, formed of tall bamboos placed close together, very like a New Zealand "Pa;" indeed, the whole country hereabouts much recals the grassy clay hills, marshy valleys, and bushy ridges of the Bay of Islands.

The hills on either side are sometimes dotted with pine-woods, sometimes conical and bare, with small clumps of pines on the summit only; while in other places are broad tracts containing nothing but young trees, resembling

plantations, but which, I am assured, are not planted; on
the other hand, however, Mr. Yule states, that the natives do
plant fir-trees, especially near the iron forges, which give
employment to all the people of Monai.

All the streams rise in flat marshy depressions amongst
the hills with which the whole country is covered; and
both these features, together with the flat clay marshes
into which the rivers expand, are very suggestive of tidal
action. Rock is hardly anywhere seen, except in the
immediate vicinity of Kollong, where are many scattered
boulders of fine-grained gneiss, of which are made the
broad stone slabs, placed as seats, and the other erections
of this singular people. We repeatedly remarked cones of
earth, clay, and pebbles, about twelve feet high, upon the
hills, which appeared to be artificial, but of which the
natives could give no explanation. Wild apple and birch are
common trees, but there is little jungle, except in the
hollows, and on the north slopes of the higher hills. Coarse
long grass, with bushes of Labiate and Composite plants,
are the prevalent features.

Kollong rock is a steep dome of red granite,* accessible
from the north and east, but almost perpendicular to the
southward, where the slope is 80° for 600 feet. The
elevation is 400 feet above the mean level of the surrounding
ridges, and 700 above the bottom of the valleys. The south
or steepest side is encumbered with enormous detached
blocks, while the north is clothed with a dense forest, containing red tree-rhododendrons and oaks; on its skirts
grew a white bushy rhododendron, which we found
nowhere else. The hard granite of the top was covered
with matted mosses, lichens, Lycopodiums, and ferns,

* This granite is highly crystalline, and does not scale or flake, nor is its
surface polished.

amongst which were many curious and beautiful air-plants.*

The view from the top is very extensive to the north-ward, but not elsewhere: it commands the Assam valley

KOLLONG ROCK.

and the Himalaya, and the billowy range of undulating grassy Khasia mountains. Few houses were visible, but the curling smoke from the valleys betrayed their lurking-places, whilst the tinkling sound of the hammers from the distant

---

* *Eria, Cœlogyne* (*Wallichii, maculata,* and *elata*), *Cymbidium, Dendrobium, Sunipia,* some of them flowering profusely; and though freely exposed to the sun and wind, dews and frosts, rain and droughts, they were all fresh, bright, green and strong, under very different treatment from that to which they are exposed in the damp, unhealthy, steamy orchid-houses of our English gardens. A wild onion was most abundant all over the top of the hill, with *Hymenopogon, Vaccinium, Ophiopogon, Anisadenia, Commelyna, Didymocarpus, Remusatia, Hedychium,* grass and small bamboos, and a good many other plants. Many of the lichens were of European kinds; but the mosses (except *Bryum argenteum*) and ferns were different. A small *Staphylinus,* which swarmed under the sods, was the only insect I remarked.

forges on all sides was singularly musical and pleasing; they
fell on the ear like "bells upon the wind," each ring being
exquisitely melodious, and chiming harmoniously with the
others. The solitude and beauty of the scenery, and the
emotions excited by the music of chimes, tended to tran-
quillise our minds, wearied by the fatigues of travel, and
the excitement of pursuits that required unremitting attention; and we rested for some time, our imaginations
wandering to far-distant scenes, brought vividly to our
minds by these familiar sounds.

CHAPTER XXX.

THE snowy Himalaya was not visible during our first stay
at Myrung, from the 5th to the 10th of July; but on three
subsequent occasions, viz., 27th and 28th of July, 13th to
17th October, and 22nd to 25th October, we saw these
magnificent mountains, and repeatedly took angular heights
and bearings of the principal peaks. The range, as seen from
the Khasia, does not form a continuous line of snowy mountains, but the loftiest eminences are conspicuously grouped
into masses, whose position is probably between the great
rivers which rise far beyond them and flow through Bhotan.
This arrangement indicates that relation of the rivers to the
masses of snow, which I have dwelt upon in the Appendix;

and further tends to prove that the snowy mountains,
seen from the southward, are not on the axis of a mountain chain, and do not even indicate its position; but
that they are lofty meridional spurs which, projecting
southward, catch the moist vapours, become more deeply
snowed, and protect the dry loftier regions behind.

The most conspicuous group of snows seen from the
Khasia bears N.N.E. from Myrung, and consists of three
beautiful mountains with wide-spreading snowy shoulders.
These are distant (reckoning from west to east) respectively
164, 170, and 172 miles from Myrung, and subtend angles
of $+ 0° 4' 0''$, $- 0° 1' 30''$, and $- 0° 2' 28''$.* From
Nunklow (940 feet lower than Myrung) they appear
higher, the western peak rising $14' 35''$ above the horizon;
whilst from Moflong (32 miles further south, and elevation 6,062 feet) the same is sunk $2'$ below the horizon.
My computations make this western mountain upwards of
24,000 feet high; but according to Col. Wilcox's angles,
taken from the Assam valley, it is only 21,600, the others
being respectively 20,720 and 21,475. Captain Thuillier
(the Deputy Surveyor General) agrees with me in considering that Colonel Wilcox's altitudes are probably much

---

* These angles were taken both at sunrise and sunset, and with an excellent
theodolite, and were repeated after two considerable intervals. The telescopes
were reversed after each observation, and every precaution used to insure
accuracy; nevertheless the mean of one set of observations of angular height
often varied 1' from that of another set. This is probably much due to atmospheric refraction, whose effect and amount it is impossible to estimate accurately
in such cases. Here the objects are not only viewed through 160 miles of
atmosphere, but through belts from between 6,000 to 20,000 feet of vertical
height, varying in humidity and transparency at different parts of the interval. If
we divide this column of atmosphere into sections parallel to those of latitude,
we have first a belt fifteen miles broad, hanging over the Khasia, 2,000 to 4,000
feet above the sea; beyond it, a second belt, seventy miles broad, hangs over the
Assam valley, which is hardly 300 feet above the level of the sea; and thirdly,
the northern part of the column, which reposes on 60 to 100 miles of the Bhotan
lower Himalaya: each of these belts has probably a different refractive
power.

under-estimated, as those of other Himalayan peaks to
the westward were by the old surveyors. It is further
evident that these mountains have (as far as can be
estimated by angles) fully 6-8,000 feet of snow on them,
which would not be the case were the loftiest only 21,600
feet high.

It is singular, that to the eastward of this group, no
snowy mountains are seen, and the lower Himalaya also
dip suddenly. This depression is no doubt partly due to
perspective; but as there is no such sudden disappearance
of the chain to the westward, where peaks are seen 35°
to the west of north, it is far more probable that the
valley of the Soobansiri river, which rises in Tibet far
behind these peaks, is broad and open; as is that of the
Dihong, still farther east, which we have every reason to
believe is the Tibetan Yaru or Burrampooter.

Supposing then the eastern group to indicate the mountain mass separating the Soobansiri from the Monass
river, no other mountains conspicuous for altitude or
dimension rise between N. N. E. and north, where there
is another immense group. This, though within 120
miles of Myrung, is below its horizon, and scarcely above
that of Nunklow (which is still nearer to it), and cannot
therefore attain any great elevation.

Far to the westward again, is a very lofty peaked mountain bearing N. N. W., which subtends an angle of
— 3′ 30″ from Myrung, and + 6′ 0″ from Nunklow. The
angles of this seem to indicate its being either Chumulari,
or that great peak which I saw due east from Bhomtso top,
and which I then estimated at ninety miles off and 23,500
feet high. From the Khasia angles, its latitude and longitude are 28° 6′ and 89° 30′, its elevation 27,000 feet, and
its distance from Myrung 200 miles. I need hardly add

that neither the position nor the elevation computed from such data is worthy of confidence.

Further still, to the extreme west, is an immense low hog-backed mass of snow, with a small peak on it; this bears north-west, both from Myrung and Nunklow, subtending an angle of — 25' from the former, and — 17' from the latter station. It is in all probability Chumulari, 210 miles distant from Nunklow. Donkia, if seen, would be distant 230 miles from the same spot in the Khasia, and Kinchinjunga 260; possibly they are visible (by refraction) from Chillong, though even further from it.

The distance from Myrung to Nunklow is ten miles, along an excellent road. The descent is at first sudden, beyond which the country is undulating, interspersed with jungle (of low trees, chiefly oaks) and marshes, with much rice cultivation. Grasses are exceedingly numerous; we gathered fifty kinds, besides twenty *Cyperaceæ* : four were cultivated, namely sugar-cane, rice, *Coix*, and maize. Most of the others were not so well suited to pasturage as those of higher localities. Dwarf Phœnix palm occurs by the roadside at 5000 feet elevation.

Gneiss (with garnets) highly inclined, was the prevalent rock (striking north-east), and scattered boulders of syenite became very frequent. In one place the latter rock is seen bursting through the gneiss, which is slaty and very crystalline at the junction.

Nunklow is placed at the northern extremity of a broad spur that over-hangs the valley of the Burrampooter river, thirty miles distant. The descent from it is very rapid, and beyond it none of the many spurs thrown out by the Khasia attain more than 1,000 feet elevation; hence, though the range does not present so abrupt a

J.D.H del.^t

John Murray, Albemarle Street, 1854.

W.L.Walton, Lith.

Bhotan Himalaya (dis.^t 120 miles) looking across the Assam Valley & Burrampooter River.

face to the Burrampooter as it does to the Jheels,
Nunklow is considered as on the brink of its north slope.
The elevation of the bungalow is 4,688 feet, and the climate
being hot, it swarms with mosquitos, fleas, and rats. It
commands a superb view to the north, of the Himalayan
snows, of the Burrampooter, and intervening malarious
Terai forest; and to the south, of the undulating Khasia,
with Kollong rock bearing south-west. All the hills
between this and Myrung look from Nunklow better
wooded than they do from Myrung, in consequence of
the slopes exposed to the south being bare of forest.

A thousand feet below the bungalow, a tropical forest
begins, of figs, birch, horse-chestnut, oak, nutmeg, *Cedrela,*
Engelhardtia, Artocarpeæ, and *Elæocarpus,* in the gullies, and
tall pines on the dry slopes, which are continued down to
the very bottom of the valley in which flows the Bor-panee,
a broad and rapid river that descends from Chillong, and
winds round the base of the Nunklow spur. Many of the
pines are eighty feet high, and three or four in diameter,
but none form gigantic trees. The quantity of balsams in
the wet ravines is very great, and tree-ferns of several
kinds are common.

The Bor-panee is about forty yards wide, and is spanned
by an elegant iron suspension-bridge, that is clamped to
the gneiss rock (strike north-east, dip north-west) on either
bank; beneath is a series of cascades, none high, but all of
great beauty from the broken masses of rocks and picturesque scenery on either side. We frequently botanised
up and down the river with great success : many curious
plants grow on its stony and rocky banks, and amongst
them *Rhododendron formosum* at the low elevation of 2000
feet. A most splendid fern, *Dipteris Wallichii,* is abundant, with the dwarf Phœnix palm and *Cycas pectinata.*

Wild animals are very abundant here, though extremely rare on the higher part of the Khasia range; tigers, however, and bears, ascend to Nunklow. We saw troops of wild dogs ("Kuleam," Khas.), deer, and immense quantities of the droppings of the wild elephant; an animal considered in Assam dangerous to meet, whereas in other parts of India it is not dreaded till provoked. There is, however, no quadruped that varies more in its native state than this: the Ceylon kind differs from the Indian in the larger size and short tusks, and an experienced judge at Calcutta will tell at once whether the newly caught elephant is from Assam, Silhet, Cuttack, Nepal, or Chittagong. Some of the differences, in size, roundness of shoulders and back, quantity of hair, length of limb, and shape of head, are very marked; and their dispositions are equally various.

The lowest rocks seen are at a considerable distance down the Bor-panee; they are friable sandstones that strike uniformly with the gneiss. From the bridge upwards the rocks are all gneiss, alternating with chert and quartz. The Nunklow spur is covered with enormous rounded blocks of syenite, reposing on clay or on one another. These do not descend the hill, and are the remains of an extensive formation which we could only find *in situ* at one spot on the road to Myrung (see p. 300), but which must have been of immense thickness.* One block within ten yards of the bungalow door was fifteen feet long, six high, and eight broad; it appeared half buried, and was rapidly decomposing from the action of the rain. Close by, to the westward, in walking amongst the masses we

---

* The tendency of many volcanic rocks to decompose in spheres is very well known: it is conspicuous in the black basalts north of Edinburgh, but I do not know any instance equal to this of Nunklow, for the extent of decomposition and dimensions of the resulting spheres.

were reminded of a moraine of most gigantic sized blocks ; one which I measured was forty feet long and eleven above the ground ; its edges were rounded, and its surface flaked off in pieces a foot broad and a quarter of an inch thick. Trees and brushwood often conceal the spaces between these fragments, and afford dens for bears and leopards, into which man cannot follow them.

Sitting in the cool evenings on one of these great blocks, and watching the Himalayan glaciers glowing with the rays of sunset, appearing to change in form and dimensions with the falling shadows, it was impossible to refrain from speculating on the possibility of these great boulders heaped on the Himalayan-ward face of the Khasia range, having been transported hither by ice at some former period ; especially as the Mont Blanc granite, in crossing the lake of Geneva to the Jura, must have performed a hardly less wonderful ice journey : but this hypothesis is clearly untenable ; and unparalleled in our experience as the results appear, if attributed to denudation and weathering alone, we are yet compelled to refer them to these causes.   The further we travel, and the longer we study, the more positive becomes the conviction that the part played by these great agents in sculpturing the surface of our planet, is as yet but half recognised.

We returned on the 7th of August to Churra, where we employed ourselves during the rest of the month in collecting and studying the plants of the neighbourhood. We hired a large and good bungalow, in which three immense coal fires * were kept up for drying plants and

---

* This coal is excellent for many purposes. We found it generally used by the Assam steamers, and were informed on board that in which we traversed the Sunderbunds, some months afterwards, that her furnaces consumed 729 lbs. per hour; whereas the consumption of English coal was 800 lbs., of Burdwan coal 840 lbs., and of Assam 900 lbs.

papers, and fifteen men were always employed, some in
changing, and some in collecting, from morning till night.
The coal was procured within a mile of our door, and cost
about six shillings a month; it was of the finest quality,
and gave great heat and few ashes.　Torrents of rain
descended almost daily, twelve inches in as many hours
being frequently registered; and we remarked that it was
impossible to judge of the quantity by estimation, an apparent deluge sometimes proving much less in amount than
much lighter but steadier falls; hence the greatest fall is
probably that in which the drops are moderately large,
very close together, and which pass through a saturated
atmosphere.　The temperature of the rain here and elsewhere in India was always a degree or two below that of
the air.

Though the temperature in August rose to 75°, we never
felt a fire oppressive, owing to the constant damp, and
absence of sun.　The latter, when it broke through the clouds,
shone powerfully, raising the thermometer 20° and 30° in
as many minutes.　On such occasions, hot blasts of damp
wind ascend the valleys, and impinge suddenly against
different houses on the flat, giving rise to extraordinary
differences between the mean daily temperatures of places
not half a mile apart.

On the 4th of September we started for the village of
Chela, which lies west from Churra, at the embouchure of
the Boga-panee on the Jheels.　The path runs by Mamloo,
and down the spur to the Jasper hill (see p. 280): the vegetation all along is very tropical, and pepper, ginger, maize,
and Betel palm, are cultivated around small cottages, which
are only distinguishable in the forest by their yellow thatch
of dry *Calamus* (Rattan) leaves.　From Jasper hill a very
steep ridge leads to another, called Lisouplang, which is

hardly so high as Mamloo ; the rocks are the same sandstone, with fragments of coal, and remains of the limestone
formation capping it.

Hot gusts of wind blow up the valleys, alternating
with clouds and mists, and it is curious to watch the
effects of the latter in stilling the voices of insects (Cicadas)
and birds. Common crows and vultures haunt the villages,
but these, and all other large birds, are very rare in the
Khasia. A very few hawks are occasionally seen, also
sparrows and kingfishers, and I once heard a cuckoo ;
pheasants are sometimes shot, but we never saw any.
Kites become numerous after the rains, and are regarded
as a sign of their cessation. More remarkable than the
rarity of birds is the absence of all animals except
domestic rats, as a more suitable country for hares and
rabbits could not be found. Reptiles, and especially
Colubridæ, are very common in the Khasia mountains,
and I procured sixteen species and many specimens.
The natives repeatedly assured us that these were all
harmless, and Dr. Gray, who has kindly examined all my
snakes, informs me of the remarkable fact (alluded to in a
note to p. 25), that whereas none of these are poisonous,
four out of the eleven species which I found in Sikkim
are so. One of the Khasia blind-worms (a new species)
belongs to a truly American genus (*Ophisaurus*), a fact as
important as is that of the Sikkim skink and *Agama* being
also American forms.

Arundina, a beautiful purple grassy-leaved orchid, was
abundantly in flower on the hill-top, and the great white
swallow-tailed moth (*Saturnia Atlas*) was extremely common, with tropical butterflies and other insects. The
curious leaf-insect (*Mantis*) was very abundant on the
orange trees, on the leaves of which the natives believe

it to feed; nor indeed could we persuade some of our
friends that its thin sharp jaws are unsuited for masti-
cating leaves, and that these and its prehensile feet
indicate its predacious nature: added to which, its singular
resemblance to a leaf is no less a provision against its
being discovered by its enemies, than an aid in deceiving
its prey.

We descended rapidly for many miles through beautiful
rocky woods, with villages nestling amongst groves of
banana and trellised climbers; and from the brow of a
hill looked down upon a slope covered with vegetation and
huts, which formed the mart of Chela, and below which the
Boga-panee flowed in a deep gorge. The view was a very
striking one: owing to the steepness of the valley below
our feet, the roofs alone of the cottages were visible, from
which ascended the sounds and smells of a dense native
population, and to which there appeared to be no way of
descending. The opposite side rose precipitously in lofty
table-topped mountains, and the river was studded with
canoes.

The descent was fully 800 feet, on a slope averaging 25°
to 35°. The cottages were placed close together, each
within a little bamboo enclosure, eight to ten yards deep;
and no two were on the same level. Each was built against
a perpendicular wall which supported a cutting in the bank
behind; and a similar wall descended in front of it, forming
the back of the compartment in which the cottage next
below it was erected. The houses were often raised on
platforms, and some had balconies in front, which overhung
the cottage below. All were mere hovels of wattle or mud,
with very high-pitched roofs: stone tanks resembling fonts,
urns, coffins, and sarcophagi, were placed near the better
houses, and blocks of stone were scattered everywhere.

We descended from hovel to hovel, alternately along the gravelled flat of each enclosure, and perpendicularly down steps cut in the sandstone or let into the walls. I counted 800 houses from the river, and there must be many

CHELA VILLAGE.

more: the inhabitants are Bengalees and Khasias, and perhaps amount to 3000 or 4000; but this is a very vague estimate.

We lodged in a curious house, consisting of one apartment, twenty feet long, and five high, raised thirty feet upon bamboos: the walls were of platted bamboo matting, fastened to strong wooden beams, and one side opened on a balcony that overhung the river. The entrance was an oval aperture reached by a ladder, and closed by folding-doors that turned on wooden pivots.

x 2

The roof was supported by tressels of great thickness, and like the rest of the woodwork, was morticed, no nails being used throughout the building. The floor was of split bamboos laid side by side.

We ascended the Boga-panee in canoes, each formed of a hollowed trunk fifty feet long and four broad; we could not, however, proceed far, on account of the rapids. The rocks in its bed are limestone, but a great bluff cliff of sandy conglomerate (strike east-south-east and dip south-south-west 70°), several hundred feet high, rises on the east bank close above the village, above which occurs amygdaloidal basalt. The pebbles in the river (which was seventy yards broad, and turbid) were of slate, basalt, sandstone, and syenite: on the opposite bank were sandstones over-lain by limestone, both dipping to the southward.

Beautiful palms, especially *Caryota urens* (by far the handsomest in India), and groves of betel-nut bordered the river, with oranges, lemons, and citrons; intermixed with feathery bamboos, horizontally-branched acacias, oaks, with pale red young leaves, and deep green foliaged figs. Prickly rattans and *Plectocomia* climbed amongst these, their enormous plumes of foliage upborne by the matted branches of the trees, and their arrowy tops shooting high above the forest.

After staying three days at Chela, we descended the stream in canoes, shooting over pebbly rapids, and amongst rocks of limestone, water-worn into fantastic shapes, till we at last found ourselves gliding gently along the still canals of the Jheels. Many of these rapids are so far artificial, that they are enclosed by gravel banks, six feet high, which, by confining the waters, give them depth; but, Chela being hardly above the level of the sea, their fall is very trifling.   We proceeded across the Jheels* to Chattuc, and then north again to Pundua, and so to Churra.

Having pretty well exhausted the botany of Churra, Dr. Thomson and I started on the 13th of September for the eastern part of the Khasia and Jyntea mountains.   On the Kala-panee road,† which we followed, we passed crowds of market people, laden with dried fish in a half-putrid state, which scented the air for many yards : they were chiefly carp, caught and dried at the foot of the hills. Large parties were bringing down baskets of bird-cherries, cinnamon-bark, iron, pine planks, fire-wood, and potatoes. Of these, the bird-cherries (like damsons) are made into an excellent preserve by the English residents, who also make capital cherry-brandy of them : the trade in cinnamon is of recent introduction, and is much encouraged by the Inglis family, to whose exertions these people are so greatly indebted ; the cinnamon is the peeled bark of a small species of *Cinnamomum* allied to that of Ceylon, and though inferior in flavour and mucilaginous (like cassia), finds a ready market at Calcutta.   It has been used to adulterate the Ceylon cinnamon ; and an extensive fraud was attempted by some Europeans at Calcutta, who sent boxes of this, with a top layer of the genuine, to England.   The smell of the cinnamon loads was as fragrant as that of the fish was offensive.

The road from Kala-panee bungalow strikes off north-easterly, and rounds the head of the deep valley to the east of Churra ; it then crosses the head-waters of the

---

* The common water-plants of the Jheels are *Vallisneria serrata, Damasonium,* 2 *Myriophylla,* 2 *Villarsiœ, Trapa,* blue, white, purple and scarlet water-lilies, *Hydrilla, Utricularia, Limnophila, Azolla, Salvinia, Ceratopteris,* and floating grasses.

† The Pea-violet (*Crotalaria occulta*) was very common by the road-side, and smelt deliciously of violets : the English name suggests the appearance of the flower, for which and for its fragrance it is well worth cultivation.

Kala-panee river, still a clear stream, the bed of which is comparatively superficial : the rocks consist of a little basalt and much sandstone, striking east by north, and dipping north by west. The Boga-panee is next reached, flowing in a shallow valley, about 200 feet below the general level of the hills, which are grassy and treeless. The river * is thirty yards across, shallow and turbid; its bed is granite, and beyond it scattered stunted pines are met with ; a tree which seems to avoid the sandstone. In the evening we arrived at Nonkreem, a large village in a broad marshy valley, where we procured accommodation with some difficulty, the people being by no means civil, and the Rajah, Sing Manuk, holding himself independent of the British Government.

Atmospheric denudation and weathering have produced remarkable effects on the lower part of the Nonkreem valley, which is blocked up by a pine-crested hill, 200 feet high, entirely formed of round blocks of granite, heaped up so as to resemble an old moraine; but like the Nunklow boulders, these are not arranged as if by glacial action. The granite is micaceous, and usually very soft, decomposing into a coarse reddish sand, that colours the Boga-panee. To procure the iron-sand, which is disseminated through it, the natives conduct water over the beds of granite sand, and as the lighter particles are washed away, the remainder is removed to troughs, where the separation of the ore is completed. The smelting is very rudely carried on in charcoal fires, blown by enormous double-action bellows, worked by two persons, who stand on the machine, raising the flaps with their hands, and expanding them with their feet, as shown in the cut at p. 312.

---

* The fall of this river, between this elevation (which may be considered that of its source) and Chela, is about 5,500 feet.

There is neither furnace nor flux used in the reduction. The fire is kindled on one side of an upright stone (like the head-stone of a grave), with a small arched hole close to the ground : near this hole the bellows are suspended ; and a bamboo tube from each of its compartments, meets in a larger one, by which the draught is directed under the

NONKREEM VILLAGE.

hole in the stone to the fire.   The ore is run into lumps as large as two fists, with a rugged surface : these lumps are afterwards cleft nearly in two, to show their purity.

The scenery about Nonkreem village is extremely picturesque, and we procured many good plants on the rocks, which were covered with the purple-flowered Orchid, *Cœlogyne Wallichii.* The country is everywhere intersected

with trenches for iron-washing, and some large marshes
were dammed up for the same purpose: in these we
found some beautiful balsams, *Hypericum* and *Parnassia;* also a diminutive water-lily, the flower of which is
no larger than a half-crown; it proves to be the *Nymphœa
pygmœa* of China and Siberia—a remarkable fact in the
geographical distribution of plants.

BELLOWS.

From Nonkreem we proceeded easterly to Pomrang,
leaving Chillong hill on the north, and again crossing the
Boga-panee, beyond which the sandstone appeared (strike

north-east and dip north-west 60°); the soil was poor in the extreme; not an inhabitant or tree was to be seen throughout the grassy landscape, and hardly a bush, save an occasional rhododendron, dwarf oak, or *Pieris*, barely a few inches high.

At Pomrang we took up our quarters in an excellent empty bungalow, built by Mr. Stainforth (Judge of Silhet), who kindly allowed us the use of it. Its elevation was 5,143 feet, and it occupied the eastern extremity of a lofty spur that overhangs the deep fir-clad valley of the Oongkot, dividing Khasia from Jyntea. The climate of Pomrang is so much cooler and less rainy than at Churra, that this place is more eligible for a station; but the soil is quite impracticable, there is an occasional scarcity of water, the pasture is wholly unsuited for cattle or sheep, and the distance from the plains is too great.

A beautiful view extends eastwards to the low Jyntea hills, backed by the blue mountains of Cachar, over the deep valley in front; to the northward, a few peaks of the Himalaya are seen, and westward is Chillong. We staid here till the 23rd September, and then proceeded south-eastward to Mooshye. The path descends into the valley of the Oongkot, passing the village of Pomrang, and then through woods of pine, *Gordonia*, and oak, the latter closely resembling the English, and infested with galls. The slopes are extensively cultivated with black awnless unirrigated rice, and poor crops of *Coix*, protected from the birds by scarecrows of lines stretched across the fields, bearing tassels and tufts of fern, shaken by boys. This fern proved to be a very curious and interesting genus, which is only known to occur elsewhere at Hong-Kong in China, and has been called *Bowringia*, after the eminent Dr. Bowring.

We crossed the river* twice, proceeding south-west to Mooshye, a village placed on an isolated, flat-topped, and very steep-sided hill, 4,863 feet above the sea, and perhaps 3,500 above the Oongkot, which winds round its base. A very steep path led up slate rocks to the top (which was of sandstone), where there is a stockaded guard-house, once occupied by British troops, of which we took possession. A Labiate plant (*Mesona Wallichiana*) grew on the ascent, whose bruised leaves smelt as strongly of patchouli, as do those of the plant producing that perfume, to which it is closely allied. The *Pogostemon Patchouli* has been said to occur in these parts of India, but we never met with it, and doubt the accuracy of the statement. It is a native of the Malay peninsula, whence the leaves are imported into Bengal, and so to Europe.

The summit commands a fine view northward of some Himalayan peaks, and southwards of the broad valley of the Oongkot, which is level, and bounded by steep and precipitous hills, with flat tops. On the 25th we left Mooshye for Amwee in Jyntea, which lies to the south-east. We descended by steps cut in the sandstone, and fording the Oongkot, climbed the hills on its east side, along the grassy tops of which we continued, at an elevation of 4,000 feet. Marshy flats intersect the hills, to which wild elephants sometimes ascend, doing much damage to the rice

---

* *Podostemon* grew on the stones at the bottom: it is a remarkable water-plant, resembling a liver-wort in its mode of growth. Several species occur at different elevations in the Khasia, and appear only in autumn, when they often carpet the bottom of the streams with green. In spring and summer no traces of them are seen; and it is difficult to conceive what becomes of the seeds in the interval, and how these, which are well known, and have no apparent provision for the purpose, attach themselves to the smooth rocks at the bottom of the torrents. All the kinds flower and ripen their seeds under water; the stamens and pistil being protected by the closed flower from the wet. This genus does not inhabit the Sikkim rivers, probably owing to the great changes of temperature to which these are subject.

crops. We crossed a stream by a bridge formed of one
gigantic block of sandstone, 20 feet long, close to the village,
which is a wretched one, and is considered unhealthy: it
stands on the high road from Jynteapore (at the foot of the
hills to the southward) to Assam : the only road that crosses
the mountains east of that from Churra to Nunklow.

Though so much lower, this country, from the barrenness of the soil, is more thinly inhabited than the Khasia.

OLD BRIDGE AT AMWEE.

The pitcher-plant (*Nepenthes*) grows on stony and
grassy hills about Amwee, and crawls along the ground ;
its pitchers seldom contain insects in the wild state, nor
can we suggest any special function for the wonderful
organ it possesses.

About eight miles south of the village is a stream,
crossed by a bridge, half of which is formed of slabs of
stone (of which one is twenty-one feet long, seven broad,
and two feet three and a half inches thick), supported on
piers, and the rest is a well turned arch, such as I have not

seen elsewhere among the hill tribes of India. It is fast
crumbling away, and is covered with tropical plants, and a
beautiful white-flowered orchis * grew in the mossy crevices
of its stones.

From Amwee our route lay north-east across the Jyntea
hills to Joowye, the hill-capital of the district. The path
gradually ascended, dipping into valleys scooped out in the
horizontal sandstone down to the basalt; and boulders of
the same rock were scattered about. Fields of rice
occupy the bottoms of these valleys, in which were placed
gigantic images of men, dressed in rags, and armed with
bows and arrows, to scare away the wild elephants ! Slate
rocks succeed the sandstone (strike north-east, dip north-
west), and with them pines and birch appear, clothing
the deep flanks of the Mintadoong valley, which we
crossed.

The situation of Joowye is extremely beautiful: it occupies the broken wooded slope of a large open flat valley,
dotted with pines; and consists of an immense number of
low thatched cottages, scattered amongst groves of bamboo,
and fields of plantain, tobacco, yams, sugar-cane, maize,
and rice, surrounded by hedges of bamboo, *Colquhounia*, and
Erythrina. Narrow steep lanes lead amongst these, shaded
with oak, birch, *Podocarpus*, Camellia, and *Araliaceæ*;
the larger trees being covered with orchids, climbing palms,
Pothos, *Scindapsus*, pepper, and *Gnetum*; while masses of
beautiful red and violet balsams grew under every hedge
and rock. The latter was of sandstone, overlying highly
inclined schists, and afforded magnificent blocks for the
natives to rear on end, or make seats of. Some erect stones

---

* *Diplomeris; Apostasia* also grew in this gulley, with a small *Arundina*, some
beautiful species of *Sonerila*, and *Argostemma*. The neighbourhood was very rich
in plants.

on a hill at the entrance are immensely large, and surround a clump of fine fig and banyan trees.*

We procured a good house after many delays, for the people were far from obliging; it was a clean, very long cottage, with low thatched eaves almost touching the ground, and was surrounded by a high bamboo paling that enclosed out-houses built on a well-swept floor of beaten earth. Within, the woodwork was carved in curious patterns, and was particularly well fitted. The old lady to whom it belonged got tired of us before two days were over, and first tried to smoke us out by a large fire of green wood at that end of the cottage which she retained; and afterwards by inviting guests, to a supper, with whom she kept up a racket all night. Her son, a tall, sulky fellow, came to receive the usual gratuity on our departure, which we made large to show we bore no ill-will: he, however, behaved so scornfully, pretending to despise it, that I had no choice but to pocket it again; a proceeding which was received with shouts of laughter, at his expense, from a large crowd of bystanders.

On the 30th of September we proceeded north-east from Joowye to Nurtiung, crossing the watershed of the Jyntea range, which is granitic, and scarcely raised above the mean level of the hills; it is about 4,500 feet elevation. To the north the descent is at first rather abrupt for 500 feet, to a considerable stream, beyond which is the village of Nurtiung. The country gradually declines hence to the north-east, in grassy hills,

---

* In some tanks we found *Hydropeltis*, an American and Australian plant allied to *Nymphœa*. Mr. Griffith first detected it here, and afterwards in Bhotan. these being the only known habitats for it in the Old World. It grows with *Typha*, *Acorus Calamus* (sweet flag), *Vallisneria*, *Potamogeton*, *Sparganium*, and other European water-plants.

which to the east become higher and more wooded : to
the west the Khasia are seen, and several Himalayan
peaks to the north.

The ascent to the village from the river is by steps cut in
a narrow cleft of the schist rocks, to a flat, elevated 4,178
feet above the sea : we here procured a cottage, and found
the people remarkably civil. The general appearance is the
same as at Joowye, but there are here extensive and very
unhealthy marshes, whose evil effects we experienced, in
having the misfortune to lose one of our servants by fever.
Except pines, there are few large trees ; but the quantity of
species of perennial woody plants contributing to form the
jungles is quite extraordinary : I enumerated 140, of which
60 were trees or large shrubs above twenty feet high.
One of these was the *Hamamelis chinensis*, a plant
hitherto only known as a native of China. This, the
Bowringia, and the little *Nymphæa*, are three out of
many remarkable instances of our approach to the eastern
Asiatic flora.

From Nurtiung we walked to the Bor-panee river, sixteen or twenty miles to the north-east (not the river of that
name below Nunklow), returning the same night ; a most
fatiguing journey in so hot and damp a climate. The path
lay for the greatest part of the way over grassy hills of
mica-schist, with boulders of granite, and afterwards of
syenite, like those of Nunklow. The descent to the river
is through noble woods of spreading oaks,* chesnuts, magnolias, and tall pines : the vegetation is very tropical, and
with the exception of there being no sal, it resembles that
of the dry hills of the Sikkim Terai. The Bor-panee is

---

* We collected upwards of fifteen kinds of oak and chesnut in these and the
Khasia mountains ; many are magnificent trees, with excellent wood, while others
are inferior as timber.

forty yards broad, and turbid; its bed, which is of basalt, is 2,454 feet above the sea: it is crossed by a raft pulled to and fro by canes.

Nurtiung contains a most remarkable collection of those sepulchral and other monuments, which form so curious a feature in the scenery of these mountains and in the habits of their savage population. They are all placed in a fine grove of trees, occupying a hollow; where several acres are covered with gigantic, generally circular, slabs of stone, from ten to twenty-five feet broad, supported five feet above the ground upon other blocks. For the most part they are buried in brushwood of nettles and shrubs, but in one place there is an open area of fifty yards encircled by them, each with a gigantic headstone behind it. Of the latter the tallest was nearly thirty feet high, six broad, and two feet eight inches in thickness, and must have been sunk at least five feet, and perhaps much more, in the ground. The flat slabs were generally of slate or horn-stone; but many of them, and all the larger ones, were of syenitic granite, split by heat and cold water with great art. They are erected by dint of sheer brute strength, the lever being the only aid. Large blocks of syenite were scattered amongst these wonderful erections.

Splendid trees of *Bombax*, fig and banyan, overshadowed them: the largest banyan had a trunk five feet in dia-meter, clear of the buttresses, and numerous small trees of *Celtis* grew out of it, and an immense flowering tuft of *Vanda cærulea* (the rarest and most beautiful of Indian orchids) flourished on one of its limbs. A small plantain with austere woolly scarlet fruit, bearing ripe seeds, was planted in this sacred grove, where trees of the most tropical genera grew mixed with the pine, birch, *Myrica*, and *Viburnum*.

The Nurtiung Stonehenge is no doubt in part religious, as the grove suggests, and also designed for cremation, the bodies being burnt on the altars. In the Khasia these upright stones are generally raised simply as memorials of great events, or of men whose ashes are not necessarily, though frequently, buried or deposited in hollow stone sarcophagi near them, and sometimes in an urn placed inside a sarcophagus, or under horizontal slabs.

The usual arrangement is a row of five, seven, or more

STONES AT NURTIUNG.

erect oblong blocks with round heads (the highest being placed in the middle), on which are often wooden discs and cones : more rarely pyramids are built. Broad slabs for seats are also common by the wayside. Mr. Yule, who first drew attention to these monuments, mentions one thirty-two feet by fifteen, and two in thickness ; and states that the sarcophagi (which, however, are rare) formed of four slabs, resemble a drawing in Bell's Circassia, and descriptions in Irby and Mangles' Travels in Syria. He adds that many villages derive their names from these stones, "mau" signifying "stone:" thus "Mausmai" is "the stone of oath," because, as his native informant said, "there was war between Churra and Mausmai, and when they made peace, they swore to it, and placed a stone as a witness;" forcibly recalling the stone Jacob set up for a pillar, and other passages in the old Testament. "Mamloo" is "the stone of salt," eating salt from a sword's point being the Khasia form of oath : "Mauflong" is "the grassy stone," &c.* Returning from this grove, we crossed a stream by a single squared block, twenty-eight feet long, five broad, and two thick, of gray syenitic granite with large crystals of felspar.

We left Nurtiung on the 4th of October, and walked to Pomrang, a very long and fatiguing day's work. The route descends north-west of the village, and turns due east along bare grassy hills of mica-schist and slate (strike east and west, and dip north). Near the village of Lernai oak woods are passed, in which *Vanda cœrulea* grows in profusion, waving its panicles of azure flowers in the wind. As this beautiful orchid is at present attracting great attention, from its high price, beauty, and difficulty of culture, I shall point out how totally at variance with its native habits, is the cultivation thought necessary for it in England.† The

* Notes on the Khasia mountains and people; by Lieutenant H. Yule, Bengal Engineers. Analogous combinations occur in the south of England and in Brittany, &c., where similar structures are found. Thus *maen, man,* or *men* is the so-called Druidical name for a stone, whence *Pen-maen-mawr,* for "the hill of the big stone," *Maen-hayr,* for the standing stones of Brittany, and *Dol-men,* "the table-stone," for a cromlech.

† We collected seven men's loads of this superb plant for the Royal Gardens

dry grassy hills which it inhabits are elevated 3000
to 4000 feet: the trees are small, gnarled, and very
sparingly leafy, so that the Vanda which grows on their
limbs is fully exposed to sun, rain, and wind. There is no
moss or lichen on the branches with the Vanda, whose
roots sprawl over the dry rough bark. The atmosphere is
on the whole humid, and extremely so during the rains ;
but there is no damp heat, or stagnation of the air, and at
the flowering season the temperature ranges between 60°
and 80°, there is much sunshine, and both air and bark
are dry during the day : in July and August, during the
rains, the temperature is a little higher than above, but
in winter it falls much lower, and hoar-frost forms on the
ground. Now this winter's cold, summer's heat, and
autumn's drought, and above all, this constant free exposure
to fresh air and the winds of heaven, are what of all things
we avoid exposing our orchids to in England. It is
under these conditions, however, that all the finer Indian
Orchideæ grow, of which we found *Dendrobium Farmeri,
Dalhousianum, Devonianum,* &c., with *Vanda cærulea;* whilst
the most beautiful species of *Cælogyne, Cymbidium,
Bolbophyllum,* and *Cypripedium,* inhabit cool climates at
elevations above 4000 feet in Khasia, and as high as
6000 to 7000 in Sikkim.

On the following day we turned out our Vanda to dress
the specimens for travelling, and preserve the flowers for
botanical purposes. Of the latter we had 360 panicles,
each composed of from six to twenty-one broad pale-blue

at Kew; but owing to unavoidable accidents and difficulties, few specimens reached
England alive. A gentleman who sent his gardener with us to be shown the
locality, was more successful : he sent one man's load to England on commission,
and though it arrived in a very poor state, it sold for 300*l.*, the individual plants
fetching prices varying from 3*l.* to 10*l.* Had all arrived alive, they would have
cleared 1000*l.* An active collector, with the facilities I possessed, might easily
clear from 2000*l.* to 3000*l.*, in one season, by the sale of Khasia orchids.

tesselated flowers, three and a half to four inches across :
and they formed three piles on the floor of the verandah,
each a yard high :—what would we not have given to have
been able to transport a single panicle to a Chiswick fête !

On the 10th of October we sent twenty-four strong
mountaineers to Churra, laden with the collections of the
previous month; whilst we returned to Nonkreem, and
crossing the shoulder of Chillong, passed through the village
of Moleem in a north-west direction to the Syong bungalow.
From this we again crossed the range to Nunklow and the
Bor-panee, and returned by Moflong and the Kala-panee
to Churra during the latter part of the month.

In November the vegetation above 4000 feet turns
wintry and brown, the weather becomes chilly, and though
the cold is never great, hoar-frost forms at Churra, and
water freezes at Moflong. We prepared to leave as these
signs of winter advanced : we had collected upwards of
2,500 species, and for the last few weeks all our diligence,
and that of our collectors, had failed to be rewarded by a
single novelty. We however procured many species in
fruit, and made a collection of upwards of 300 kinds of
woods, many of very curious structure. As, however, we
projected a trip to Cachar before quitting the neighbourhood,
we retained our collectors, giving orders for them to meet
us at Chattuc, on our way down the Soormah in December,
with their collections, which amounted to 200 men's loads,
and for the conveyance of which to Calcutta, Mr. Inglis
procured us boats.

Before dismissing the subject of the Khasia mountains,
it will be well to give a slight sketch of their prominent
geographical features, in connection with their geology.
The general geological characters of the chain may be
summed up in a few words. The nucleus or axis is of

highly inclined stratified metamorphic rocks, through which the granite has been protruded, and the basalt and syenite afterwards injected. After extensive denudations of these, the sandstone, coal, and limestone were successively deposited. These are altered and displaced along the southern edge of the range, by black amygdaloidal trap, and have in their turn been extensively denuded; and it is this last operation that has sculptured the range, and given the mountains their present aspect; for the same gneisses, slates, and basalts in other countries, present rugged peaks, domes, or cones, and there is nothing in their composition or arrangement here that explains the tabular or rounded outline they assume, or the uniform level of the spurs into which they rise, or the curious steep sides and flat floors of the valleys which drain them.

All these peculiarities of outline are the result of denudation, of the specific action of which agent we are very ignorant. The remarkable difference between the steep cliffs on the south face of the range, and the rounded outline of the hills on the northern slopes, may be explained on the supposition that when the Khasia was partially submerged, the Assam valley was a broad bay or gulf; and that while the Churra cliffs were exposed to the full sweep of the ocean, the Nunklow shore was washed by a more tranquil sea.

The broad flat marshy heads of all the streams in the central and northern parts of the chain, and the rounded hills that separate them, indicate the levelling action of a tidal sea, acting on a low flat shore ; * whilst the steep flat-

* Since our return to England, we have been much struck with the similarity in contour of the Essex and Suffolk coasts, and with the fact that the tidal coast sculpturing of this surface is preserved in the very centre of High Suffolk, twenty to thirty miles distant from the sea, in rounded outlines and broad flat marshy valleys.

floored valleys of the southern watershed may be attributed to the scouring action of higher tides on a boisterous rocky coast. These views are confirmed by an examination of the east shores of the Bay of Bengal, and particularly by a comparison of the features of the country about Silhet, now nearly 200 miles distant from the sea, with those of the Chittagong coast, with which they are identical.

The geological features of the Khasia are in many respects so similar to those of the Vindhya, Kymore, Behar, and Rajmahal mountains, that they have been considered by some observers as an eastern prolongation of that great chain, from which they are geographically separated by the delta of the Ganges and Burrampooter. The general contour of the mountains, and of their sandstone cliffs, is the same, and the association of this rock with coal and lime is a marked point of similarity; there is, however, this difference between them, that the coal-shales of Khasia and limestone of Behar are non-fossiliferous, while the lime of Khasia and the coal-shales of Behar contain fossils.

The prevalent north-east strike of the gneiss is the same in both, differing from the Himalaya, where the stratified rocks generally strike north-west. The nummulites of the limestone are the only known means we have of forming an approximate estimate of the age of the Khasia coal, which is the most interesting feature in the geology of the range : these fossils have been examined by MM. Archiac and Jules Haines,* who have pronounced the species collected by Dr. Thomson and myself to be the same as those found in the nummulite rocks of north-west India, Scinde, and Arabia.

* "Description des Animaux Fossiles des Indes Orientales," p. 178. These species are *Nummulites scabra*, Lamarck, *N. obtusa*, Sowerby, *N. Lucusana*, Deshayes, and *N. Beaumonti*, d'Arch. and Haines.

CHAPTER XXXI.

WE left Churra on the 17th of November, and taking boats
at Pundua, crossed the Jheels to the Soormah, which we
ascended to Silhet.   Thence we continued our voyage 120
miles up the river in canoes, to Silchar, the capital of the
district of Cachar: the boats were such as I described
at Chattuc, and though it was impossible to sit upright in
them, they were paddled with great swiftness.   The river
at Silhet is 200 yards broad; it is muddy, and flows with a
gentle current of two to three miles an hour, between banks
six to twelve feet high.   As we glided up its stream, villages
became rarer, and eminences more frequent in the Jheels.
The people are a tall, bold, athletic Mahometan race, who
live much on the water, and cultivate rice, sesamum, and

radishes, with betel-pepper in thatched enclosures as in Sikkim : maize and sugar are rarer, bamboos abound, and four palms (*Borassus, Areca,* cocoa-nut, and *Caryota*) are planted, but there are no date-palms.

The Teelas (or hillocks) are the haunts of wild boars, tigers, and elephants, but not of the rhinoceros; they are 80 to 200 feet high, of horizontally stratified gravel and sand slates, and clay conglomerates, with a slag-like honey-combed sandstone; they are covered with oaks, figs, *Heretiera,* and bamboos, and besides a multitude of common Bengal plants, there are some which, though generally considered mountain or cold country genera, here descend to the level of the sea ; such are *Kadsura, Rubus, Camellia,* and *Sabia* ; *Aerides* and *Saccolabia* are the common orchids, and rattan-canes and *Pandani* render the jungles impenetrable.

A very long sedge (*Scleria*) grows by the water, and is used for thatching : boatloads of it are collected for the Calcutta market, for which also were destined many immense rafts of bamboo, 100 feet long. The people fish much, using square and triangular drop-nets stretched upon bamboos, and rude basket-work weirs, that retain the fish as the river falls. Near the villages we saw fragments of pottery three feet below the surface of the ground, shewing that the bank, which is higher than the surrounding country, increases from the annual overflow.

About seventy miles up the river, the mountains on the north, which are east of Jyntea, rise 4000 feet high in forest-clad ranges like those of Sikkim. Swamps extend from the river to their base, and penetrate their valleys, which are extremely malarious : these forests are frequented by timber-cutters, who fell jarool (*Lagerstroemia Reginae*), a magnificent tree with red wood, which, though soft, is durable under water, and therefore in universal use for boat-building. The toon is also cut, with red sandal-wood (*Adenanthera pavonina*) ; also Nageesa,* *Mesua ferrea*, which is highly valued for its weight, strength, and durability: *Aquilaria agallocha*, the eagle-wood, a tree yielding uggur oil, is also much sought for its fragrant wood, which is carried to Silhet and Azmerigunj, where it is broken up and distilled. Neither teak, sissoo, sal, nor other *Dipterocarpi*, are found in these forests.

Porpoises, and both the long and the short-nosed alligator, ascend the Soormah for 120 miles, being found beyond Silchar, which place we reached on the 22nd, and were most hospitably received by Colonel Lister, the political agent commanding the Silhet Light Infantry, who was inspecting the Cookie levy, a corps of hill-natives which had lately been enrolled.

The station is a small one, and stands about forty feet above the river, which however rises half that height in the rains. Long low spurs of tertiary rocks stretch from the Tipperah hills for many miles north, through the swampy Jheels to the river; and there are also hills on the opposite or north side, but detached from the Cookie hills, as the lofty blue range twelve miles north of the Soormah is called. All these mountains swarm with tigers, wild buffalos, and boars, which also infest the long grass of the Jheels.

The elevation of the house we occupied at Silchar was

---

* There is much dispute amongst oriental scholars about the word Nageesa; the Bombay philologists refer it to a species of *Garcinia*, whilst the pundits on the Calcutta side of India consider it to be *Mesua ferrea*. Throughout our travels in India, we were struck with the undue reliance placed on native names of plants, and information of all kinds; and the pertinacity with which each linguist adhered to his own crotchet as to the application of terms to natural objects, and their pronunciation. It is a very prevalent, but erroneous, impression, that savage and half-civilised people have an accurate knowledge of objects of natural history, and a uniform nomenclature for them.

116 feet above the sea.   The bank it stood on was of clay, with soft rocks of conglomerate, which often assume the appearance of a brown sandy slag.

During the first Birmese war, Colonel Lister was sent with a force up to this remote corner of Bengal, when the country was an uninhabited jungle, so full of tigers that not a day passed without one or more of his grass or wood-cutters being carried off.   Now, thousands of acres are cultivated with rice, and during our stay we did not see a tiger. The quantity of land brought into cultivation in this part of Bengal, and indeed throughout the Gangetic delta, has probably been doubled during the last twenty years, and speaks volumes for the state of the peasant under the Indian Company's sway, as compared with his former condition.   The Silchar rice is of admirable quality, and much is imported to Silhet, the Jheels not producing grain enough for the consumption of the people.   Though Silchar grows enough for ten times its population, there was actually a famine six weeks before our arrival, the demand from Silhet being so great.

The villages of Cachar are peopled by Mahometans, Munniporees, Nagas, and Cookies; the Cacharies themselves being a poor and peaceful jungle tribe, confined to the mountains north of the Soormah.   The Munniporees* are emigrants from the kingdom of that name, which lies beyond the British possessions, and borders on Assam and

---

* The Munnipore valley has never been explored by any naturalist, its mountains are said to be pine-clad, and to rise 8000 feet above the level of the sea. The Rajah is much harassed by the Birmese, and is a dependant of the British, who are in the very frequent dilemma of supporting on the throne a sovereign opposed by a strong faction of his countrymen, and who has very dubious claims to his position.   During our stay at Silchar, the supposed rightful Rajah was prevailing over the usurper; a battle had been fought on the hills on the frontier, and two bodies floated past our bungalow, pierced with arrows.

Birmah. Low ranges of forest-clad mountains at the head
of the Soormah, separate it from Silchar, with which it is
coterminous; the two chief towns being seven marches
apart. To the south-east of Silchar are interminable jungles,
peopled by the Cookies, a wild Indo-Chinese tribe, who live
in a state of constant warfare, and possess the whole hill-
country from this, southward to beyond Chittagong. Two
years ago they invaded and ravaged Cachar, carrying many
of the inhabitants into slavery, and so frightening the
people, that land previously worth six rupees a biggah, is
now reduced to one and a half. Colonel Lister was
sent with a strong party to rescue the captives, and
marched for many days through their country without
disturbing man or beast; penetrating deep forests of
gigantic trees and tall bamboos, never seeing the sun
above, or aught to the right and left, save an occasional
clearance and a deserted village. The incursion, however, had its effects, and the better inclined near the
frontier have since come forward, and been enrolled as the
Cookie levy.

The Munnipore emigrants are industrious settlers for a
time, but never remain long in one place: their religion is
Hindoo, and they keep up a considerable trade with their
own country, whence they import a large breed of buffalos,
ponies, silks, and cotton cloths dyed with arnotto (*Bixa*),
and universally used for turbans. They use bamboo
blowing-tubes and arrows for shooting birds, make excellent shields of rhinoceros hide (imported from Assam), and
play at hockey on horseback like the Western Tibetans.
A fine black varnish from the fruit of *Holigarna longifolia*,
is imported from Munnipore, as is another made from
Sesuvium Anacardium (marking-nut), and a remarkable black
pigment resembling that from *Melanorhœa usitatissima*,

which is white when fresh, and requires to be kept under water.*

One fine moonlight night we went to see a Munnipore dance. A large circular area was thatched with plantain leaves, growing on their trunks, which were stuck in the ground; and round the enclosure was a border neatly cut from the white leaf-sheaths of the same tree. A double enclosure of bamboo, similarly ornamented, left an inner circle for the performers, and an outer for the spectators: the whole was lighted with oil lamps and Chinese paper lanterns. The musicians sat on one side, with cymbals, tomtoms, and flutes, and sang choruses.

The performances began by a copper-coloured Cupid entering and calling the virgins with a flute; these appeared from a green-room, to the number of thirty or forty, of all ages and sizes. Each had her hair dressed in a topknot, and her head covered with a veil; a scarlet petticoat loaded with tinsel concealed her naked feet, and over this was a short red kirtle, and an enormous white shawl was swathed round the body from the armpits to the waist. A broad belt passed over the right shoulder and under the left arm, to which hung gold and silver chains, corals, &c., with tinsel and small mirrors sewed on everywhere: the arms and hands were bare, and decorated with bangles and rings.

Many of the women were extremely tall, great stature being common amongst the Munniporees. They commenced with a prostration to Cupid, around whom they danced very slowly, with the arms stretched out, and the

* This turns of a beautiful black colour when applied to a surface, owing, according to Sir D. Brewster, to the fresh varnish consisting of a congeries of minute organised particles, which disperse the rays of light in all directions; the organic structure is destroyed when the varnish dries, and the rays of light are consequently transmitted.

hands in motion; at each step the free foot was swung backwards and forwards. Cupid then chose a partner, and standing in the middle went through the same motions, a compliment the women acknowledged by curtseying and whirling round, making a sort of cheese with their petti-coats, which, however, were too heavy to inflate properly.

The Nagas are another people found on this frontier, chiefly on the hills to the north : they are a wild, copper-coloured, uncouth jungle tribe, who have proved troublesome on the Assam frontier. Their features are more Tartar than those of the Munniporees, especially amongst the old men. They bury their dead under the threshold of their cottages. The men are all but naked, and stick plumes of hornbills' feathers in their hair, which is bound with strips of bamboo: tufts of small feathers are passed through their ears, and worn as shoulder lappets. A short blue cotton cloth, with a fringe of tinsel and tufts of goat's hair dyed red, is passed over the loins in front only : they also wear brass armlets, and necklaces of cowries, coral, amber, ivory, and boar's teeth. The women draw a fringed blue cloth tightly across the breast, and wear a checked or striped petticoat. They are less ornamented than the men, and are pleasing looking; their hair is straight, and cut short over the eyebrows.

The Naga dances are very different from those of the Munniporees; being quick, and performed in excellent time to harmonious music. The figures are regular, like quadrilles and country-dances : the men hold their knives erect during the performance, the women extend their arms only when turning partners, and then their hands are not given, but the palms are held opposite. The step is a sort of polka and balancez, very graceful and lively. A bar of music is always played first, and at the end the spectators applaud with two short shouts.   Their ear for music, and the nature of their dance, are as Tibetan as their countenances, and different from those of the Indo-Chinese tribes of the frontier.

We had the pleasure of meeting Lieutenant Raban at Silchar, and of making several excursions in the neighbourhood with him; for which Colonel Lister here, as at Churra, afforded us every facility of elephants and men. Had we had time, it was our intention to have visited Munnipore, but we were anxious to proceed to Chittagong. I however made a three days' excursion to the frontier, about thirty miles distant, proceeding along the north bank of the Soormah.   On the way my elephant got bogged in crossing a deep muddy stream : this is sometimes an alarming position, as should the animal become terrified, he will seize his rider, or pad, or any other object (except his driver), to place under his knees to prevent his sinking.   In this instance the driver in great alarm ordered me off, and I had to flounder out through the black mud. The elephant remained fast all night, and was released next morning by men with ropes.

The country continued a grassy level, with marshes and rice cultivation, to the first range of hills, beyond which the river is unnavigable; there also a forest commences, of oaks, figs, and the common trees of east Bengal.   The road hence was a good one, cut by Sepoys across the dividing ranges, the first of which is not 500 feet high. On the ascent bamboos abound, of the kind called Tuldah or Dulloah, which has long very thin-walled joints; it attains no great size, but is remarkably gregarious.   On the east side of the range, the road runs through soft shales and beds of clay, and conglomerates,

descending to a broad valley covered with gigantic scattered
timber-trees of jarool, acacia, *Diospyros, Urticeæ,* and
Bauhiniæ, rearing their enormous trunks above the bamboo
jungle : immense rattan-canes wound through the forest,
and in the gullies were groves of two kinds of tree-fern,
two of *Areca, Wallichia* palm, screw-pine, and *Dracæna.*
Wild rice grew abundantly in the marshes, with tall
grasses ; and *Cardiopteris* * covered the trees for upwards
of sixty feet, like hops, with a mass of pale-green foliage,
and dry white glistening seed-vessels.   This forest differed
from those of the Silhet and Khasia mountains, especially in
the abundance of bamboo jungle, which is, I believe, the
prevalent feature of the low hills in Birmah, Ava, and
Munnipore ; also in the gigantic size of the rattans, larger
palms, and different forest trees, and in the scanty
undergrowth of herbs and bushes.   I only saw, however,
the skirts of the forest ; the mountains further east, which
I am told rise several thousand feet in limestone cliffs, are
doubtless richer in herbaceous plants.

The climate of Cachar partakes of that of the Jheels
in its damp equable character : during our stay the weather
was fine, and dense fogs formed in the morning : the
mean maximum was 80°, minimum 58° 4.†

The annual rain-fall in 1850 was 211·60 inches, according
to a register kindly given me by Captain Verner.   There
are few mosquitos, which is one of the most curious facts
in the geographical distribution of these capricious blood-
suckers ; for the locality is surrounded by swamps, and

---

* A remarkable plant of unknown affinity; see Brown and Bennett, "Flora
Javæ:" it is found in the Assam valley and Chittagong.

† The temperature does not rise above 90° in summer, nor sink below 45° or
50° in January: forty-seven comparative observations with Calcutta showed the
mean temperature to be 1° 8 lower at Silchar, and the air damper, the saturation-
point being, at Calcutta 0·3791, at Silchar 0·4379.

they swarm at Silhet, and on the river lower down.   Both on the passage up and down, we were tormented in our canoes by them for eighty or ninety miles above Silhet, and thence onwards to Cachar we were free.

On the 30th of November, we were preparing for our return to Silhet, and our canoes were loading, when we were surprised by a loud rushing noise, and saw a high wave coming down the river, swamping every boat that remained on its banks, whilst most of those that pushed out into the stream, escaped with a violent rocking. It was caused by a slip of the bank three quarters of a mile up the stream, of no great size, but which propagated a high wave.   This appeared to move on at about the rate of a mile in three or four minutes, giving plenty of time for our boatmen to push out from the land on hearing the shouts of those first overtaken by the calamity; but they were too timid, and conse-quently one of our canoes, full of papers, instruments, and clothes, was swamped.   Happily our dried collections were not embarked, and the hot sun repaired much of the damage.

We left in the evening of the 2nd of December, and proceeded to Silhet, where we were kindly received by Mr. Stainforth, the district judge.   Silhet, the capital of the district of the same name, is a large Mahometan town, occupying a slightly raised part of the Jheels, where many of the Teelas seem joined together by beds of gravel and sand.   In the rains it is surrounded by water, and all communication with other parts is by boats : in winter, Jynteapore and Pundua may be reached by land, crossing creeks innumerable on the way.   Mr. Stainforth's house, like those of most of the other Europeans, occupies the top of one of the Teelas, 150 feet high, and is surrounded by fine spreading oaks,* *Garcinia*, and *Diospyros* trees. The rock of which the hill is composed, is a slag-like ochreous sandstone, covered in most places with a shrubbery of rose-flowered *Melastoma*, and some peculiar plants.†

Broad flat valleys divide the hills, and are beautifully clothed with a bright green jungle of small palms, and many kinds of ferns. In sandy places, blue-flowered *Burmannia*, *Hypoxis*, and other pretty tropical annuals, expand their blossoms, with an inconspicuous *Stylidium*, a plant belonging to a small natural family, whose limits are so confined to New Holland, that this is almost the only kind that does not grow in that continent. Where the ground is swampy, dwarf *Pandanus* abounds, with the gigantic nettle, *Urtica crenulata* ("Mealum-ma" of Sikkim, see p. 189).

The most interesting botanical ramble about Silhet is to the tree-fern groves on the path to Jynteapore, following the bottoms of shallow valleys between the Teelas, and along clear streams, up whose beds we waded for some miles, under an arching canopy of tropical shrubs, trees, and climbers, tall grasses, screw-pines, and *Aroideæ*. In the narrower parts of the valleys the tree-ferns are numerous on the slopes, rearing their slender brown trunks forty feet

---

* It is not generally known that oaks are often very tropical plants; not only abounding at low elevations in the mountains, but descending in abundance to the level of the sea. Though unknown in Ceylon, the Peninsula of India, tropical Africa, or South America, they abound in the hot valleys of the Eastern Himalaya, East Bengal, Malay Peninsula, and Indian islands; where perhaps more species grow than in any other part of the world. Such facts as this disturb our preconceived notions of the geographical distribution of the most familiar tribes of plants, and throw great doubt on the conclusions which fossil plants are supposed to indicate.

† *Gelonium, Adelia, Moacurra, Linostoma, Justicia, Trophis, Connarus, Ixora, Congea, Dalhousiea, Grewia, Myrsine, Büttneria;* and on the shady exposures a *Calamus, Briedelia,* and various ferns.

high, with feathery crowns of foliage, through which the sun-beams trembled on the broad shining foliage of the tropical herbage below.

Silhet, though hot and damp, is remarkably healthy, and does not differ materially in temperature from Silchar, though it is more equable and humid.* It derives some interest from having been first brought into notice by the enterprise of one of the Lindsays of Balcarres, at a time when the pioneers of commerce in India encountered great hardships and much personal danger. Mr. Lindsay, a writer in the service of the East India Company, established a factory at Silhet, and commenced the lime trade with Calcutta,† reaping an enormous fortune himself, and laying the foundation of that prosperity amongst the people which has been much advanced by the exertions of the Inglis family, and has steadily progressed under the protecting rule of the Indian government.

From Silhet we took large boats to navigate the Burrampooter and Megna, to their embouchure in the Bay of Bengal at Noacolly, a distance of 250 miles, whence we were to proceed across the head of the bay to Chittagong, about 100 miles further. We left on the 7th of December, and arrived at Chattuc on the 9th, where we met our Khasia collectors with large loads of plants, and paid them off. The river was now low, and presented a busy scene, from the numerous trading boats being confined to its fewer and deeper channels. Long grasses and sedges

---

* During our stay of five days the mean maximum temperature was 74°, minimum 64° 8 : that of thirty-two observations compared with Calcutta show that Silhet is only 1°·7 cooler, though Mr. Stainforth's house is upwards of 2° further north, and 150 feet more elevated. A thermometer sunk two feet seven inches, stood at 73° 5. The relative saturation-points were, Calcutta ·633, Silhet ·821.

† For an account of the early settlement of Silhet, see " Lives of the Lindsays," by Lord Lindsay.

(*Arundo, Saccharum* and *Scleria*), were cut, and stacked along the water's edge, in huge brown piles, for export and thatching.

On the 13th December, we entered the broad stream of the Megna. Rice is cultivated along the mud flats left by the annual floods, and the banks are lower and less defined than in the Soormah, and support no long grasses or bushes. Enormous islets of living water-grasses (*Oplismenus stagninus*) and other plants, floated past, and birds became more numerous, especially martins and egrets. The sun was hot, but the weather otherwise cool and pleasant: the mean temperature was nearly that of Calcutta, 69°·7, but the atmosphere was more humid.*

On the 14th we passed the Dacca river; below which the Megna is several miles wide, and there is an appearance of tide, from masses of purple *Salvinia* (a floating plant, allied to ferns), being thrown up on the beach like sea-weed. Still lower down, the vegetation of the Sunderbunds commences; there is a narrow beach, and behind it a mud bank several feet high, supporting a luxuriant green jungle of palms (*Borassus* and *Phœnix*), immense fig-trees, covered with *Calami*, and tall betel palms, clothed with the most elegant drapery of *Acrostichum scandens*, a climbing fern with pendulous fronds.

Towards the embouchure, the banks rise ten feet high, the river expands into a muddy sea, and a long swell rolls

---

* The river-water was greenish, and a little cooler (73°·8) than that of the Soormah (74°·3), which was brown and muddy. The barometer on the Soormah stood 0·028 inch higher than that of Calcutta (on the mean of thirty-eight observations), whereas on the Megna the pressure was 0·010 higher. As Calcutta is eighteen feet above the level of the Bay of Bengal, this shows that the Megna (which has no perceptible current) is at the level of the sea, and that either the Soormah is upwards of thirty feet above that level, or that the atmospheric pressure there, and at this season, is less than at Calcutta, which, as I have hinted at p. 259, is probably the case.

in, to the disquiet of our fresh-water boatmen. Low islands
of sand and mud stretch along the horizon: which,
together with the ships, distorted by extraordinary refraction, flicker as if seen through smoke. Mud is the all-
prevalent feature; and though the water is not salt, we do
not observe in these broad deltas that amount of animal
life (birds, fish, alligators, and porpoises), that teems in the
narrow creeks of the western Sunderbunds.

We landed in a canal-like creek at Tuktacolly,* on the
17th, and walked to Noacolly, over a flat of hard mud or
dried silt, covered with turf of *Cynodon Dactylon*. We
were hospitably received by Dr. Baker, a gentleman who
has resided here for twenty-three years; and who communicated to us much interesting information respecting the
features of the Gangetic delta.

Noacolly is a station for collecting the revenue and
preventing the manufacture of salt, which, with opium, are
the only monopolies now in the hands of the East India
Company. The salt itself is imported from Arracan, Ceylon,
and even Europe, and is stored in great wooden buildings
here and elsewhere. The ground being impregnated with
salt, the illicit manufacture by evaporation is not easily
checked; but whereas the average number of cases brought
to justice used to be twenty and thirty in a week, they are
now reduced to two or three. It is remarkable, that
though the soil yields such an abundance of this mineral,
the water of the Megna at Noacolly is only brackish, and it
is therefore to repeated inundations and surface evapora-
tions that the salt is due. Fresh water is found at a very
few feet depth everywhere, but it is not good.

When it is considered how comparatively narrow the
sea-board of the delta is, the amount of difference in the

* "Colly" signifies a muddy creek, such as intersect the delta.

physical features of the several parts, will appear most extraordinary. I have stated that the difference between the northern and southern halves of the delta is so great, that, were all depressed and their contents fossilised, the geologist who examined each by itself, would hardly recognise the two parts as belonging to one epoch ; and the difference between the east and west halves of the lower delta is equally remarkable.

The total breadth of the delta is 260 miles, from Chittagong to the mouth of the Hoogly, divided longitudinally by the Megna : all to the west of that river presents a luxuriant vegetation, while to the east is a bare muddy expanse, with no trees or shrubs but what are planted. On the west coast the tides rise twelve or thirteen feet, on the east, from forty to eighty. On the west, the water is salt enough for mangroves to grow for fifty miles up the Hoogly; on the east, the sea coast is too fresh for that plant for ten miles south of Chittagong. On the west, fifty inches is the Cuttack fall of rain; on the east, 90 to 120 at Noacolly and Chittagong, and 200 at Arracan. The east coast is annually visited by earthquakes, which are rare on the west; and lastly, the majority of the great trees and shrubs carried down from the Cuttack and Orissa forests, and deposited on the west coast of the delta, are not only different in species, but in natural order, from those that the Fenny and Chittagong rivers bring down from the jungles.*

We were glad to find at Noacolly that our observations

---

* The Cuttack forests are composed of teak, Sal, Sissoo, ebony, *Pentaptera*, *Buchanania*, and other trees of a dry soil, and that require a dry season alternating with a wet one. These are unknown in the Chittagong forests, which have Jarool (*Lagerstrœmia*) *Mesua*, *Dipterocarpi*, nutmegs, oaks of several kinds, and many other trees not known in the Cuttack forests, and all typical of a perennially humid atmosphere.

on the progression westwards of the Burrampooter (see p. 253) were confirmed by the fact that the Megna also is gradually moving in that direction, leaving much dry land on the Noacolly side, and forming islands opposite that coast; whilst it encroaches on the Sunderbunds, and is cutting away the islands in that direction. This advance of the fresh waters amongst the Sunderbunds is destructive to the vegetation of the latter, which requires salt; and if the Megna continues its slow course westwards, the obliteration of thousands of square miles of a very peculiar flora, and the extinction of many species of plants and animals that exist nowhere else, may ensue. In ordinary cases these plants, &c., would take up their abode on the east coast, as they were driven from the west; but such might not be the case in this delta; for the sweeping tides of the east coast prevent any such vegetation establishing itself there, and the mud which the eastern rivers carry down, becomes a caking dry soil, unsuited to the germination of seeds.

On our arrival at Calcutta in the following February, Dr. Falconer showed us specimens of very modern peat, dug out of the banks of the Hoogly a few feet below the surface of the soil, in which were seeds of the *Euryale ferox*: * this plant is not now known to be found nearer than Dacca (sixty miles north-east, see p. 255), and indicates a very different state of the surface at Calcutta at the date of its deposition than that which exists now, and also shows that the estuary was then much fresher.

The main land of Noacolly is gradually extending seawards, and has advanced four miles within twenty-

---

* This peat Dr. Falconer also found to contain bones of birds and fish, seeds of *Cucumis Madraspatana* and another Cucurbitaceous plant, leaves of *Saccharum Sara* and *Ficus cordifolia*. Specks of some glistening substance were scattered through the mass, apparently incipient carbonisation of the peat.

three years : this seems sufficiently accounted for by the recession of the Megna. The elevation of the surface of the land is caused by the overwhelming tides and south-west hurricanes in May and October : these extend thirty miles north and south of Chittagong, and carry the waters of the Megna and Fenny back over the land, in a series of tremendous waves, that cover islands of many hundred acres, and roll three miles on to the main land. On these occasions, the average earthy deposit of silt, separated by micaceous sand, is an eighth of an inch for every tide; but in October, 1848, these tides covered Sundeep island, deposited six inches on its level surface, and filled ditches several feet deep. These deposits become baked by a tropical sun, and resist to a considerable degree denudation by rain. Whether any further rise is caused by elevation from below is doubtful; there is no direct evidence of it, though slight earthquakes annually occur; and even when they have not been felt, the water of tanks has been seen to oscillate for three-quarters of an hour without intermission, from no discernible cause.*

Noacolly is considered a healthy spot, which is not the case with the Sunderbund stations west of the Megna. The climate is uniformly hot, but the thermometer never rises above 90°, nor sinks below 45°; at this temperature hoar-frost will form on straw, and ice on water placed in porous pans, indicating a powerful radiation.†

---

* The natives are familiar with this phenomenon, of which Dr. Baker remembers two instances, one in the cold season of 1834-5, the other in that of 1830-1. The earthquakes do not affect any particular month, nor are they accompanied by any meteorological phenomena.

† The winds are north-west and north in the cold season (from November to March), drawing round to west in the afternoons. North-west winds and heavy hailstorms are frequent from March to May, when violent gales set in from the southward. The rains commence in June, with easterly and southerly winds, and the temperature from 82° to 84°; May and October are the hottest months. The

We left Noacolly on the 19th for Chittagong ; the state of the tide obliging us to go on board in the night.   The distance is only 100 miles, but the passage is considered dangerous at this time (during the spring-tides) and we were therefore provided with a large vessel and an experienced crew.   The great object in this navigation is to keep afloat and to make progress towards the top of the tide and during its flood, and to ground during the ebb in creeks where the bore (tidal wave) is not violent ; for where the channels are broad and open, the height and force of this wave rolls the largest coasting craft over and swamps them.

Our boatmen pushed out at 3 in the morning, and brought up at 5, in a narrow muddy creek on the island of Sidhee.   The waters retired along channels scooped several fathoms deep in black mud, leaving our vessel aground six or seven feet below the top of the bank, and soon after- wards there was no water to be seen ; as far as the eye could reach, all was a glistening oozy mud, except the bleak level surfaces of the islands, on which neither shrub nor tree grew.   Soon after 2 P.M. a white line was seen on the low black horizon, which was the tide- wave, advancing at the rate of five miles an hour, with a hollow roar; it bore back the mud that was gradually slipping along the gentle slope, and we were afloat an hour after: at night we grounded again, opposite the mouth of the Fenny.

By moonlight the scene was oppressively solemn : on all sides the gurgling waters kept up a peculiar sound that filled the air with sullen murmurs ; the moonbeams slept

rains cease in the end of October (on the 8th of November in 1849, and 12th of November in 1850, the latest epoch ever remembered) : there is no land or sea breeze along any part of the coast. During our stay we found the mean tempe- rature for twelve observations to be precisely that of Calcutta, but the humidity was more, and the pressure 0·040 lower.

upon the slimy surface of the mud, and made the dismal landscape more ghastly still. Silence followed the ebb, broken occasionally by the wild whistle of a bird like the curlew, of which a few wheeled through the air : till the harsh roar of the bore was heard, to which the sailors seemed to waken by instinct. The waters then closed in on every side, and the far end of the reflected moonbeam was broken into flashing light, that approached and soon danced beside the boat.

We much regretted not being able to obtain any more accurate data than I have given, as to the height of the tide at the mouth of the Fenny ; but where the ebb some-times retires twenty miles from high-water mark, it is obviously impossible to plant any tide-gauge.

On the 21st we were ashore at daylight on the Chitta-gong coast far north of the station, and were greeted by the sight of hills on the horizon : we were lying fully twenty feet below high-water mark, and the tide was out for several miles to the westward. The bank was covered with flocks of white geese feeding on short grass, upon what appeared to be detached islets on the surface of the mud. These islets, which are often an acre in extent, are composed of stratified mud ; they have perpendicular sides several feet high, and convex surfaces, owing to the tide washing away the earth from under their sides ; and they were further slipping seawards, along the gently sloping mud-beach. Few or no shells or seaweed were to be seen, nor is it possible to imagine a more lifeless sea than these muddy coasts present.

We were three days and nights on this short voyage, without losing sight of mud or land. I observed the baro-meter whenever the boat was on the shore, and found the mean of six readings (all reduced to the same level) to be identical with that at Calcutta. These being all taken at elevations lower than that of the Calcutta observatory, show either a diminished atmospheric pressure, or that the mean level of high-water is not the same on the east and west coasts of the Bay of Bengal: this is quite possible, considering the widely different direction of the tides and currents on each, and that the waters may be banked up, as it were, in the narrow channels of the western Sunderbunds. The temperature of the air was the same as at Calcutta, but the atmosphere was damper. The water was always a degree warmer than the air.

We arrived at Chittagong on the 23rd of December, and became the guests of Mr. Sconce, Judge of the district, and of Mr. Lautour; to both of whom we were greatly indebted for their hospitality and generous assistance in every way.

Chittagong is a large town of Mahometans and Mugs, a Birmese tribe who inhabit many parts of the Malay peninsula, and the coast to the northward of it. The town stands on the north shore of an extensive delta, formed by rivers from the lofty mountains separating this district from Birma. These mountains are fine objects on the horizon, rising 4,000 to 8,000 feet; they are forest-clad, and inhabited by turbulent races, who are coterminous with the Cookies of the Cachar and Tipperah forests; if indeed they be not the same people. The mountains abound with the splendid timber-trees of the Cachar forests, but like these are said to want teak, Sal, and Sissoo; they have, besides many others, magnificent Gurjun trees (*Dipterocarpi*), the monarchs of the forests of these coasts.

The natives of Chittagong are excellent shipbuilders and active traders, and export much rice and timber to Madras and Calcutta. The town is large and beautifully situated, interspersed with trees and tanks; the hills resemble those of Silhet, and are covered with a similar vegetation: on these the European houses are built. The climate is very healthy, which is not remarkable, considering how closely it approximates in character to that of Silhet and other places in Eastern Bengal, but very extraordinary, if it be compared with Arracan, only 200 miles further south, which is extremely unhealthy. The prominent difference between the physical features of Chittagong and Arracan, is the presence of mangrove swamps at the latter place, for which the water is too fresh at the former.

The hills about the station are not more than 150 or 200 feet high, and are formed of stratified gravel, sand, and clay, that often becomes nodular, and is interstratified with slag-like iron clay. Fossil wood is found; and some of the old buildings about Chittagong contain nummulitic limestone, probably imported from Silhet or the peninsula of India, with which countries there is no such trade now. The views are beautiful, of the blue mountains forty to fifty miles distant, and the many-armed river, covered with sails, winding amongst groves of cocoa-nuts, Areca palm, and yellow rice fields. Good European houses surmount all the eminences, surrounded by trees of *Acacia* and *Cæsalpinia*. In the hollows are native huts amidst vegetation of every hue, glossy green *Garciniæ* and figs, broad plantains, feathery *Cassia* and Acacias, dark *Mesua*, red-purple *Terminalia*, leafless scarlet-flowered *Bombax*, and grey *Casuarina*.* Seaward the tide leaves immense flats, called churs, which stretch for many miles on either side the offing.

---

* This, which is almost exclusively an Australian genus, is not indigenous at Chittagong: to it belongs an extra-Australian species common in the Malay islands, and found wild as far north as Arracan.

We accompanied Mr. Sconce to a bungalow which he has built at the telegraph station at the south head of the harbour : its situation, on a hill 100 feet above the sea, is exposed, and at this season the sea-breeze was invigorating, and even cold, as it blew through the mat-walls of the bungalow.* To the south, undulating dunes stretch along the coast, covered with low bushes, of which a red-flowered *Melastoma* is the most prevalent,† and is considered a species of *Rhododendron* by by many of the residents ! The flats along the beach are several miles broad, intersected with tidal creeks, and covered with short grass, while below high-water mark all is mud, coated with green *Conferva*. There are no leafy seaweeds or mangroves, nor any seaside shrub but *Dilivaria ilicifolia*. Animal life is extremely rare; and a *Cardium*-like shell and small crab are found sparingly.

Coffee has been cultivated at Chittagong with great success; it is said to have been introduced by Sir W. Jones, and Mr. Sconce has a small plantation, from which his table is well supplied. Both Assam and Chinese teas flourish, but Chinamen are wanted to cure the leaves. Black pepper succeeds admirably, as do cinnamon, arrow-root, and ginger.

Early in January we accompanied Mr. Lautour on an excursion to the north, following a valley separated from the coast by a range of wooded hills, 1,000 feet high. For several marches the bottom of this valley was broad, flat, and full of villages. At Sidhee, about twenty-five miles

* The mean temperature of the two days (29th and 30th) we spent at this bungalow was 66°·5, that of Calcutta being 67°·6; the air was damp, and the barometer 0·144 lower at the flagstaff hill, but it fell and rose with the Calcutta instrument.

† *Melastoma*, jasmine, *Calamus*, *Ægle Marmelos*, *Adelia*, *Memecylon*, *Ixora*, *Linostoma*, *Congea*, climbing *Cæsalpinia*, and many other plants; and along their bases large trees of *Amoora*, *Gaurea*, figs, *Mesua*, and *Micromelon*.

from Chittagong, it contracts, and spurs from the hills on either flank project into the middle: they are 200 to 300 feet high, formed of red clay, and covered with brushwood. At Kajee-ke-hath, the most northern point we reached, we were quite amongst these hills, and in an extremely picturesque country, intersected by long winding flat valleys, that join one another : some are full of copse-wood, while others present the most beautiful park-like scenery, and a third class expand into grassy marshes or lake-beds, with wooded islets rising out of them. The hill-sides are clothed with low jungle, above which tower magnificent Gurjun trees (wood-oil). The whole contour of this country is that of a low bay, whose coast is raised above the sea, and over which a high tide once swept for ages.

The elevation of Hazari-ke-hath is not 100 feet above the level of the sea. It is about ten miles west of the mouth of the Fenny, from which it is separated by hills 1,000 feet high; its river falls into that at Chittagong, thirty miles south. Large myrtaceous trees (*Eugenia*) are common, and show a tendency to the Malayan flora, which is further demonstrated by the abundance of Gurjun (*Dipterocarpus turbinatus*). This is the most superb tree we met with in the Indian forests : we saw several species, but this is the only common one here; it is conspicuous for its gigantic size, and for the straightness and graceful form of its tall unbranched pale grey trunk, and small symmetrical crown : many individuals were upwards of 200 feet high, and fifteen in girth. Its leaves are broad, glossy, and beautiful; the flowers (then falling) are not conspicuous ; the wood is hard, close-grained, and durable, and a fragrant oil exudes from the trunk, which is extremely valuable as pitch and varnish, &c., besides being a good medicine. The natives procure it by cutting transverse holes in the trunk, pointing

downwards, and lighting fires in them, which causes the
oil to flow.*

On the 8th of January we experienced a sharp earth-
quake, preceded by a dull thumping sound; it lasted about

GURJUN TREE.

twenty seconds, and seemed to come up from the southward; the water of a tank by which we were seated was

* The other trees of these dry forests are many oaks, *Henslowia, Gordonia,
Engelhardtia, Duabanga, Adelia, Byttneria, Bradleia,* and large trees of *Pongamia,*
whose seeds yield a useful oil.

smartly agitated. The same shock was felt at Mymen-
sing and at Dacca, 110 miles north-west of this.*

We crossed the dividing ridge of the littoral range on
the 9th, and descended to Seetakoond bungalow, on the
high road from Chittagong to Comilla. The forests at the
foot of the range were very extensive, and swarmed with
large red ants that proved very irritating: they build
immense pendulous nests of dead and living leaves at the
ends of the branches of trees, and mat them with a white
web. Tigers, leopards, wild dogs, and boars, are numerous;
as are snipes, pheasants, peacocks, and jungle-fowl, the
latter waking the morn with their shrill crows; and in
strange association with them, common English woodcock,
is occasionally found.

The trees are of little value, except the Gurjun, and
" Kistooma," a species of *Bradleia*, which was stacked
extensively, being used for building purposes. The papaw†
is abundantly cultivated, and its great gourd-like fruit is
eaten (called " Papita " or " Chinaman "); the flavour is
that of a bad melon, and a white juice exudes from the
rind. The *Hodgsonia heteroclita* (*Trichosanthes* of Roxburgh), a magnificent Cucurbitaceous climber, grows in
these forests; it is the same species as the Sikkim one (see
p. 7). The long stem bleeds copiously when cut, and like
almost all woody climbers, is full of large vessels; the
juice does not, however, exude from these great tubes,
which hold air, but from the close woody fibres. A
climbing *Apocyneous* plant grows in these forests, the

* Earthquakes are extremely common, and sometimes violent, at Chittagong,
and doubtless belong to the volcanic forces of the Malayan peninsula.
† The Papaw tree is said to have the curious property of rendering tough meat
tender, when hung under its leaves, or touched with the juice; this hastening the
process of decay. With this fact, well-known in the West Indies, I never found
a person in the East acquainted.

milk of which flows in a continuous stream, resembling caoutchouc (it is probably the *Urceola elastica*, which yields Indian-rubber).

The subject of bleeding is involved in great obscurity, and the systematic examination of the motions in the juices of tropical climbers by resident observers, offers a fertile field to the naturalist. I have often remarked that if a climbing stem, in which the circulation is vigorous, be cut across, it bleeds freely from both ends, and most copiously from the lower, if it be turned down-wards; but that if a truncheon be severed, there will be no flow from either of its extremities. This is the case with all the Indian watery-juiced climbers, at whatever season they may be cut. When, however, the circulation in the plant is feeble, neither end of a simple cut will bleed much, but if a truncheon be taken from it, both the extremities will.

The ascent of the hills, which are densely wooded, was along spurs, and over knolls of clay; the rocks were sandy and slaty (dip north-east 60°). The road was good, but always through bamboo jungle, and it wound amongst the low spurs, so that there was no defined crest or top of the pass, which is about 800 feet high. There were no tall palms, tree-ferns, or plantains, no *Hymeno-phylla* or *Lycopodia*, and altogether the forest was smaller and poorer in plants than we had expected. The only palms (except a few rattans) were two kinds of *Wallichia*.

From the summit we obtained a very extensive and singular view. At our feet was a broad, low, grassy, alluvial plain, intersected by creeks, bounding a black expanse of mud which (the tide being out) appeared to stretch almost continuously to Sundeep Island, thirty miles distant; while beyond, the blue hills of Tipperah rose on

352   CHITTAGONG.   Chap. XXXI.

the north-west horizon. The rocks yielded a dry poor soil,
on which grew dwarf *Phœnix* and cycas-palm (*Cycas
circinalis* or *pectinata*).

Descending, we rode several miles along an excellent
road, that runs to Tipperah, and stopped at the bungalow
of Seetakoond, twenty-five miles north of Chittagong. The
west flank of the range which we had crossed is much
steeper than the east, often precipitous, and presents the
appearance of a sea-worn cliff towards the Bay of Bengal.
Near Seetakoond (which is on the plain) a hill on the range,
bearing the same name, rises 1,136 feet high, and being
damper and more luxuriantly wooded, we were anxious to
explore it, and therefore spent some days at the bungalow.
Fields of poppy and sun (*Crotalaria juncea*), formed most
beautiful crops; the latter grows from four to six feet
high, and bears masses of laburnum-like flowers, while the
poppy fields resembled a carpet of dark-green velvet,
sprinkled with white stars; or, as I have elsewhere
remarked, a green lake studded with water-lilies.

The road to the top of Seetakoond leads along a most
beautiful valley, and then winds up a cliff that is in many
places almost precipitous, the ascent being partly by steps
cut in the rock, of which there are 560. The mountain is
very sacred, and there is a large Brahmin temple on its
flank; and near the base a perpetual flame bursts out of
the rock. This we were anxious to examine, and were
extremely disappointed to find it a small vertical hole in a
slaty rock, with a lateral one below for a draught, and that
it is daily supplied by pious pilgrims and Brahmins with
such enormous quantities of ghee (liquid butter), that it
is to all intents and purposes an artificial lamp; no trace
of natural phenomena being discoverable.

On the dry but wooded west face of the mountain,

W.L.Walton,lith.

J.D.H.delt

John Murray, Albemarle Street, 1854.

Seetakoond Hill.

grows *Falconeria*, a curious Euphorbiaceous tree, with an acrid milky juice that affects the eyes when the wood is cut. Beautiful *Cycas* palms are also common, with *Terminalia*, *Bignonia*, *Sterculia*, dwarf *Phœnix* palm, and Gurjun trees. The east slope of the mountain is damper, and much more densely wooded; we there found two wild species of nutmeg trees, whose wood is full of a brown acrid oil, seven palms, tree-ferns, and many other kinds of ferns, several kinds of oak, *Dracæna*, and figs. The top is 1,136 feet above the sea, and commands an extensive view to all points of the compass; but the forests, in which the ashy bark of the Gurjun trees is conspicuous, and the beautiful valley on the west, are the only attractive features.

The weather on the east side of the range differs at this season remarkably from that on the west, where the vicinity of the sea keeps the atmosphere more humid and warm, and at the same time prevents the formation of the dense fogs that hang over the valleys to the eastward every morning at sunrise. We found the mean temperature at the bungalow, from January 9th till the 13th, to be 70° 2.

We embarked again at Chittagong on the 16th of January, at 10 P.M., for Calcutta, in a very large vessel, rowed by twelve men : we made wretchedly slow progress, for the reasons mentioned above (p. 343), being for four days within sight of Chittagong ! On the 20th we only reached Sidhee, and thence made a stretch to Hattiah, an island which may be said to be moving bodily to the westward, the Megna annually cutting many acres from the east side, and the tide-wave depositing mud on the west. The surface is flat, and raised four feet above mean high-water level; the tide rises about 14 feet up the bank, and then retires for miles; the total rise and fall is, however, much less here than in the Fenny, higher up the gulf. The turf is composed of *Cynodon* and a *Fimbristylis;* and the earth being impregnated with salt, supports different kinds of *Chenopodium*. Two kinds of tamarisk, and a thorny *Cassia* and *Excœcaria,* are the only shrubs on the eastern islands; on the central ones a few dwarf mangroves appear, with the holly-leaved *Dilivaria,* dwarf screw-pine (*Pandanus*), a shrub of *Compositæ,* and a curious fern, a variety of *Acrostichum aureum*. Towards the northern end of Hattiah, Talipot, cocoa-nut and date-palms appear.

On the 22nd we entered the Sunderbunds, rowing amongst narrow channels, where the tide rises but a few feet. The banks were covered with a luxuriant vegetation, chiefly of small trees, above which rose stately palms. On the 25th, we were overtaken by a steamer from Assam, a novel sight to us, and a very strange one in these creeks, which in some places seemed hardly broad enough for it to pass through. We jumped on board in haste, leaving our boat and luggage to follow us. She had left Dacca two days before, and this being the dry season, the route to Calcutta, which is but sixty miles in a straight line, involved a détour of three hundred.

From the masts of the steamer we obtained an excellent *coup-d'œil* of the Sunderbunds; its swamps clothed with verdure, and intersected by innumerable inosculating channels, with banks a foot or so high. The amount of tide, which never exceeds ten feet, diminishes in proceeding westwards into the heart of these swamps, and the epoch, direction, and duration of the ebb and flow vary so much in every canal, that at times, after stemming a powerful current, we found ourselves, without materially changing our course, suddenly swept along with a favouring stream

This is owing to the complex ramifications of the creeks, the flow of whose waters is materially influenced by the most trifling accidents of direction.

Receding from the Megna, the water became salter, and *Nipa fruticans* appeared, throwing up pale yellow-green tufts of feathery leaves, from a short thick creeping stem, and bearing at the base of the leaves its great head of nuts, of which millions were floating on the waters, and vegetating in the mud. Marks of tigers were very frequent, and the footprints of deer, wild boars, and enormous crocodiles : these reptiles were extremely common, and glided down the mud banks on the approach of the steamer, leaving between the footmarks a deep groove in the mud made by their tail. The *Phœnix paludosa*, a dwarf slender-stemmed date-palm, from six to eight feet high, is the all-prevalent feature, covering the whole landscape with a carpet of feathery fronds of the liveliest green. The species is eminently gregarious, more so than any other Indian palm, and presents so dense a mass of foliage, that when seen from above, the stems are wholly hidden.[*]

The water is very turbid, and only ten to twenty feet deep, which, we were assured by the captain, was not increased during the rains : it is loaded with vegetable matter, but the banks are always muddy, and we never saw any peat. Dense fogs prevented our progress in the morning, and we always anchored at dusk. We did not see a village or house in the heart of the Sunderbunds (though such do occur), but we saw canoes, with fishermen, who use the tame otter in fishing ; and the banks were covered with piles of firewood, stacked for the

---

[*] *Sonneratia, Heritiera littoralis,* and *Careya,* form small gnarled trees on the banks, with deep shining green-leaved species of *Carallia, Rhizophora,* and other Mangroves. Occasionally the gigantic reed-mace (*Typha elephantina*) is seen, and tufts of tall reeds (*Arundo*).

Calcutta market. As we approached the Hoogly, the water became very salt and clear; the Nipa fruits were still most abundant, floating out to sea, but no more of the plant itself was seen. As the channels became broader, sand-flats appeared, with old salt factories, and clumps of planted *Casuarina*.

On the 28th of January we passed Saugor island, and entered the Hoogly, steamed past Diamond Harbour, and landed at the Botanic Garden Ghat, where we received a hearty welcome from Dr. Falconer. Ten days later we bade farewell to India, reaching England on the 25th of March, 1851.

APPENDIX.

---◆---

A.

METEOROLOGICAL OBSERVATIONS IN BEHAR, AND IN THE VALLEYS
OF THE SOANE AND GANGES.

Most of the instruments which I employed were constructed by Mr. Newman, and with considerable care: they were in general accurate, and always extremely well guarded, and put up in the most portable form, and that least likely to incur damage; they were further frequently carefully compared by myself. These are points to which too little attention is paid by makers and by travellers in selecting instruments and their cases. This remark applies particularly to portable barometers, of which I had five at various times. Although there are obvious defects in the system of adjustment, and in the method of obtaining the temperature of the mercury, I found that these instruments invariably worked well, and were less liable to derangement and fracture than any I ever used; the best proof I can give of this is that I preserved three uninjured during nearly all my excursions, left two in India, and brought a third home myself that had accompanied me almost throughout my journey.

In very dry climates these and all other barometers are apt to leak, from the contraction of the box-wood plug through which the tube passes into the cistern. This must, in portable barometers, in very dry weather, be kept moist with a sponge. A small iron bottle of pure mercury to supply leakage should be supplied with every barometer, as also a turnscrew. The vernier plate and scale should be screwed, not soldered on the metal sheath, as if an escape occurs in the barometer-case the solder is acted upon at once. A table of

corrections for capacity and capillarity should accompany every
instrument, and simple directions, &c., in cases of trifling derangement, and alteration of neutral point.

The observations for temperature were taken with every precaution
to avoid radiation, and the thermometers were constantly compared
with a standard, and the errors allowed for. The maximum thermometer with a steel index, I found to be extremely liable to
derangement and very difficult to re-adjust. Negretti's maximum
thermometer was not known to me during my journey. The spirit
minimum thermometers again, are easily set to rights when out of
order, but in every one (of six or seven) which I took to India,
by several makers, the zero point receded, the error in some
increasing annually, even to—6° in two years. This seems due to a
vaporisation of the spirit within the tube. I have seen a thermometer of this description in India, of which the spirit seemed to
have retired wholly into the bulb, and which I was assured had
never been injured. In wet-bulb observations, distilled water or
rain, or snow water was used, but I never found the result to differ
from that obtained by any running fresh water, except such as was
polluted to the taste and eye.

The hours of observation selected were at first sunrise, 9 A. M.,
3 P. M., sunset, and 9 P. M., according to the instructions issued to
the Antarctic expedition by the Royal Society. In Sikkim, however,
I generally adopted the hours appointed at the Surveyor General's
office, Calcutta; viz., sunrise, 9h. 50m. A.M., noon, 2h. 40m. P.M.,
4 P.M., and sunset, to which I added a 10 P.M. observation, besides
many at intermediate hours as often as possible. Of these the 9h.
50m. A.M. and 4 P.M. have been experimentally proved to be those of
the maximum and minimum of atmospheric pressure at the level of the
sea in India, and I did not find any great or marked deviation from
this at any height to which I attained, though at 15,000 or 16,000
feet the morning maximum may occur rather earlier.

The observations for nocturnal (terrestrial) radiation were made
by freely suspending thermometers with naked bulbs, or by laying
them on white cotton, wool, or flannel; also by means of a thermometer placed in the focus of a silvered parabolic reflector. I
did not find that the reflector possessed any decided advantage over

the white cotton: the means of a number of observations taken by each approximated closely, but the difference between individual observations often amounted to 2°.

Observations again indicative of the radiation from grass, whether dewed or dry, are not strictly comparable; not only does the power of radiation vary with the species, but much more with the luxuriance and length of the blades, with the situation, whether on a plane surface or raised, and with the subjacent soil. Of the great effect of the soil I had frequent instances; similar tufts of the same species of grass radiating more powerfully on the dry sandy bed of the Soane, than on the alluvium on its banks; the exposure being equal in both instances. Experiments for the surface-temperature of the soil itself, are least satisfactory of any :—adjoining localities being no less affected by the nature, than by the state of disintegration of the surface, and by the amount of vegetation in proximity to the instrument.

The power of the sun's rays in India is so considerable, and protracted through so long a period of the day, that I did not find the temperature of springs, or of running water, even of large deep rivers, so constant as was to be expected.

The temperature of the earth was taken by sinking a brass tube a yard long in the soil.

A thermometer with the bulb blackened affords the only means the traveller can generally compass, of measuring the power of the sun's rays. It should be screened or put in a blackened box, or laid on black wool.

A good Photometer being still a desideratum, I had recourse to the old wedge of coloured glass, of an uniform neutral tint, the distance between whose extremes, or between transparency and total opacity, was one foot. A moveable arm carrying a brass plate with a slit and a vernier, enables the observer to read off at the vanishing point of the sun's limb, to one five-hundredth of an inch. I generally took the mean of five readings as one, and the mean of five of these again I regarded as one observation; but I place little dependence upon the results. The causes of error are quite obvious. As far as the effects of the sun's light on vegetation are concerned, I am inclined to think that it is of more importance to register the number of hours or rather of parts of each hour, that the sun shines, and its clearness during the time. To secure valuable results this should be done repeatedly, and the strength of the rays by the black-bulb thermometer registered at each hour. The few actino-meter observations will be found in another part of the Appendix.

The dew-point has been calculated from the wet-bulb, by Dr. Apjohn's formula, or, where the depression of the barometer is considerable, by that as modified by Colonel Boileau. * The saturation-point was obtained by dividing the tension at the dew-point by that at the ordinary temperature, and the weight of vapour, by Daniell's formula.

The following summary of meteorological observations is alluded to at vol. i., p. 15.

I.—*Table-land of Birbhoom and Behar, from Taldanga to Dunwah. Average elevation* 1,135 *feet.*

It is evident from these observations, that compared with Calcutta, the dryness of the atmosphere is the most remarkable feature of this table-land, the temperature not being high; and to this, combined with the sterility of the soil over a great part of the surface, must be attributed the want of a vigorous vegetation. Though so favourably exposed to the influence of nocturnal radiation, the amount of the latter is small. The maximum depression of a thermometer laid on grass never exceeded 10°, and averaged 7°; whereas the average depression of the dew-point at the same hour amounted to 25° in the morning. Of course no dew was deposited even in the clearest star-light night.

* Journal of Asiatic Society, No. 147 (1844), p. 135.

February. 1848. Hour.	TEMPERATURE.				WET-BULB.			Elasticity of Vapour.	DEW-POINT.					Weight of Vapour in cubic feet.	SATURATION.			Number of observations.
	Mean.	Max.	Min.	Range.	Mean.	Max. Depression.	Min. Depression.		Mean.	Max.	Min.	Max Depression.	Min. Depression.		Mean.	Max.	Min.	
Sun-rise...	56·6	65·2	46·3	18·9	48·2	12·5	6·0	·276	39·5	52·0	23·3	31·7	10·4	3·088	·550	·680	·330	7
9 A.M. ...	70·1	77·0	61·2	15·8	53·7	19·3	14·3	·264	37·9	52·7	24·5	39·2	24·3	2·875	·330	·450	·260	7
3 P.M. ...	75·5	81·7	65·2	16·5	55·3	22·5	16·7	·248	36·0	46·8	24·3	48·4	34·9	2·674	·260	·320	·190	7
9 P.M. ...	61·7	66·2	55·5	10·7	49·3	20·5	9·0	·248	36·1	50·0	*9·1	56·9	16·2	2·745	·410	·590	·140	10

Extreme variations of Temperature . . . . . 35·4°

   ,,   of relative humidity . . . . ·540

   ,,   diff. between Solar and Nocturnal Radiation . . 96·5°

* Taken during a violent N. W. dust-storm.

SOLAR RADIATION.

		MORNING.					AFTERNOON.		
Hour.	Th.	Black Bulb.	Diff.	Phot.	Hour.	Th.	Black Bulb.	Diff.	Phot.
9½ A.M.	77·0	130	53·0	...	3½ P.M.	81·7	109	27·3	...
10 . .	69·5	124	54·5	10·320	3 . .	80·5	120	39·5	10·320
10 . .	77·0	137	60·0	...	3 . .	81·5	127	45·5	10·330
9 . .	63·5	94	30·5	10·230	3½ . .	72·7	105	32·3	10·230
9 . .	61·2	106	44·8	...	3 . .	72·5	110	37·5	10·390
9 . .	67·0	114	47·0	10·350	...	...	...	...	...
Mean .	69·2	117·5	48·1	10·300	...	77·8	114·2	36·4	10·318

NOCTURNAL RADIATION.

	SUNRISE.				NINE P.M.			
	Temperature.	Mean Diff. from Air.	Max. Diff. from Air.	Number of observations.	Temperature.	Mean Diff. from Air.	Max. Diff. from Air.	Number of observations.
Exposed Th	51·1	4·0	9·0	6	56·4	5·3	7·5	7
On Earth	48·3	2·5	3·7	3	53·8	4·9	5·5	6
On Grass	46·6	6·2	9·0	5	54·4	7·2	10·0	7

On one occasion, and that at night, the dew-point was as low as 11° 5, with a temperature of 66°, a depression rarely equalled at so low a temperature: this phenomenon was transient, and caused by the passage of a current of air loaded with dust, whose particles possibly absorbed the atmospheric humidity. From a comparison of the night and morning observations of thermometers laid on grass, the earth, and freely exposed, it appears that the grass parts with its heat much more rapidly than the earth, but that still the effect of radiation is slight, lowering its temperature but 2° below that of the freely exposed thermometer.

As compared with the climate of Calcutta, these hills present a remarkable contrast, considering their proximity in position and moderate elevation.

The difference of temperature between Calcutta and Birbhoom, deduced from the sunrise, morning and afternoon observations, amounts to 4°, which, if the mean height of the hills where crossed by the road, be called 1135 feet, will be equal to a fall of one degree for every 288 feet.

In the dampness of its atmosphere, Calcutta contrasts very remarkably with these hills ; the dew-point on the Hoogly averaging 51°·3, and on these hills 38°, the corresponding saturation-points being 0·559 and 0·380.

The difference between sunrise, forenoon and afternoon dew-points at Calcutta and on the hills, is 13°·6 at each observation ; but the atmosphere at Calcutta is relatively drier in the afternoon than that of the hills; the difference between the Calcutta sunrise and afternoon saturation-point being 0·449, and that between the hill sunrise and afternoon, 0·190. The march of the dew-point is thus the same in both instances, but owing to the much higher temperature of Calcutta, and the greatly increased tension of the vapour there, the relative humidity varies greatly during the day.

In other words, the atmosphere of Calcutta is loaded with moisture in the early morning of this season, and is relatively dry in the afternoon : in the hills again, it is scarcely more humid at sunrise than at 3 P.M. That this dryness of the hills is partly due to elevation, appears from the disproportionately moister state of the atmosphere below the Dunwah pass.

II.—*Abstract of the Meteorological observations taken in the Soane Valley (mean elevation 422 feet).*

The difference in mean temperature (partly owing to the sun's more northerly declination) amounts to 2°·5 of increase in the Soane valley, above that of the hills. The range of the thermometer from day to day was considerably greater on the hills (though fewer observations were there recorded) : it amounted to 17°·2 on the hills, and only 12°·8 in the valley. The range from the maximum to the minimum of each day amounts to the same in both, above 20°. The extreme variations in temperature too coincide within 1°·4.

The hygrometric state of the atmosphere of the valley differs most

decidedly from that of the hills.   In the valley dew is constantly
formed, which is owing to the amount of moisture in the air, for
nocturnal radiation is more powerful on the hills.   The sunrise and
9 P.M. observations in the valley, give a mean depression of the dew-
point below the air of 12°·3, and those at the upper level of 21°·2,
with no dew on the hills and a copious deposit in the valley.   The
corresponding state of the atmosphere as to saturation is 0·480 on
the hills and 0·626 in the valley.

The vegetation of the Soane valley is exposed to a less extreme
temperature than that of the hills; the difference between solar
and nocturnal radiation amounting here only to 80°·5, and on the
hills to 96°·5.   There is no material difference in the power of the
sun's rays at the upper and lower levels, as expressed by the black-
bulb thermometer, the average rise of which above one placed in the
shade, amounted to 48° in both cases, and the maximum occurred
about 11 A.M.   The decrease of the power of the sun's rays in the
afternoon is much the most rapid in the valley, coinciding with a
greater reduction of the elasticity of vapour and of humidity in the
atmosphere.

The photometer observations show a greater degree of sun's light
on the hills than below, but there is not at either station a decided
relation between the indications of this instrument and the black-
bulb thermometer.   From observations taken elsewhere, I am inclined
to attribute the excess of solar light on the hills to their elevation;
for at a far greater elevation I have met with much stronger solar
light, in a very damp atmosphere, than I ever experienced in the
drier plains of India.   In a damp climate the greatest intensity may
be expected in the forenoon, when the vapour is diffused near the
earth's surface; in the afternoon the lower strata of atmosphere are
drier, but the vapour is condensed into clouds aloft which more
effectually obstruct the sun's rays.   On the Birbhoom and Behar
hills, where the amount of vapour is so small that the afternoon
is but little drier than the forenoon, there is little difference
between the solar light at each time.   In the Soane valley again,
where a great deal of humidity is removed from the earth's
surface and suspended aloft, the obstruction of the sun's light is
very marked.

DUNWAH TO SOANE RIVER, AND UP SOANE TO TURA, FEBRUARY 10-19TH.

	TEMPERATURE.				WET-BULB.				DEW-POINT.						SATURATION.			Number of observations.
	Mean.	Max.	Min.	Range.	Mean.	Max. Depression.	Min. Depression.	Elasticity of Vapour.	Mean.	Max.	Min.	Max. Depression.	Min. Depression.	Weight of Vapour in cubic feet.	Mean.	Max.	Min.	
Sunrise ...	57·6	62·0	53·5	8·5	51·7	8·5	3·8	0·352	46·1	53·6	40·6	16·9	7·0	3·930	·680	·787	·566	10
9 A.M.......	74·0	81·0	63·5	17·5	59·5	18·5	4·0	0·382	48·5	56·7	38·0	33·5	6·8	4·066	·460	·818	·338	8
3 P.M.......	77·6	87·5	71·0	16·5	59·9	26·0	6·8	0·357	46·4	60·0	36·0	44·2	11·0	3·658	·352	·703	·237	9
9 P.M.......	64·5	68·7	60·0	8·7	55·5	12·5	2·5	0·370	47·5	55·6	41·0	24·1	4·4	4·014	·572	·860	·452	10

Extreme variation of Temperature . . . = 34·0°

„   of relative humidity . . . = ·623

„   diff. between Solar and Nocturnal Radiation . = 80·5°

NOCTURNAL RADIATION.

	SUNRISE.				NINE P.M.			
	Temperature.	Mean Diff from Air.	Max. Diff from Air.	Number of observations.	Temperature.	Mean Diff from Air.	Max. Diff from Air.	Number of observations.
Exposed Th.......	53·2	4·5	8·5	9	59·9	4·6	11·5	10
On Earth	54·0	3·7	9·0	9	60·7	3·8	10·5	10
On Grass	51·5	6·2	7·5	8	56·4	8·1	13·5	10

SOLAR RADIATION.

	MORNING.					AFTERNOON.			
Time.	Temp.	Black bulb.	Diff.	Phot.	Time.	Temp.	Black bulb.	Diff.	Phot.
9 P. M...	70·0	125	55·0	10·300	4 P.M.	76·5	90	13·5	...
11	81·0	119	38·0	10·230	3	80·0	105	25·0	10·210
10½......	71·5	126	54·5	10·300	3	76·0	102	26·0	10·170
10	72·0	117	45·0	10·220	3	87·5	126	38·5	...
10	80·0	122	42·0	...	...	...	...	...	...
10½......	78·0	128	50·0	...	...	...	...	...	...
Mean ...	75·4	122·8	47·4	10·262	..	80·0	105·7	25·7	10·190

NOCTURNAL RADIATION FROM PLANTS.

	SUNRISE.				NINE P.M.				
Air Temp.	Calotropis.	Diff.	Argemone.	Diff.	Temp.	Calotropis.	Diff.	Argemone.	Diff.
59·5	...	...	57·0	2·5	67·5	...	...	53·0	14·0
55·0	49·5	5·5	47·0	8·0	67·0	...	...	56·0	11·0
					64·3	58·5	5·8	57·0	7·3

III. VALLEY OF SOANE RIVER, TURA TO SULKUN (MEAN ELEV. 517 FEET), FEBRUARY 20TH TO MARCH 3RD.

	TEMPERATURE.				WET-BULB.			Elasticity of Vapour.	DEW-POINT.					Vapour in cubic feet.	SATURATION.			Number of Observations.
	Mean.	Max.	Min.	Range.	Mean.	Max. Depression.	Min. Depression.		Mean.	Max.	Min.	Max. Depression.	Min. Depression.		Mean.	Max	Min.	
Sun-rise...	56·8	70·0	50·0	20·0	52·5	10·0	1·5	·380	48·3	53·1	41·1	17·3	5·4	4·240	·754	·831	·570	12
9 A.M.......	82·0	89·0	69·0	20·0	61·2	24·3	12·0	·385	48·7	60·2	40·3	45·2	22·0	4·097	·342	·488	·226	11
3 P.M.......	88·6	94·7	81·5	13·2	62·4	30·2	14·5	·289	40·8	50·9	32·3	57·2	25·1	2·975	·211	·598	·154	11
9 P.M.......	68·0	74·0	61·0	13·0	56·8	15·0	6·0	·369	47·4	51·8	42·6	27·1	10·2	3·933	·511	·703	·415	11

Extreme variation of Temperature . . . 44·7°

" of relative humidity . . . ·677°

diff. between Solar and Nocturnal Radiation . 100°

NOCTURNAL RADIATION.

	SUNRISE.				NINE P.M.			
	Temperature.	Mean Diff. from Air.	Max. Diff. from Air.	Number of Observations.	Temperature.	Mean Diff. from Air.	Max. Diff. from Air.	Number of Observations.
Exposed Th.......	51·7	4·1	8·0	9	61·2	6·8	10·5	10
On Earth	52·4	3·4	7·0	9	64·3	4·6	8·5	9
On Grass	48·8	7·0	11·5	9	55·8	11·8	17·0	9

SOLAR RADIATION.

MORNING.					AFTERNOON.				
Time.	Temp.	Black Bulb.	Diff.	Phot.	Time.	Temp.	Black Bulb.	Diff.	Phot.
11½ A.M.	85·5	129	44·5	...	3 P.M.	85·5	116	30·5	...
10½ . .	89·0	132	43·0	...	...	92·5	128	35·5	...
Noon .	90·0	132	42·0	10·140	...	92·0	120	28·0	...
,,	85·0	130	45·0	...	...	89·5	128	38·5	...
,,	86·0	138	52·0	...	...	93·5	144	50·5	...
,,	90·0	138	48·0	...	...	...	...	...	...
Mean.	87·6	133	45·8	10·140	...	90·6	127	36·6	...

NOCTURNAL RADIATION FROM PLANTS.

SUNRISE.							NINE P.M.						
Temp. Air.	Barley.	Diff.	Calotropis.	Diff.	Argemone.	Diff.	Temp. Air.	Barley.	Diff.	Calotropis.	Diff.	Argemone.	Diff.
61·0	56	5·0	56·5	4·5	57·0	4·0	68·5	...	...	...	...	56·0	12·5
57·0	46	11·0	48·0	9·0	50·0	7·0	70·0	...	...	65·0	5·0	67·0	3·0
57·0	52	5·0	...	...	50·0	7·0	69·0	...	...	57·0	12·0	57·0	12·0
58·5	52	6·5	...	...	...	...	74·0	...	...	59·0	15·0	...	...
57·0	52	5·0	...	...	...	...	62·5	51·5	11·0	...	...	...	...
50·0	45	5·0	45·5	4·5	...	...	67·5	67·5	10·0	62·5	5·0	...	...
50·5	43	7·5	...	...	...	...	61·0	50·0	11·0	...	...	...	...
56·0	...	...	...	...	49·0	7·0	...	...	...	...	...	...	...
55·9	49·4	6·4	50·0	6·0	51·5	6·2	67·5	56·3	10·7	60·9	9·3	60·0	9·2

The upper course of the Soane being in some places confined, and exposed to furious gusts from the gullies of the Kymore hills, and at others expanding into a broad and flat valley, presents many fluctuations of temperature. The mean temperature is much above that of the lower parts of the same valley (below Tura), the excess amounting to 5°4. The nights and mornings are cooler, by 1°2, the days hotter by 10°. There were also 10° increase of range during the thirteen days spent there; and the mean range from day to day was nearly as great as it was on the hills of Bengal.

There being much exposed rock, and the valley being swept by violent dust-storms, the atmosphere is drier, the mean saturation-point being ·454, whereas in the lower part of the Soane's course it was ·516.

A remarkable uniformity prevails in the depression of thermometers exposed to nocturnal radiation, whether laid on the earth, grass, or freely exposed; both the mean and maximum indication coincide very nearly with those of the lower Soane valley and of the hills. The temperature of tufts of green barley laid on the ground is one degree higher than that of short grass; *Argemone* and *Calotropis* leaves maintain a still warmer temperature; from the previous experiments the *Argemone* appeared to be considerably the cooler, which I was inclined to attribute to the smoother and more shining surface of its leaf, but from these there would seem to be no sensible difference between the radiating powers of the two plants.

IV. TABLE-LAND OF KYMORE HILLS (MEAN ELEV. 979 FEET), MARCH 3RD TO 8TH, 1848.

	TEMPERATURE				WET-BULB			Elasticity of Vapour.	DEW-POINT					Vapour in cubic feet.	SATURATION			Number of Observations.
	Mean	Max.	Min.	Range.	Mean.	Max Depression.	Min. Depression.		Mean.	Max.	Min.	Max. Depression.	Min. Depression.		Mean.	Max.	Min.	
Sun-rise	65·3	69·0	57·5	11·5	57·7	8·0	6·0	·428	52·0	55·5	45·9	14·1	11·6	4·710	·647	·741	·648	4
9 A.M........	81·6	83·5	79·5	4·0	65·3	19·0	14·0	·468	54·5	57·9	49·0	33·0	12·9	5·000	·421	·479	·344	3
3 P.M........	88·1	90·0	84·5	5·5	63·3	26·5	21·5	·324	43·7	47·8	37·9	46·6	42·2	3·417	·240	·295	·214	3
9 P.M........	71·1	76·0	68·0	8·0	60·3	13·0	8·3	·433	52·3	56·7	46·8	21·9	13·8	4·707	·542	·643	·491	4

Extreme variation of Temperature   .   .   .   32·5°

     „   of relative humidity   .   .   .   „   ·527°

     „   diff. between Nocturnal and Solar Radiation   .   110·5°

NOCTURNAL RADIATION.

	SUN-RISE.				NINE P.M.			
	Temperature.	Mean Diff. from Air.	Max Diff. from Air.	Number of Observations.	Temperature.	Mean Diff. from Air.	Max. Diff. from Air.	Number of Observations.
Exposed Th.......	59·5	3·5	3·5	2	71·5	3·3	7·0	3
On Earth	56·0	1·5	1·5	1	62·5	5·5	5·5	1
On Grass	54·7	8·2	8·5	2	61·0	8·2	11·0	2

The rapid drying of the lower strata of the atmosphere during the day, as indicated by the great decrease in the tension of the vapour from 9 A.M. to 3 P.M., is the effect of the great violence of the north-west winds.

From the few days' observations taken on the Kymore hills, the temperature of their flat tops appeared 5° higher than that of the Soane valley, which is 500 feet below their mean level. I can account for this anomaly only on the supposition that the thick bed of alluvium, freely exposed to the sun (not clothed with jungle), absorbs the sun's rays and parts with its heat slowly. This is indicated by the increase of temperature being due to the night and morning observations, which are 3°.1 and 8°.5 higher here than below, whilst the 9 A.M. and 3 P.M. temperatures are half a degree lower.

The variations of temperature too are all much less in amount, as are those of the state of the atmosphere as to moisture, though the climate is rather damper.

On the subject of terrestrial radiation the paucity of the observations precludes my dwelling. Between 9 P.M. and sunrise the following morning I found the earth to have lost but 6°.5 of heat, whereas a mean of nine observations at the same hours in the valley below indicated a loss of 12°.

Though the mean temperature deduced from the few days I spent on this part of the Kymore is so much above that of the upper Soane valley, which it bounds, I do not suppose that the whole hilly range

partakes of this increase. When the alluvium does not cover the
rock, as at Rotas and many other places, especially along the
southern and eastern ridges of the ghats, the nights are considerably
cooler than on the banks of the Soane ; and at Rotas itself, which
rises almost perpendicularly from the river, and is exposed to no such
radiation of heat from a heated soil as Shahgunj is, I found the
temperature considerably below that of Akbarpore on the Soane,
which however is much sheltered by an amphitheatre of rocks.

V.—*Mirzapore on the Ganges.*

During the few days spent at Mirzapore, I was surprised to find
the temperature of the day cooler by nearly 4° than that of the hills
above, or of the upper part of the Soane valley, while the nights on the
other hand were decidedly warmer. The dew-point was even lower
in proportion, 7°.6, and the climate consequently drier. The following
is an abstract of the observations taken at Mr. Hamilton's house on
the banks of the Ganges (page 373).

It is remarkable that nocturnal radiation as registered at sunrise
is much more powerful at Mirzapore than on the more exposed
Kymore plateau ; the depression of the thermometer freely
exposed being 3° greater, that laid on bare earth 6°, and that on
the grass 1°.4 greater, on the banks of the Ganges.

During my passage down the Ganges the rise of the dew-point
was very steady, the maximum occurring at the lowest point on the
river, Bhaugulpore, which, as compared with Mirzapore, showed an
increase of 8° in temperature, and of 30°.6 in the rise of the dew-
point.   The saturation-point at Mirzapore was ·331, and at the corresponding hours at Bhaugulpore ·742.

MIRZAPORE (ELEV. 362 FEET). MARCH 9TH TO 13TH, 1848.

	TEMPERATURE.				WET-BULB.			Elasticity of Vapour.	DEW-POINT.					Vapour in cubic foot of atmosp.	SATURATION.			Number of Observations.
	Mean.	Max.	Min.	Range.	Mean.	Max. Diff.	Min. Diff.		Mean.	Max.	Min.	Max. Diff.	Min. Diff.		Mean.	Max.	Min.	
Sun-rise...	61·1	63·0	58·0	5·0	48·8	51·5	47·0	·236	34·3	39·7	29·7	32·8	23·8	2·574	·405	·450	·327	3
9 A.M.......	76·1	83·0	71·0	12·0	58·5	56·5	51·7	·302	41·9	...	...	52·3	15·7	3·271	·324	·603	·176	3
3 P.M.......	86·0	...	...	...	61·7	24·3	...	·295	41·3	...	...	44·7	...	3·089	·264	...	...	1
9 P.M.......	76·0	...	...	...	63·5	12·5	...	·480	55·2	...	...	20·8	...	5·127	·511	...	...	1

TERRESTRIAL RADIATION.

Air in Shade. Sunrise.	Exposed Th.	Diff	Exposed on earth.	Diff.	Exposed on grass.	Diff.
60·0	55·0	5·0	...	...	52·0	8·0
62·5	54·5	8·0	56·0	6·5	52·5	10·0
63·0	55·5	7·5	50·5	12·5	50·5	12·5
58·0	53·0	5·0	54·0	4·0	50·0	8·0
Mean, 60·9	54·6	6·4	53·5	7·7	51·3	9·6

B.

ON THE MINERAL CONSTITUENTS AND ALGÆ OF THE HOT-SPRINGS
OF BEHAR, THE HIMALAYA, AND OTHER PARTS OF INDIA, ETC.,
INCLUDING NOTES ON THE FUNGI OF THE HIMALAYA.

(By Dr. R. D. Thomson and the Rev. M. J. Berkeley, M.A., F.L.S.)

The following remarks, for which I am indebted to the kindness
of the able chemist and naturalist mentioned above, will be highly
valued, both by those who are interested in the many curious
physiological questions involved in the association of the most
obscure forms of vegetable life with the remarkable phenomena of
mineral springs; or in the exquisitely beautiful microscopic structure
of the lower Algæ, which has thrown so much light upon a branch
of natural history, whose domain, like that of astronomy, lies to a
great extent beyond the reach of the unassisted eye.—J. D. H.

1. Mineral water, Soorujkoond, Behar (vol. i., p. 27), contains
chloride of sodium and sulphate of soda.

2. Mineral water, hot springs, Yeumtong, altitude 11,730 feet
(see vol. ii., p. 117). Disengages sulphuretted hydrogen when fresh.—
This water was inodorous when the bottle was opened. The saline
matter in solution was considerably less than in the Soorujkoond
water, but like that consisted of chloride of sodium and sulphate of
soda. Its alkaline character suggests the probability of its containing carbonate of soda, but none was detected.

The rocks decomposed by the waters of the spring consist of
granite impregnated with sulphate of alumina. It appears that in
this case the sulphurous waters of Yeumtong became impregnated
in the air with sulphuric acid, which decomposed the felspar,*
and united with its alumina. I found traces only of potash in
the salt.

Sulphuretted hydrogen waters appear to give origin to sulphuric
acid, when the water impregnated with the gas reaches the surface;

* I have, in my journal, particularly alluded to the garnets (an aluminous
mineral) being thus entirely decomposed.—J. D. H.

and I have fine fibrous specimens of sulphate of lime accompanied with sulphur, from the hot springs of Pugha in west Tibet, brought by Dr. T. Thomson.

3. Mineral water, Momay hot springs, (vol. ii., p. 133).—When the bottle was uncorked, a strong smell of sulphuretted hydrogen was perceived. The water contains about twenty-five grains per imp. gallon, of chloride of sodium, sulphate and carbonate of soda; the reaction being strongly alkaline when the solution was concentrated.

4. Effloresced earth from Behar (vol. i., p. 13), consists of granite sand, mixed with sesquicarbonate of soda.

On the Indian Algæ which occur principally in different parts of the Himalayan Range, in the hot-springs of Soorujkoond in Bengal, Pugha in Tibet, and Momay in Sikkim; and on the Fungi of the Himalayas. By the Rev. M. J. Berkeley, M.A.

It is not my intention in the present appendix to give specific characters or even accurately determined specific names to the different objects within its scope, which have come under investigation, as collected by Dr. Hooker and Dr. Thomson. To do so would require far more time than I have at present been able to devote to the subject, for though every species has been examined microscopically, either by myself or Mr. Broome, and working sketches secured at the same time, the specific determination of fresh water Algæ from Herbarium specimens is a matter which requires a very long and accurate comparison of samples from every available locality, and in the case of such genera as *Zygnema*, *Tyndaridea*, and *Conferva*, is, after all, not a very satisfactory process.

The object in view is merely to give some general notion of the forms which presented themselves in the vast districts visited by the above-mentioned botanists, comprising localities of the greatest possible difference as regards both temperature and elevation; but more especially in the hot-springs which occur in two distant parts of the Himalayas and in Behar, and these again under very different degrees of elevation and of extrinsic temperature.

The Algæ from lower localities are but few in number, and some of these of very common forms. We have for instance from the Ganges, opposite Bijnour, a *Batrachospermum* and *Conferva crispata*, the former purple below, with specimens of *Chantransia*, exactly as they might occur in the Thames. The *Conferva*, or more properly *Cladophora*, which occurs also under various forms, at higher elevations, as in the neighbourhood of Simla and Iskardo, swarms with little parasites, but of common or uninteresting species. In the Bijnour specimens, these consist of common forms of *Synedra, Meridion circulare,* and a *Cymbella*, on others from Dacca, there are about three species of *Synedra*,* a minute *Navicula* and *Gomphonema curvatum*. Nothing, in fact, can well be more European. One splendid Alga, however, occurs at Fitcoree, in Behar, on the banks of nullahs, which are dry in hot weather, forming a purple fleece of coarse woolly hairs, which are singularly compressed, and of extreme beauty under the microscope, from the crystalline green of the articulated string which threads the bright red investing sheath. This curious Alga calls to mind in its colouring *Cænocoleus Smithii*, figured in English Botany, t. 2940, but it has not the common sheath of that Alga, and is on a far larger scale. One or two other allied forms, or species, occur in East Nepal, to which I purpose giving, together with the Behar plant, the generic name of *Erythronema*. From the Soane River, also, is an interesting Alga, belonging to the curious genus *Thwaitesia*, in which the division of the endochrome in the fertile cells into four distinct masses, sometimes entirely free, is beautifully marked. In some cases, indeed, instead of the ordinary spores, the whole mass is broken up into numerous bodies, as in the fertile joints of *Ulothrix*, and probably, as in that case, the resultant corpuscles are endowed with active motion. In Silhet, again, is a magnificent *Zygnema*, allied to *Z. nitidum*, with large oval spores, about $\frac{1}{285}$ part of an inch long, and a dark golden brown colour, and containing a spiral green endochrome.

Leaving, however, the lower parts of India, I shall first take the species which occur in Khasia, Sikkim, Eastern Nepal, and the adjoining parts of Tibet.

---

* Two of these appear to be *S. Vaucheriæ* and *S. inæqualis*.

In the hot valleys of the Great Rungeet, at an elevation of about 2000 feet, we have the *Erythronema*, but under a slightly different form; at Nunklow, at about the same height; in Khasia, again, at twice that elevation; in Eastern Nepal, at 12,000; and, finally, at Momay, reaching up to 16,000 feet. In water, highly impregnated with oxide of iron, at 4,000 feet in Sikkim, a *Leptothrix* occurred in great abundance, coloured with the oxide, exactly as is the case with Algæ which grow in iron springs in Europe. At elevations between 5000 and 7000 feet, several European forms occur, consisting of *Ulothrix*, *Zygnema*, *Oscillatoria*, *Lyngbya*, *Sphærozyga*, *Scytonema*, *Conferva*, and *Cladophora*. The species may indeed not be identical with European species, but they are all more or less closely allied to well-known Hydrophytes. One very interesting form, however, either belonging to the genus *Zygnema*, or possibly constituting a distinct genus, occurs in streams at 5000 feet in Sikkim, consisting of highly gelatinous threads of the normal structure of the *Zygnema*, but forming a reticulated mass. The threads adhere to each other laterally, containing only a single spiral endochrome, and the articulations are very long. Amongst the threads are mixed those of some species of *Tyndaridea*. There is also a curious *Hormosiphon*, at a height of 7000 feet, forming anastomosing gelatinous masses. A fine new species of *Lyngbya* extends up as high as 11,000 feet. At 13,000 feet occurs either some simple *Conferva* or *Zygnema*, it is doubtful which from the condition of the specimens; and at the same elevation, in the nearly dry bed of the stream which flows from the larger lake at Momay, amongst flat cakes, consisting of felspathic silt from the glaciers above, and the debris of Algæ, and abounding in Diatomaceæ, some threads of a *Zygnema*. At 17,000 feet, an *Oscillatoria*, attached or adherent to *Zannichellia;* and, finally, on the bare ground, at 18,000 feet, on the Donkia mountains, an obscure species of *Cænocoleus*. On the surface of the glaciers at Kinchinjhow, on silt, there is a curious *Palmella*, apparently quite distinct from any European form.

Amongst the greater part of the Algæ, from 4000 feet to 18,000 feet, various Diatomaceæ occur, which will be best noticed in a tabular form, as follows; the specific name, within brackets, merely indicating the species to which they bear most resemblance:—

Himantidium (*Soleirolii*)	. .	4000 to 7000 feet.	Sikkim.
Odontidium (*hiemale*, forma minor)		5000 to 7000 ,,	,,
Epithemia, *n. sp.*	. . . .	7000 ,,	,,
Cymbella	. . . . .	— ,,	,,
Navicula, *n. sp.*	. . . .	— ,,	,,
Tabillaria (*flocculosa*)	. . .	6000 to 7000 ,,	,,
Odontidium (*hiemale*)	. .	11,000 ,,	,,
Himantidium	. . .	16,000 ,,	Momay.
Odontidium (*turgidulum*)	. . .	17,000 ,,	,,
Epithemia (*ocellata*)	. . .	— ,,	Tibet.
Fragillaria	. . . .	18,000 ,,	Momay.
Odontidium (*turgidulum*)	. .	— ,,	,,
Dictyocha (*gracilis*)	. . .	— ,,	,,
Odontidium (*hiemale*)	. . .	— ,,	Kinchinjhow.

We now turn to those portions of Tibet or the neighbouring
regions, explored by Dr. Thomson and Captain Strachey. The
principal feature in the Algology is the great prevalence of species
of *Zygnema* and *Tyndaridea*, which occur under a variety of forms,
sometimes with very thick gelatinous coats. In not a single instance,
however, is there the slightest tendency to produce fructification.
Conferva crispata again, as mentioned above, occurs in several localities; and in one locality a beautiful unbranched *Conferva*, with torulose
articulations.   At Iskardo, Dr. Thomson gathered a very gelatinous
species of *Draparnaldia*, or more properly, a *Stygeoclonium*, if we
may judge from a little conglomeration of cells which appeared
amongst the threads.   A *Tetraspora* in Piti, an obscure *Tolypothrix*,
and one or two *Oscillatoriæ*, remarkable for their interrupted mode
of growth, complete the list of Algæ, with the exception of one, to be
mentioned presently ; as also of *Diatomaceæ*, and of the species of
Nostoc and *Hormosiphon*, which occurred in great profusion, and
under several forms, sometimes attaining a very large size (several
inches across), especially in the districts of Le and Piti, and where
the soil or waters were impregnated with saline matters.   It is well
known that some species of *Nostoc* form an article of food in China,
and one was used for that purpose in a late Arctic expedition, as
reported by Dr. Sutherland; but it does not seem that any use is
made of them in Tibet, though probably all the large species would
form tolerable articles of food, and certainly, from their chemical
composition, prove very nutritious.   One species is mentioned by
Dr. Thomson as floating, without any attachment, in the shallow

water of the pools scattered over the plains, on the Parang River, separated only by a ridge of mountains from Piti, broad and foliaceous, and scarcely different from the common *Nostoc*, which occurs in all parts of the globe. I must not, however, neglect to record a very singular new genus, in which the young threads have the characters of *Tyndaridea*, but, after a time, little swellings occur on their sides, in which a distinct endochrome is formed, extending backwards into the parent endochrome, separated from it by a well defined membrane, and producing, either by repeated pullulation, a compound mass like that of *Calothrix*, or simply giving rise to a forked thread. In the latter case, however, there is no external swelling, but a lateral endochrome is formed, which, as it grows, makes its way through an aperture, whose sides are regularly inflected. I have given to this curious production the name of *Cladozygia Thomsoni*.

The whole of the above Algæ occurred at heights varying from 10,000 to 15,500 feet. As in the Southern Himalayan Algæ, the specimens were infested with many Diatomaceæ, amongst which the most conspicuous were various *Cymbellæ* and *Epithemiæ*. The following is a list of the species observed.

Cymbella (*gastroides*).	Epithemia *n. sp.*
— (*gracilis*).	Synedra (*arcus*).
— (*Ehrenbergii*)	— (*tenuis*).
and three others.	— (*æqualis*).
Odontidium (*hiemale*).	Denticula (*obtusa*).
— (*mesodon*).	Gomphonema (*abbreviatum*).
— *n. sp.*	Meridion circulare.
Epithemia *n. sp.*	

There is very little identity between this list and that before given from the Southern Himalayas, as is the case also with the other Algæ. Till the species, however, have been more completely studied, a very accurate comparison cannot be made.

In both instances the species which grow in hot springs have been reserved in order to make their comparison more easy. I shall begin in an inverse order, with those of the springs of Pugha in Tibet, which attain a temperature of 174°. Two *Confervæ* only occur in the specimens which have been preserved, viz., an *Oscillatoria* allied to that which I have called *O. interrupta*, and a true *Conferva* extremely delicate with very long articulations, singularly swollen at the commissures. The *Diatomaceæ* are :—

Odontidium (*hiemale*).	Denticula (*obtusa*).
— (*mesodon*).	Navicula.
— *n. sp.*, same as at Piti on	Cymbella, three species.
Conferva.	Epithemia.

Scarcely any one of these except the *Navicula* is peculiar to the locality. A fragment apparently of some *Closterium*, the only one which I have met with in the collection, accompanies one of the specimens.

The hot springs of Momay, (temp. 110°) at 16,000 feet, produce a golden brown *Cænocoleus* representing a small form of *C. cirrhosus*, and a very delicate *Sphærozyga*, an *Anabaina*, and *Tolypothrix;* and at 17,000 feet, a delicate green *Conferva* with long even articulations. With the latter is an *Odontidium* allied to, or identical with *O. turgidulum*, and with the former a fine species of *Epithemia* resembling in form, but not in marking, *E. Faba, E.* (*Zebra*) a fine *Navicula*, perhaps the same with *N. major* and *Fragilaria* (*virescens*).* In mud from one of the Momay springs (*a*), I detected *Epithemia* (*Broomeii n. s.*), and two small *Naviculæ*, and in the spring (*c*) two species of *Epithemia* somewhat like *E. Faba*, but different from that mentioned above.

The hot springs of Soorujkoond, of the vegetation of which very numerous specimens have been preserved, are extremely poor in species. In the springs themselves and on their banks, at temperatures varying from 80° to 158°, at which point vegetation entirely ceases, a minute *Leptothrix* abounds everywhere, varying a little in the regularity of the threads in different specimens, but scarcely presenting two species. Between 84° and 112° there is an imperfect *Zygnema* with very long articulations, and where the green scum passes into brown, there is sometimes an *Oscillatoria*, or a very minute stellate *Scytonema*, probably in an imperfect state. *Epithemia ocellata* also contributes often to produce the tint. An *Anabaina* occurs at a temperature of 125°, but the same species was found also in the stream from the springs where the water had become cold, as was also the case with the *Zygnema*.

---

* Mr. Thomas Brightwell finds in a portion of the same specimen *Epithemia alpestris, Surirella splendida, S. linearis,* Smith, *Pinnularia viridis,* Smith, *Navicula* (*lanceolata*) and *Himantidium* (*arcus*).

The Diatomaceæ consisted of :—

Epithemia Broomeii, *n. s.*	Epithemia inæqualis, *n. sp.*
— thermalis, *n. sp.*	Navicula Beharensis, *n. sp.*

The vegetation in the three sets of springs was very different. As regards the *Confervæ*, taking the word in its older sense, the species in the three are quite different, and even in respect of genera there is little identity, but amongst the *Diatomaceæ* there is no striking difference, except in those of the Behar springs where three out of the four did not occur elsewhere. In the Pugha and Momay springs, the species were either identical with, or nearly allied to those found in neighbouring localities, where the water did not exceed the ordinary temperature. A longer examination will doubtless detect more numerous forms, but those which appear on a first examination are sure to give a pretty correct general notion of the vegetation. The species are certainly less numerous than I had expected, or than might be supposed from the vegetation of those European hot springs which have been most investigated.

In conclusion, I shall beg to add a few words on the Fungi of the Himalayas, so far as they have at present been investigated. As regards these there is a marked difference, as might be anticipated from the nature of the climates between those parts of Tibet investigated by Dr. Thomson, and the more southern regions. The fungi found by Dr. Thomson were but few in number, and for the most part of very ordinary forms, differing but little from the produce of an European wood. Some, however, grow to a very large size, as for instance, *Polyporus fomentarius* on poplars near Iskardo, exceeding in dimensions anything which this species exhibits in Europe. A very fine *Æcidium* also infests the fir trees (*Abies Smithiana*), a figure of which has been given in the "Gardeners' Chronicle," 1852, p. 627, under the name of *Æcidium Thomsoni*. This is allied to the Hexenbesen of the German forests, but is a finer species and quite distinct. *Polyporus oblectans, Geaster limbatus, Geaster mammosus, Erysiphe taurica,* a *Boletus* infested with *Sepedonium mycophilum, Scleroderma verrucosum,* an *Æcidium,* and a *Uromyces,* both on *Mulgedium Tataricum,* about half-a-dozen Agarics, one at an altitude of 16,000 feet above the Nubra river, a *Lycoperdon,* and *Morchella semilibera,* which is eaten in Kashmir, and exported when dry to the plains of India, make up the list of fungi.

The region of Sikkim is perhaps the most productive in fleshy fungi of any in the world, both as regards numbers and species, and Eastern Nepal and Khasia yield also an abundant harvest. The forms are for the most part European, though the species are scarcely ever quite identical. The dimensions of many are truly gigantic, and many species afford abundant food to the natives. Mixed with European forms a few more decidedly tropical occur, and amongst those of East Nepal is a *Lentinus* which has the curious property of staining every thing which touches it of a deep rhubarb yellow, and is not exceeded in magnificence by any tropical species. The *Polypori* are often identical with those of Java, Ceylon, and the Philippine Isles, and the curious *Trichocoma paradoxum* which was first found by Junghuhn in Java, and very recently by Dr. Harvey in Ceylon, occurs abundantly on the decayed trunks of laurels, as it does in South Carolina. The curious genus *Mitremyces* also is scattered here and there, though not under the American form, but that which occurs in Java. Though *Hymenomycetes* are so abundant, the *Discomycetes* and *Ascomycetes* are comparatively rare, and very few species indeed of *Sphæria* were gathered. One curious matter is, that amongst the very extensive collections which have been made there is scarcely a single new genus. The species moreover in Sikkim are quite different, except in the case of some more or less cosmopolite species from those of Eastern Nepal and Khasia: scarcely a single *Lactarius* or *Cortinarius* for instance occurs in Sikkim, though there are several in Khasia. The genus *Boletus* through the whole district assumes the most magnificent forms, which are generally very different from anything in Europe.

C.*

ON THE SOILS OF SIKKIM.

THERE is little variety in the soil throughout Sikkim, and, as far as vegetation is concerned, it may be divided into vegetable mould and stiff clay—each, as they usually occur, remarkably characteristic in composition of such soils. Bog-earth is very rare, nor did I find peat at any elevation.

The clay is uniformly of great tenacity, and is, I believe, wholly due to the effect of the atmosphere on crumbling gneiss and other rocks. It makes excellent bricks, is tenacious, seldom friable, and sometimes accumulated in beds fourteen feet thick, although more generally only about two feet. In certain localities, beds or narrow seams of pure felspathic clay and layers of vegetable matter occur in it, probably wholly due to local causes. An analysis of that near Dorjiling gives about 30 per cent. of alumina, the rest being silica, and a fraction of oxide of iron. Lime is wholly unknown as a constituent of the soil, and only occasionally seen as a stalactitic deposit from a few springs.

A layer of vegetable earth almost invariably covers the clay to the depth of from three to twelve or fourteen inches. It is a very rich black mould, held in its position on the slopes of the hills by the dense vegetation, and accumulated on the banks of small streams to a depth at times of three and four feet. The following is an analysis of an average specimen of the surface-soil of Dorjiling, made for me by my friend C. J. Muller, Esq., of that place :—

a.—DRY EARTH.

Anhydrous	83·84
Water	16·16
	100·00

* The tables referred to, at v. i. p. 31, as under Appendix C., will be found under Appendix A.

b.—ANHYDROUS EARTH.

Humic acid.	3·89
Humine .	4·61
Undecomposed vegetable matter.	20·98
Peroxide of iron and manganese	7·05
Alumina	8·95
Siliceous matter, insoluble in dilute hydrochloric acid.	54·52
Traces of soda and muriatic acid.	..
	100·00

c.—Soluble in water, gr. 1·26 per cent., consisting of soda, muriatic acid, organic matter, and silica.

The soil from which this example was taken was twelve inches deep; it abounded to the eye in vegetable matter, and was siliceous to the touch. There were no traces of phosphates or of animal matter, and doubtful traces of lime and potash. The subsoil of clay gave only 5·7 per cent. of water, and 5.55 of organic matter. The above analysis was conducted during the rainy month of September, and the sample is an average one of the surface-soil at 6000 to 10,000 feet. There is, I think, little difference anywhere in the soils at this elevation, except where the rock is remarkably micaceous, or where veins of felspathic granite, by their decomposition, give rise to small beds of kaolin.

D.

(Vol. i., p. 37.)

AN AURORA SEEN FROM BAROON ON THE EAST BANK OF THE SOANE RIVER.

Lat. 24° 52′ N.; Long. 84° 22′ E. ; Alt. 345 feet.

THE following appearances are as noted in my journal at the time. They so entirely resembled auroral beams, that I had no hesitation in pronouncing them at the time to be such. This opinion has, however, been dissented from by some meteorologists, who consider that certain facts connected with the geographical distribution of auroras (if I may use the term), are opposed to it. I am well aware of the force of these arguments, which I shall not attempt to controvert; but for the information of those who may be interested in the matter, I may remark, that I am very familiar with the Aurora borealis in the northern temperate zone, and during the Antarctic expedition was in the habit of recording in the log-book the appearance presented by the Aurora australis. The late Mr. Williams, Mr. Haddon, and Mr. Theobald, who were also witnesses of the appearances on this occasion, considered it a brilliant display of the aurora.

Feb. 14th, 9 P.M.—Bar . Corr. 29·751; temp. 62°; D. P. 41·0°; calm, sky clear; moon three-quarters full, and bright.

Observed about thirty lancet beams rising in the north-west from a low luminous arch, whose extremes bore W. 20° S., and N. 50° E. ; altitude of upper limb of arch 20°, of the lower 8°. The beams crossed the zenith, and converged towards S. 15° E. The extremity of the largest was forked, and extended to 25° above the horizon in the S.E. by S. quarter. The extremity of the centre one bore S. 50° E., and was 45° above the horizon. The western beams approached nearest the southern horizon. All the beams moved and flashed slowly, occasionally splitting and forking, fading and brightening; they were brightly defined, though the milky way and zodiacal light could not be discerned, and the stars and planets, though clearly discernible, were very pale.

At 10 P.M., the luminous appearance was more diffused; upper limb of the arch less defined; no beams crossed the zenith; but occasionally beams appeared there and faded away.

Between 10 and 11, the beams continued to move and replace one another, as usual in auroras, but disappeared from the south-east quarter, and became broader in the northern hemisphere; the longest beams were near the north and north-east horizon.

At half-past 10, a dark belt, 4° broad, appeared in the luminous arch, bearing from N. 55° W. to N. 10° W.; its upper limb was 10° above the horizon : it then gradually dilated, and thus appeared to break up the arch. This appeared to be the commencement of the dispersion of the phenomenon.

At 10.50, P.M. the dark band had increased so much in breadth that the arch was broken up in the north-west, and no beams appeared there. Eighteen linear beams rose from the eastern part of the arch, and bore from north to N. 20° E.

Towards 11 P.M., the dark band appeared to have replaced the luminous arch; the beams were all but gone, a few fragments appearing in the N.E. A southerly wind sprang up, and a diffused light extended along the horizon.

At midnight, I saw two faint beams to the north-east, and two well defined parallel ones in the south-west.

E.

PHYSICAL GEOGRAPHY OF THE SIKKIM HIMALAYA, EAST NEPAL, AND ADJACENT PROVINCES OF TIBET.

SIKKIM is included in a section of the Himalaya, about sixty miles broad from east to west, where it is bounded respectively by the mountain states of Bhotan and Nepal. Its southern limits are easily defined, for the mountains rise abruptly from the plains of Bengal, as spurs of 6000 to 10,000 feet high, densely clothed with forest to their summits. The northern and north-eastern frontier of Sikkim is beyond the region of much rain, and is not a natural, but a political line, drawn between that country and Tibet. Sikkim lies nearly due north of Calcutta, and only four hundred miles from the Bay of Bengal; its latitude being 26° 40′ to 28° N., and longitude 88° to 89° E.

The main features of Sikkim are Kinchinjunga, the loftiest hitherto measured mountain, which lies to its north-west, and rises 28,178 feet above the level of the sea; and the Teesta river, which flows throughout the length of the country, and has a course of upwards of ninety miles in a straight line. Almost all the sources of the Teesta are included in Sikkim; and except some comparatively insignificant streams draining the outermost ranges, there are no rivers in this country but itself and its feeders, which occupy the largest of the Himalayan valleys between the Tambur in East Nepal, and the Machoo in Western Bhotan.

An immense spur, sixty miles long, stretches south from Kinchin to the plains of India; it is called Singalelah, and separates Sikkim from East Nepal; the waters from its west flank flow into the Tambur, and those from the east into the Great Rungeet, a feeder of the Teesta.   Between these two latter rivers is a second spur from Kinchinjunga, terminating in Tendong.

The eastern boundary of Sikkim, separating it from Bhotan, is formed for the greater part by the Chola range, which stretches south from the immense mountain of Donkia, 23,176 feet high, situated fifty miles E.N.E. of Kinchinjunga: where the frontier approaches the plains of India, the boundary line follows the course of the Teesta, and of the Rinkpo, one of its feeders, flowing from the Chola range.   This range is much more lofty than that of Singalelah, and the drainage from its eastern flank is into the Machoo river, the upper part of whose course is in Tibet, and the lower in Bhotan.

The Donkia mountain, though 4000 feet lower than Kinchin, is the culminant point of a much more extensive and elevated mountain mass.   It throws off an immense spur from its north-west face, which runs west, and then south-west, to Kinchin, forming the watershed of all the remote sources of the Teesta.   This spur has a mean elevation of 18,000 to 19,000 feet, and several of its peaks (of which Chomiomo is one) rise much higher.   The northern boundary of Sikkim is not drawn along this, but runs due west from Donkia, following a shorter, but stupendous spur, called Kinchinjhow; whence it crosses the Teesta to Chomiomo, and is continued onwards to Kinchinjunga.

Though the great spur connecting Donkia with Kinchin is in Tibet, and bounds the waters that flow directly south into the Teesta, it is far from the true Himalayan axis, for the rivers that rise on its northern slope do not run into the valley of the Tsampu, or Tibetan Burrampooter, but into the Arun of Nepal, which rises to the north of Donkia, and flows south-west for many miles in Tibet, before entering Nepal and flowing south to the Ganges.

Sikkim, thus circumscribed, consists of a mass of mountainous spurs, forest-clad up to 12,000 feet; there are no flat valleys or plains in the whole country, no lakes or precipices of any consequence below that elevation, and few or no bare slopes, though the latter are uniformly steep.   The aspect of Sikkim can only be understood by a reference to its climate and vegetation, and I shall therefore take these together, and endeavour, by connecting these phenomena,

c c 2

to give an intelligible view of the main features of the whole country.*

The greater part of the country between Sikkim and the sea is a dead level, occupied by the delta of the Ganges and Burrampooter, above which the slope is so gradual to the base of the mountains, that the surface of the plain from which the Himalayas immediately rise is only 300 feet above the sea.   The most obvious effect of this position is, that the prevailing southerly wind reaches the first range of hills, loaded with vapour.   The same current, when deflected easterly to Bhotan, or westerly to Nepal and the north-west Himalaya, is intercepted and drained of much moisture, by the Khasia and Garrow mountains (south of Assam and the Burrampooter) in the former case, and the Rajmahal hills (south of the Ganges) in the latter.   Sikkim is hence the dampest region of the whole Himalaya.

Viewed from a distance on the plains of India, Sikkim presents the appearance—common to all mountainous countries—of consecutive parallel ridges, running east and west: these are all wooded, and backed by a beautiful range of snowy peaks, with occasional breaks in the foremost ranges, through which the rivers debouch.   Any view of the Himalaya, especially at a sufficient distance for the remote snowy peaks to be seen overtopping the outer ridges, is, however, rare, from the constant deposition of vapours over the forest-clad ranges during the greater part of the year, and the haziness of the dry atmosphere of the plains in the winter months.   At the end of the rains, when the south-east monsoon has ceased to blow with constancy, views are obtained, sometimes from a distance of nearly two hundred miles.   From the plains, the highest peaks subtend so small an angle, that they appear like white specks very low on the horizon, tipping the black lower and outer wooded ranges, which always rise out of a belt of haze, and from the density, probably, of the lower strata of atmosphere, are never seen to rest on the visible horizon.   The remarkable lowness on the horizon of the whole stupendous mass is always a disappointing feature to the new comer, who expects to see dazzling peaks towering in the air.   Approaching nearer, the snowy mountains

* This I did with reference especially to the cultivation of Rhododendrons, in a paper which the Horticultural Society of London did me the honour of printing.   Quarterly Journ. of Hort. Soc., vol. vii., p. 82.

sink behind the wooded ones, long before the latter have assumed gigantic proportions; and when they do so, they appear a sombre, lurid grey-green mass of vegetation, with no brightness or variation of colour. There is no break in this forest caused by rock, precipices, or cultivation; some spurs project nearer, and some valleys appear to retire further into the heart of the foremost great chain that shuts out all the country beyond.

From Dorjiling the appearance of parallel ridges is found to be deceptive, and due to the inosculating spurs of long tortuous ranges that run north and south throughout the whole length of Sikkim, dividing deep wooded valleys, which form the beds of large rivers. The snowy peaks here look like a long east and west range of mountains, at an average distance of thirty or forty miles. Advancing into the country, this appearance proves equally deceptive, and the snowy range is resolved into isolated peaks, situated on the meridional ridges; their snow-clad spurs, projecting east and west, cross one another, and being uniformly white, appear to connect the peaks into one grand unbroken range. The rivers, instead of having their origin in the snowy mountains, rise far beyond them; many of their sources are upwards of one hundred miles in a straight line from the plains, in a very curious country, loftier by far in mean elevation than the meridional ridges which run south from it, yet comparatively bare of snow. This rearward part of the mountain region is Tibet, where all the Sikkim, Nepal, and Bhotan rivers rise as small streams, increasing in size as they receive the drainage from the snowed parts of the ridges that bound them in their courses. Their banks, between 8000 and 14,000 feet, are generally clothed with rhododendrons, sometimes to the almost total exclusion of other woody vegetation, especially near the snowy mountains—a cool temperature and great humidity being the most favourable conditions for the luxuriant growth of this genus.

The source of this humidity is the southerly or sea wind which blows steadily from May till October in Sikkim, and prevails throughout the rest of the year, if not as the monsoon properly so called, as a current from the moist atmosphere above the Gangetic delta. This rushes north to the rarefied regions of Sikkim, up the great valleys, and does not appear materially disturbed by the north-west wind, which blows during the afternoons of the winter months over the plains, and along the flanks of the outer range, and is a dry surface current, due to the diurnal heating of the soil. When it is considered that this wind, after passing lofty mountains on the outer range, has to traverse eighty or one hundred miles of alps before it has watered all the forest region, it will be evident that its moisture must be expended before it reaches Tibet.

Let the figures in the accompanying woodcut, the one on the true scale, the other with the heights exaggerated, represent two of these long meridional ridges, from the watershed to the plains of India, following in this instance the course of the Teesta river, from its source at 19,000 feet to where it debouches from the Himalaya at 300. The lower rugged outline represents one meridional ridge, with all its most prominent peaks (whether exactly or not on the line of section); the upper represents the parallel ridge of 'Singalelah (D.E.P.), of greater mean elevation, further west, introduced to show the maximum elevation of the Sikkim mountains, Kinchinjunga (28,178 feet), being represented on it. A deep valley is interposed between these two ridges, with a feeder of the Teesta in it (the Great Rungeet), which runs south from Kinchin, and turning west enters the Teesta at R. The position of the bed of the Teesta river is indicated by a dotted line from its source at T to the plains at S; of Dorjiling, on the north flank of the outer range, by d; of the first point where perpetual snow is met with, by P; and of the first indications of a Tibetan climate, by C.

A warm current of air, loaded with vapour, will deposit the bulk of its moisture on the ridge of Sinchul, which rises above Dorjiling (d), and is 8,500 feet high. Passing on, little will be precipitated on e, whose elevation is the same as that of Sinchul; but much at f (11,000 feet), where the current, being further cooled, has less capacity for holding vapour, and is further exhausted. When it ascends to P (15,000 feet) it is sufficiently cooled to deposit snow in the winter and spring months, more of which falling than can be melted during the summer, it becomes perennial. At the top of Kinchin very little falls, and it is doubtful if the southerly current ever reaches that prodigiously elevated isolated summit. The amount of surface above 20,000 feet is, however, too limited and

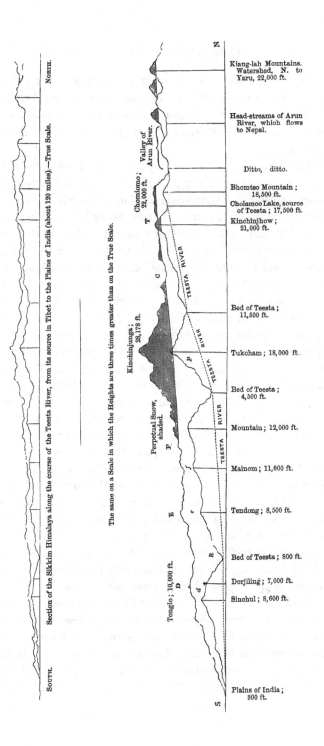

SOUTH.   Section of the Sikkim Himalaya along the course of the Teesta River, from its source in Tibet to the Plains of India (about 120 miles).—True Scale.   NORTH.

The same on a Scale in which the Heights are three times greater than on the True Scale.

N

Kiang-lah Mountains. Watershed, N. to Yaru, 22,000 ft.

Head-streams of Arun River, which flows to Nepal.

Valley of Arun River.

Chomiomo; 22,000 ft.

T

Ditto, ditto.

Bhomtso Mountain; 18,500 ft.

Cholamoo Lake, source of Teesta; 17,500 ft.

Kinchinjhow; 21,000 ft.

TEESTA RIVER

C

Bed of Teesta; 11,500 ft.

Kinchinjunga; 28,178 ft.

RIVER

P

Tukcham; 18,000 ft.

TEESTA

Bed of Teesta; 4,500 ft.

Perpetual Snow, shaded.

P

RIVER

Mountain; 12,000 ft.

TEESTA

Mainom; 11,000 ft.

f

B

e

Tendong; 8,500 ft.

R

Bed of Teesta; 800 ft.

Tongio; 10,000 ft.

D

d

Dorjiling; 7,000 ft.

Sinchul; 8,600 ft.

Plains of India; 800 ft.

S

broken into isolated peaks to drain the already nearly exhausted current, whose condensed vapours roll along in fog beyond the parallel of Kinchin, are dissipated during the day over the arid mountains of Tibet, and deposited at night on the cooled surface of the earth.

Other phenomena of no less importance than the distribution of vapour, and more or less depending on it, are the duration and amount of solar and terrestrial radiation. Towards D the sun is rarely seen during the rainy season, as well from the constant presence of nimbi aloft, as from fog on the surface of the ground. An absence of both light and heat is the result south of the parallel of Kinchin; and at C low fogs prevail at the same season, but do not intercept either the same amount of light or heat; whilst at T there is much sunshine and bright light. During the night, again, there is no terrestrial radiation between S and P; the rain either continues to pour—in some months with increased violence—or the saturated atmosphere is condensed into a thick white mist, which hangs over the redundant vegetation. A bright starlight night is almost unknown in the summer months at 6000 to 10,000 feet, but is frequent in December and January, and at intervals between October and May, when, however, vegetation is little affected by the cold of nocturnal radiation. In the regions north of Kinchin, starlight nights are more frequent, and the cold produced by radiation, at 14,000 feet, is often severe towards the end of the rains in September. Still the amount of clear weather during the night is small; the fog clears off for an hour or two at sunset as the wind falls, but the returning cold north current again chills the air soon afterwards, and rolling masses of vapour are hence flying overhead, or sweeping the surface of the earth, throughout the summer nights. In the Tibetan regions, on the other hand, bright nights and even sharp frosts prevail throughout the warmest months.

Referring again to the cut, it must be borne in mind that neither of the two meridional ridges runs in a straight line, but that they wind or zigzag as all mountain ranges do; that spurs from each ridge are given off from either flank alternately, and that the origin of a spur on one side answers to the source of a river (*i.e.*, the head of a valley) on the other. These rivers are feeders of the main

stream, the Teesta, and run at more or less of an angle to the latter.
The spurs from the east flank of one ridge cross, at their ends, those
from the west flank of another; and thus transverse valleys are
formed, presenting many modifications of climate with regard to
exposure, temperature, and humidity.

The roads from the plains of India to the watershed in Tibet
always cross these lateral spurs. The main ridge is too winding and
rugged, and too lofty for habitation throughout the greater part of
its length, while the river-channel is always very winding, unhealthy
for the greater part of the year below 4000 feet, and often narrow,
gorge-like, and rocky. The villages are always placed above the
unhealthy regions, on the lateral spurs, which the traveller repeatedly
crosses throughout every day's march; for these spurs give off
lesser ones, and these again others of a third degree, whence the
country is cut up into as many spurs, ridges, and ranges, as there
are rills, streams, and rivers amongst the mountains.

Though the direction of the main atmospheric current is to the
north, it is in reality seldom felt to be so, except the observer be on
the very exposed mountain tops, or watch the motions of the upper
strata of atmosphere. Lower currents of air rush up both the main
and lateral valleys, throughout the day; and from the sinuosities in
the beds of the rivers, and the generally transverse directions of their
feeders, the current often becomes an east or west one. In the
branch valleys draining to the north the wind still ascends; it is, in
short, an ascending warm, moist current, whatever course be pursued
by the valleys it follows.

The sides of each valley are hence equally supplied with moisture,
though local circumstances render the soil on one or the other flank
more or less humid and favourable to a luxuriant vegetation: such
differences are a drier soil on the north side, with a too free exposure
to the sun at low elevations, where its rays, however transient, rapidly
dry the ground, and where the rains, though very heavy, are of shorter
duration, and where, owing to the capacity of the heated air for retaining moisture, day fogs are comparatively rare. In the northern
parts of Sikkim, again, some of the lateral valleys are so placed that
the moist wind strikes the side facing the south, and keeps it very
humid, whilst the returning cold current from the neighbouring

Tibetan mountains impinges against the side facing the north, which is hence more bare of vegetation. An infinite number of local peculiarities will suggest themselves to any one conversant with physical geography, as causing unequal local distribution of light, heat, and moisture in the different valleys of so irregular a country; namely, the amount of slope, and its power of retaining moisture and soil; the composition and hardness of the rocks; their dip and strike; the protection of some valleys by lofty snowed ridges; and the free southern exposures of others at great elevations.

The position and elevation of the perpetual snow * vary with those of the individual ranges, and their exposure to the south wind. The expression that the perpetual snow lies lower and deeper on the southern slopes of the Himalayan mountains than on the northern,

---

* It appears to me, as I have asserted in the pages of my Journal, that the limit of perpetual snow is laid down too low in all mountain regions, and that accumulations in hollows, and the descent of glacial ice, mask the phenomenon more effectually than is generally allowed. In this work I define the limit, as is customary, in general terms only, as being that where the accumulations are very great, and whence they are continuous upwards, on gentle slopes. All perpetual snow, however, becomes ice, and, as such, obeys the laws of glacial motion, moving as a viscous fluid; whence it follows that the lower edge of a snow-bed placed on a slope is, in one sense, the termination of a glacier, and indicates a position below that where all the snow that falls melts. I am well aware that it is impossible to define the limit required with any approach to accuracy. Steep and broken surfaces, with favourable exposures to the sun or moist winds, are bare much above places where snow lies throughout the year; but the occurrence of a gentle slope, free of snow, and covered with plants, cannot but indicate a point below that of perpetual snow. Such is the case with the " Jardin " on the Mer de Glace, whose elevation is 9,500 feet, whereas that of perpetual snow is considered by Professor J. Forbes, our best authority, to be 8,500 feet. Though limited in area, girdled by glaciers, presenting a very gentle slope to the east, and screened by surrounding mountains from a considerable proportion of the sun's rays, the Jardin is clear, for fully three months of the year, of all but sporadic falls of snow, that never lie long; and so are similar spots placed higher on the neighbouring slopes; which facts are quite at variance with the supposition that the perpetual snow-line is below that point in the Mont Blanc Alps. On the Monte Rosa Alps, again, Dr. Thomson and I gathered plants in flower, above 12,000 feet, on the steep face of the Weiss-thor Pass, and at 10,938 feet on the top of St. Théodule; but in the former case the rocks are too steep for any snow to lie, they are exposed to the south-east, and overhang a gorge 8000 feet deep, up which no doubt warm currents ascend; while at St. Théodule the plants were growing on a slope which, though gentle, is black and stony, and exposed to warm ascending currents, as on the Weiss-thor; and I do not consider either of these as evidences of the limit of perpetual snow being higher than their position.

conveys a false impression. It is better to say that the snow lies
deeper and lower on the southern faces of the individual mountains and spurs that form the snowy Himalaya. The axis itself of the
chain is generally far north of the position of the spurs that catch all
the snow, and has comparatively very little snow on it, most of what
there is lying upon north exposures.

A reference to the woodcut will show that the same circumstances
which affect the distribution of moisture and vegetation, determine
the position, amount, and duration of the snow. The principal fall
will occur, as before shown, where the meridional range first attains
a sufficiently great elevation, and the air becomes consequently cooled
below 32° ; this is at a little above 14,000 feet, sporadic falls occurring
even in summer at that elevation : these, however, melt immediately,
and the copious winter falls also are dissipated before June. As the
depth of rain-fall diminishes in advancing north to the higher parts
of the meridional ranges, so does that of the snow-fall. The permanence of the snow, again, depends on—1. The depth of the accumulation ; 2. The mean temperature of the spot ; 3. The melting power of
the sun's rays ; 4. The prevalence and strength of evaporating winds.
Now at 14,000 feet, though the accumulation is immense, the amount
melted by the sun's rays is trifling, and there are no evaporating
winds; but the mean temperature is so high, and the corroding
powers of the rain (which falls abundantly throughout summer) and
of the warm and humid ascending currents are so great, that the
snow is not perennial. At 15,500 feet, again, it becomes perennial,
and its permanence at this low elevation (at P) is much favoured by
the accumulation and detention of fogs over the rank vegetation
which prevails from S nearly to P ; and by the lofty mountains
beyond it, which shield it from the returning dry currents from the
north. In proceeding north all the circumstances that tend to the
dispersion of the snow increase, whilst the fall diminishes. At P the
deposition is enormous and the snow-line low—16,000 feet ; whilst
at T little falls, and the limit of perpetual snow is 19,000 and 20,000
feet. Hence the anomaly, that the snow-line ascends in advancing
north to the coldest Himalayan regions. The position of the greatest
peaks and of the greatest mass of perpetual snow being generally
assumed as indicating a ridge and watershed, travellers, arguing from

single mountains alone, on the meridional ridges, have at one time supported and at another denied the assertion, that the snow lies longer and deeper on the north than on the south slope of the Himalayan ridge.

The great accumulation of snow at 15,000 feet, in the parallel of P, exercises a decided influence on the vegetation. The alpine rhododendrons hardly reach 14,000 feet in the broad valleys and round-headed spurs of the mountains of the Tunkra and Chola passes; whilst the same species ascend to 16,000, and one to 17,000 feet, at T. Beyond the latter point, again, the great aridity of the climate prevents their growth, and in Tibet there are generally none even as low as 12,000 and 14,000 feet. Glaciers, again, descend to 15,000 feet in the tortuous gorges which immediately debouch from the snows of Kinchinjunga, but no plants grow on the débris they carry down, nor is there any sward of grass or herbage at their base, the atmosphere immediately around being chilled by enormous accumulations of snow, and the summer sun rarely warming the soil. At T, again, the glaciers do not descend below 16,000 feet, but a greensward of vegetation creeps up to their bases, dwarf rhododendrons cover the moraines, and herbs grow on the patches of earth carried down by the latter, which are thawed by the more frequent sunshine, and by the radiation of heat from the unsnowed flanks of the valleys down which these ice-streams pour.

Looking eastward or westward on the map of India, we perceive that the phenomenon of perpetual snow is regulated by the same laws. From the longitude of Upper Assam in 95° E to that of Kashmir in 75° E, the lowest limit of perpetual snow is 15,500 to 16,000 feet, and a shrubby vegetation affects the most humid localities near it, at 12,000 to 14,000 feet. Receding from the plains of India and penetrating the mountains, the climate becomes drier, the snow-line rises, and vegetation diminishes, whether the elevation of the land increases or decreases; plants reaching 17,000 and 18,000 feet, and the snow-line, 20,000 feet. To mention extreme cases; the snow-level of Sikkim in 27° 30 is at 16,000 feet, whereas in latitude 35° 30 Dr. Thomson found the snow line 20,000 feet on the mountains near the Karakoram Pass, and vegetation up to 18,500 feet—features I found to be common also to Sikkim in latitude 28°.

The Himalaya, north of Nepal, and thence eastward to the bend of the Yaru-Tsampu (or Tibetan Burrampooter) has for its geographical limits the plains of India to the south, and the bed of the Yaru to the north. All between these limits is a mountain mass, to which Tibet (though so often erroneously called a plain)* forms no exception. The waters from the north side of this chain flow into the Tsampu, and those from the south side into the Burrampooter of Assam, and the Ganges. The line, however tortuous, dividing the heads of these waters, is the watershed, and the only guide we have to the axis of the Himalaya. This has never been crossed by Europeans, except by Captain Turner's embassy in 1798, and Captain Bogle's in 1779, both of which reached the Yaru river. In the account published by Captain Turner, the summit of the watershed is not rigorously defined, and the boundary of Tibet and Bhotan is sometimes erroneously taken for it; the boundary being at that point a southern spur of Chumulari.† Eastwards from the sources of the Tsampu, the watershed of the Himalaya seems to follow a very winding course, and to be everywhere to the north of the snowy peaks seen from the plains of India. It is by a line through these snowy peaks that the axis of the Himalaya is represented in all our maps; because they *seem* from the plains to be situated on an east and west ridge, instead of being placed on subsidiary meridional ridges, as explained above. It is also across or along the subsidiary ridges that the boundary line between the Tibetan provinces and those of Nepal, Sikkim, and Bhotan, is usually drawn; because the enormous accumulations of snow form a more

---

* The only true account of the general features of eastern Tibet is to be found in MM. Huc and Gabet's travels. Their description agrees with Dr. Thomson's account of western Tibet, and with my experience of the parts to the north of Sikkim, and the information I everywhere obtained. The so-called *plains* are the flat floors of the valleys, and the terraces on the margins of the rivers, which all flow between stupendous mountains. The term "maidan," so often applied to Tibet by the natives, implies, not a plain like that of India, but simply an open, dry, treeless country, in contrast to the densely wooded wet regions of the snowy Himalaya, south of Tibet.

† Between Donkia and Chumulari lies a portion of Tibet (including the upper part of the course of the Machoo river) bounded on the east by Bhotan, and on the west by Sikkim (see p. 110). Turner, when crossing the Simonang Pass, descended westwards into the valley of the Machoo, and was still on the Indian watershed.

efficient natural barrier than the greater height of the less snowed central part of the chain beyond them.

Though, however, our maps draw the axis through the snowy peaks, they also make the rivers to rise beyond the latter, on the northern slopes as it were, and to flow southwards through gaps in the axis. Such a feature is only reconcilable with the hypothesis of the chain being double, as the Cordillera of Peru and Chili is said to be, geographically, and which in a geological sense it no doubt is: but to the Cordillera the Himalaya offers no parallel. The results of Dr. Thomson's study of the north-west Himalaya and Tibet, and my own of the north-east extreme of Sikkim and Tibet, first gave me an insight into the true structure of this chain. Donkia mountain is the culminant point of an immensely elevated mass of mountains, of greater mean height than a similarly extensive area around Kinchinjunga: It comprises Chumulari, and many other mountains much above 20,000 feet, though none equalling Kinchinjunga, Junnoo, and Kubra. The great lakes of Ramchoo and Cholamoo are placed on it, and the rivers rising on it flow in various directions; the Painomchoo north-west into the Yaru; the Arun west to Nepal; the Teesta south-west through Sikkim; the Machoo south, and the Pachoo south-east, through Bhotan. All these rivers have their sources far beyond the great snowed mountains, the Arun most conspicuously of all, flowing completely at the back or north of Kinchinjunga. Those that flow southwards, break through no chain, nor do they meet any contraction as they pass the snowy parts of the mountains which bound the valleys in which they flow, but are bound by uniform ranges of lofty mountains, which become more snowy as they approach the plains of India. These valleys, however, gradually contract as they descend, being less open in Sikkim and Nepal than in Tibet, though there bounded by rugged mountains, which from being so bare of snow and of vegetation, do not give the same impression of height as the isolated sharper peaks which rise out of a dense forest, and on which the snow limit is 4,000 or 5,000 feet lower.

The fact of the bottom of the river valleys being flatter towards the watershed, is connected with that of their fall being less rapid at that part of their course; this is the consequence of the great extent in breadth of the most elevated portion of the chain. If we select the Teesta as an example, and measure its fall at three points of its course, we shall find the results very different. From its principal source at Lake Cholamoo, it descends from 17,000 to 15,000 feet, with a fall of 60 feet to the mile; from 15,000 to 12,000 feet, the fall is 140 feet to the mile; in the third part of its course it descends from 12,000 to 5,000 feet, with a fall of 160 feet to the mile; and in the lower part the descent is from 5,000 feet to the plains of India at 300 feet, giving a fall of 50 feet to the mile. There is, however, no marked limit to these divisions; its valley gradually contracts, and its course gradually becomes more rapid. It is worthy of notice that the fall is at its maximum through that part of its valley of which the flanks are the most loaded with snow; where the old moraines are very conspicuous, and where the present accumulations from landslips, &c., are the most extensive.*

With reference to Kinchinjunga, these facts are of importance, as showing that mere elevation is in physical geography of secondary importance. That lofty mountain rises from a spur of the great range of Donkia, and is quite removed from the watershed or axis of the Himalaya, the rivers which drain its northern and southern flanks alike flowing to the Ganges. Were the Himalaya to be depressed 18,000 feet, Kubra, Junnoo, Pundim, &c., would form a small cluster of rocky islands 1,000 to 7,000 feet high, grouped near Kinchinjunga, itself a cape 10,000 feet high, which would be connected by a low, narrow neck, with an extensive and mountainous tract of land to its north-east; the latter being represented by Donkia. To the north of Kinchin a deep bay or inlet would occupy the present valley of the Arun, and would be bounded on the north by the axis of the Himalaya, which would form a continuous tract of land beyond it. Since writing the above, I have seen Professor J. Forbes's beautiful work on the glaciers of Norway: it fully justifies a comparison of the Himalaya to Norway, which has long been a familiar subject of

* It is not my intention to discuss here the geological bearings of this curious question; but I may state that as the humidity of the climate of the middle region of the river-course tends to increase the fall in a given space, so I believe the dryness of the climate of the loftier country has the opposite effect, by preserving those accumulations which have raised the floors of the valleys and rendered them level.

theoretical enquiry with Dr. Thomson and myself. The deep narrow valleys of Sikkim admirably represent the Norwegian fiords; the lofty, rugged, snowy mountains, those more or less submerged islands of the Norwegian coast; the broad rearward watershed, or axis of the chain, with its lakes, is the same in both, and the Yaru-tsampu occupies the relative position of the Baltic.

Along the whole chain of the Himalaya east of Kumaon. there are, I have no doubt, a succession of such lofty masses as Donkia, giving off stupendous spurs such as that on which Kinchin forms so conspicuous a feature. In support of this view we find every river rising far beyond the snowy peaks, which are separated by continuously unsnowed ranges placed between the great white masses that these spurs present to the observer from the south.* From the Khasia mountains (south-east of Sikkim) many of these groups or spurs were seen by Dr. Thomson and myself, at various distances (80 to 210 miles); and these groups were between the courses of the great rivers the Soobansiri, Monass, and Pachoo, all east of Sikkim. Other masses seen from the Gangetic valley probably thus mark the relative positions of the Arun, Cosi, Gunduk, and Gogra rivers.

Another mass like that of Chumulari and Donkia, is that around the Mansarowar lakes, so ably surveyed by the brothers Captains R. and H. Strachey, which is evidently the centre of the Himalaya. From it the Gogra, Sutlej, Indus, and Yaru rivers all flow to the Indian side of Asia; and from it spring four chains, two of which are better known than the others. These are:—1. The eastern Himalaya, whose axis runs north of Nepal, Sikkim, and Bhotan, to the bend of the Yaru, the valley of which it divides from the plains of India. 2. The north-west Himalaya, which separates the valley of the Indus from the plains of India. Behind these, and probably parallel to them, lie two other chains. 3. The Kouenlun or Karakoram chain, dividing the Indus from the Yarkand river. 4. The chain north of the Yaru, of which nothing is known. All the waters from the two first of these chains, flow into the Indian Ocean, as do those from

* At vol. i. p. 185, I have particularly called attention to the fact, that west of Kinchinjunga there is no continuation of a snowy Himalaya, as it is commonly called. So between Donkia and Chumulari there is no perpetual snow, and the valley of the Machoo is very broad, open, and comparatively flat.

the south faces of the third and fourth; those from the north side
of the Kouenlun, and of the chain north of the Yaru, flow into the
great valley of Lake Lhop, which may once have been continuous
with the Amoor river.*

For this view of the physical geography of the western Himalaya
and central Asia, I am indebted to Dr. Thomson. It is more
consonant with nature, and with what we know of the geography
of the country and of the nature of mountain chains, than that of
the illustrious Humboldt, who divides central Asia by four parallel
chains, united by two meridional ones; one at each extremity of
the mountain district. It follows in continuation and conclusion of
our view that the mountain mass of Pamir or Bolor, between the
sources of the Oxus and those of the Yarkand river, may be regarded
as a centre from which spring the three greatest mountain systems of
Asia. These are:—1. A great chain, which runs in a north-easterly
direction as far as Behring's Straits, separating all the rivers of
Siberia from those which flow into the Pacific Ocean. 2. The
Hindoo Koosh, continued through Persia and Armenia into Taurus.
And, 3. The Muztagh or Karakorum, which probably extends due
east into China, south of the Hoang-ho, but which is broken up
north of Mansarowar into the chains which have been already
enumerated.

F.

ON THE CLIMATE OF SIKKIM.

THE meteorology of Sikkim, as of every part of the Himalayan
range, is a subject of growing interest and importance ; as it becomes
yearly more necessary for the Government to afford increased
facilities for a residence in the mountains to Europeans in search of
health, or of a salubrious climate for their families, or for themselves on
retirement from the exhausting service of the plains. I was therefore
surprised to find no further register of the weather at Dorjiling, than

* The Chinese assert that Lake Lhop once drained into the Hoang-ho; the
statement is curious, and capable of confirmation when central Asia shall have
been explored.

an insufficient one of the rain-fall, kept by the medical officer in charge of the station; who, in this, as in all similar cases,* has neither the time nor the opportunity to give even the minimum of required attention to the subject of meteorology. This defect has been in a measure remedied by Dr. Chapman, who kept a twelve-months' register in 1837, with instruments carefully compared with Calcutta standards by the late James Prinsep, Esq., one of the most accomplished men in literature and science that India ever saw.

The annual means of temperature, rain-fall, &c., vary greatly in the Himalaya; and apparently slight local causes produce such great differences of temperature and humidity, that one year's observations taken at one spot, however full and accurate they may be, are insufficient : this is remarkably the case in Sikkim, where the rain-fall is great, and where the difference between those of two consecutive years is often greater than the whole annual London fall. My own meteorological observations necessarily form but a broken series, but they were made with the best instruments, and with a view to obtaining results that should be comparable *inter se*, and with those of Calcutta; when away from Dorjiling too, in the interior of Sikkim, I had the advantage of Mr. Muller's services in taking observations at hours agreed upon previous to my leaving, and these were of the greatest importance, both for calculating elevations, and for ascertaining the differences of temperature, humidity, diurnal atmospheric tide, and rain-fall; all of which vary with the elevation, and the distance from the plains of India.

Mr. Hodgson's house proved a most favourable spot for an observatory, being placed on the top of the Dorjiling spur, with its broad verandah facing the north, in which I protected the instruments from

---

* The government of India has gone to an immense expense, and entailed a heavy duty upon its stationary medical officers, in supplying them with some-times admirable, but more often very inaccurate, meteorological instruments, and requiring that daily registers be made, and transmitted to Calcutta. In no case have I found it to be in the officer's power to carry out this object; he has never time, seldom the necessary knowledge and experience, and far too often no inclination. The majority of the observations are in most cases left to personal native or other servants, and the laborious results I have examined are too frequently worthless:

radiation * and wind. Broad grass-plots and a gravel walk surrounded the house, and large trees were scattered about; on three sides the ground sloped away, while to the north the spur gently rose behind.

Throughout the greater part of the year the prevailing wind is from the south-east, and comes laden with moisture from the Bay of Bengal: it rises at sunrise, and its vapours are early condensed on the forests of Sinchul; billowy clouds rapidly succeed small patches of vapour, which rolling over to the north side of the mountain, are carried north-west, over a broad intervening valley, to Dorjiling. There they bank on the east side of the spur, and this being partially clear of wood, the accumulation is slow, and always first upon the clumps of trees. Very generally by 9 A.M., the whole eastern sky, from the top of Dorjiling ridge, is enveloped in a dense fog, while the whole western exposure enjoys sunshine for an hour or two later. At 7 or 8 A.M., very small patches are seen to collect on Tonglo, which gradually dilate and coalesce, but do not shroud the mountain for some hours, generally not before 11 A.M. or noon. Before that time, however, masses of mist have been rolling over Dorjiling ridge to the westward, and gradually filling up the valleys, so that by noon, or 1 P.M., every object is in cloud. Towards sunset it falls calm, when the mist rises, first from Sinchul, or if a south-east wind sets in, from Tonglo first.

The temperature is more uniform at Mr. Hodgson's bungalow, which is on the top of the Dorjiling ridge, than on either of its flanks; this is very much because a good deal of wood is left upon it, whose cool foliage attracts and condenses the mists. Its mean temperature is lower by nearly 2½° than that of Mr. Muller's and Dr. Campbell's houses, both situated on the slopes, 400 feet below. This I ascertained by numerous comparative observations of the temperature of the air, and by burying thermometers in the earth:

---

* This is a most important point, generally wholly neglected in India, where I have usually seen the thermometer hung in good shade, but exposed to reflected heat from walls, gravel walks, or dry earth. I am accustomed from experience to view all extreme temperatures with great suspicion, on this and other accounts. It is very seldom that the temperature of the free shaded air rises much above 100°, except during hot winds, when the lower stratum only of atmosphere (often loaded with hot particles of sand), sweeps over the surface of a soil scorched by the direct rays of the sun.

it is chiefly to be accounted for by the more frequent sunshine at the lower stations, the power of the sun often raising the thermometer in shade to 80°, at Mr. Muller's ; whereas during the summer I spent at Mr. Hodgson's it never rose much above 70°, attaining that height very seldom and for a very short period only. The nights, again, are uniformly and equally cloudy at both stations, .so that there is no corresponding cold of nocturnal radiation to reduce the temperature.

The mean decrease of temperature due to elevation, I have stated (Appendix I.) to be about 1° for every 300 feet of ascent ; according to which law Mr. Hodgson's should not be more than 1½° colder than Mr. Muller's. These facts prove how difficult it is to choose unexceptionable sites for meteorological observatories in mountainous countries ; discrepancies of so great an amount being due to local causes, which, as in this case, are perhaps transient ; for should the top of the spur be wholly cleared of timber, its temperature would be materially raised ; at the expense, probably, of a deficiency of water at certain seasons. Great inequalities of temperature are also produced by ascending currents of heated air from the Great Rungeet valley, which affect certain parts of the station only ; and these raise the thermometer 10° (even when the sun is clouded) above what it indicates at other places of equal elevation.

The mean temperature of Dorjiling (elev. 7,430 feet) is very nearly 50°, or 2° higher than that of London, and 26° below that of Calcutta (78°,[*] or 78° 5 in the latest published tables [†]) ; which, allowing 1° of diminution of temperature for every degree of latitude leaves 1° due to every 300 feet of ascent above Calcutta to the height of Dorjiling, agreeably to my own observations. This diminution is not the same for greater heights, as I shall have occasion to show in a separate chapter of this Appendix, on the decrement of heat with elevation.

A remarkable uniformity of temperature prevails throughout the year at Dorjiling, there being only 22° difference between the mean temperatures of the hottest and coldest months ; whilst in London,

* Prinsep, in As. Soc. Journ., Jan. 1832, p. 30.
† Daniell's Met. Essays, vol. ii. p. 341.

with a lower mean temperature, the equivalent difference is 27°. At 11,000 feet this difference is equal to that of London. In more elevated regions, it is still greater, the climate becoming excessive at 15,000 feet, where the difference amounts to 30° at least. * The accompanying table is the result of an attempt to approximate to the mean temperatures and ranges of the thermometer at various elevations.

Altitude.	Mean Shade.	Mean Warmest Month.	Mean Coldest Month.	Mean Daily Range of Temperature.	Rain-fall in inches.	
11,000 feet...	40·9	50·0	24·0	20·0	40·0	1° = 320 feet.
15,000 feet...	29·8	40·0	11·0	27·0	20·0	1° = 350 feet.
19,000 feet...	19·8	32·0	0·0	35·0	10·0	1° = 400 feet.

Supposing the same formula to apply (which I exceedingly doubt) to heights above 19,000 feet, 2° would be the mean annual temperature of the summit of Kinchinjunga, altitude 28,178 feet, the loftiest known spot on the globe: this is a degree or two higher than the temperature of the poles of greatest cold on the earth's surface, and about the temperature of Spitzbergen and Melville island.

The upper limit of phenogamic vegetation coincides with a mean temperature of 30° on the south flank of Kinchinjunga, and of 22° in Tibet; in both cases annuals and perennial-rooted herbaceous plants are to be found at elevations corresponding to these mean temperatures, and often at higher elevations in sheltered localities. I have assumed the decrease of temperature for a corresponding

* This is contrary to the conclusions of all meteorologists who have studied the climate of the Alps, and is entirely due to the local disturbances which I have so often dwelt upon, and principally to the unequal distribution of moisture in the loftier rearward regions, and the aridity of Tibet. Professor James Forbes states (Ed. Phil. Trans., v. xiv. p. 489):—1. That the decrement of temperature with altitude is most rapid in summer: this (as I shall hereafter show) is not the case in the Himalaya, chiefly because the warm south moist wind then prevails. 2. That the annual range of temperature diminishes with the elevation: this, too, is not the case in Sikkim, because of the barer surface and more cloudless skies of the rearward loftier regions. 3. That the diurnal range of temperature diminishes with the height: that this is not the case follows from the same cause. 4. That radiation is least in winter: this is negatived by the influence of the summer rains.

amount of elevation to be gradually less in ascending ($1° = 320$ feet at 6000 to 10,000 feet, $1° = 400$ feet at 14,000 to 18,000 feet). My observations appear to prove this, but I do not regard them as conclusive; supposing them to be so, I attribute it to a combination of various causes, especially to the increased elevation and yet unsnowed condition of the mass of land elevated above 16,000 feet, and consequent radiation of heat; also to the greater amount of sunshine there, and to the less dense mists which obstruct the sun's rays at all elevations. In corroboration of this I may mention that the decrease of temperature with elevation is much less in summer than in winter, $1°$ of Fahr. being equivalent to only 250 feet in January between 7000 and 13,000 feet, and to upwards of 400 feet in July. Again, at Dorjiling (7,430 feet) the temperature hardly ever rises above 70° in the summer months, yet it often rises even higher in Tibet at 12,000 to 14,000 feet. On the other hand, the winters, and the winter nights especially, are disproportionately cold at great heights, the thermometer falling upwards of 40° below the Dorjiling temperature at an elevation only 6000 feet higher.

The diurnal distribution of temperature is equally and similarly affected by the presence of vapour at different altitudes. The lower and outer ranges of 6000 to 10,000 feet, first receive the diurnal charge of vapour-loaded southerly winds; those beyond them get more of the sun's rays, and the rearward ones more still. Though the summer days of the northern localities are warmer than their elevation would indicate, the nights are not proportionally cold; for the light mist of 14,000 feet, which replaces the dense fog of 7000 feet, effectually obstructs nocturnal radiation, though it is less an obstacle to solar radiation. Clear nights, be it observed, are as rare at Momay (15,300 feet) as at Dorjiling, the nights if windy being rainy; or, if calm, cold currents descend from the mountains, condensing the moist vapours of the valleys, whose narrow floors are at sunrise bathed in mist at all elevations in Sikkim. The rise and dispersion of these dense mists, and their collection and recondensation on the mountains in the morning, is one of the most magnificent phenomena of the Himalaya, when viewed from a proper elevation; it commences as soon as the sun appears on the horizon.

The mean daily range of the thermometer at 7000 feet is 13° in cleared spots, but considerably less in wooded, and certainly one-third less in the forest itself. At Calcutta, which has almost an insular climate, it amounts to 17°; at Delhi, which has a continental one, to 24° 6; and in London to 17° 5. At 11,000 feet it amounts to about 20°, and at 15,000 feet to 27°. These values vary widely in the different months, being much less in the summer or rainy months. The following is probably a fair approximation :—

```
At 7,000 feet it amounts to 8°-9° in Aug. and Sept., and 17° in Dec.
   11,000   „         „       12°      „          „      30°   „
   15,000   „         „       15°      „          „      40°   „
   London             „       20°      „          „      10°   „
```

The distribution of temperature throughout the day and year varies less at Dorjiling than in most mountainous countries, owing to the prevailing moisture, the effect of which is analogous to that of a circumambient ocean to an island : the difference being, that in the case of the island the bulk of water maintains an uniform temperature ; in that of Dorjiling the quantity of vapour acts directly by interfering with terrestrial and solar radiation, and indirectly by nurturing a luxuriant vegetation. The result in the latter case is a climate remarkable for its equability, and similar in many features to that of New Zealand, South-west Chili, Fuegia, and the damp west coasts of Scotland and Ireland, and other countries exposed to moist sea winds.

The mean temperature of the year at Dorjiling, as taken by maxima and minima thermometers * by Dr. Chapman, is nearly the same as that of March and October : January, the coldest month, is more than 13° 4 colder than the mean of the year; but the hottest month is only 8° 3' warmer than the same mean : at Calcutta the months vary less from the mean; at Delhi more; and in London the distribution is wholly different; there being no rains to modify the summer heat, July is 13° hotter, and January 14° colder than the mean of the year.

---

* The mean of several of the months, thus deduced, often varies a good deal from the truth, owing to the unequal diurnal distribution of heat; a very few minutes' sunshine raises the temperature 10° or 15° above the mean of the day ; which excessive heat (usually transient) the maximum thermometer registers, and consequently gives too high a mean.

This distribution of the seasons has a most important effect upon vegetation, to which sufficient attention has not been paid by cultivators of alpine Indian plants; in the first place, though English winters are cold enough for such, the summers are too hot and dry; and, in the second place, the great accession of temperature, causing the buds to burst in spring, occurs in the Himalaya in March, when the temperature at 7000 feet rises 8° above that of February, raising the radiating thermometer always above the freezing point, whence the young leaves are never injured by night frost: in England the corresponding rise is only 3°, and there is no such accession of temperature till May, which. is 8° warmer than April; hence, the young foliage of many Himalayan plants is cut off by night frosts in English gardens early in the season, of which *Abies Webbiana* is a conspicuous example.

The greatest heat of the day occurs at Dorjiling about noon, owing to the prevalent cloud, especially during the rainy months, when the sun shines only in the mornings, if at all, and the clouds accumulate as the day advances. According to hourly observations of my own, it occured in July at noon, in August at 1 P.M., and in September (the most rainy month) there was only four-tenths of a degree difference between the means of noon, 1 P.M., and 2 P.M., but I must refer to the abstracts at the end of this chapter for evidence of this, and of the wonderful uniformity of temperature during the rainy months. In the drier season again, after September, the greatest heat occurs between 2 and 3 P.M.; in Calcutta the hottest hour is about 2·45 P.M., throughout the year; and in England also about 3 P.M.

The hour whose temperature coincides with the mean of the day necessarily varies with the distribution of cloud and sunshine; it is usually about 7 A.M. and 7 P.M.; whereas in Calcutta the same coincidence occurs at a little before 10 A.M., and in England at about 8 A.M.

Next to the temperature of the air, observations on that of the earth are perhaps of the greatest value; both from their application to horticulture, and from the approximation they afford to the mean temperature of the week or month in which they are taken. These form the subject of a separate chapter.

Nocturnal and solar radiation, the one causing the formation of

dew and hoar-frost when the air in the shade is above freezing, and
killing plants by the rapid abstraction of heat from all their surfaces
which are exposed to the clear sky, and the other scorching the skin
and tender plants during the day, are now familiar phenomena, and
particularly engaged my attention during my whole Indian journey.
Two phenomena particularly obstruct radiation in Sikkim—the
clouds and fog from the end of May till October, and the haze from
February till May. Two months alone are usually clear; one before
and one after the rains, when the air, though still humid, is transparent. The haze has never been fully explained, though a well-
known phenomenon. On the plains of India, at the foot of the hills,
it begins generally in the forenoon of the cold season, with the rise of
the west wind; and, in February especially, obscures the sun's disc
by noon; frequently it lasts throughout the twenty-four hours, and
is usually accompanied by great dryness of the atmosphere. It
gradually diminishes in ascending, and I have never experienced it at
10,000 feet; at 7000, however, it very often, in April, obscures the
snowy ranges 30 miles off, which are bright and defined at sunrise,
and either pale away, or become of a lurid yellow-red, according to
the density of this haze, till they disappear at 10 A.M. I believe it
always accompanies a south-west wind (which is a deflected current
of the north-west) and dry atmosphere in Sikkim.

The observations for solar radiation were taken with a black-bulb
thermometer, and also with actinometers, but the value of the data
afforded by the latter not being fixed or comparative, I shall give the
results in a separate section. (See Appendix K.) From a multitude
of desultory observations, I conclude that at 7,400 feet, 125° 7,
or + 67° above the temperature of the air, is the average maximum
effect of the sun's rays on a black-bulb thermometer * throughout the
year, amounting rarely to + 70° and + 80° in the summer months,
but more frequently in the winter or spring. These results, though
greatly above what are obtained at Calcutta, are not much, if at all,
above what may be observed on the plains of India. This effect is

---

* From the mean of very many observations, I find that 10° is the average
difference at the level of the sea, in India, between two similar thermometers,
with spherical bulbs (½-inch diam.), the one of black, and the other of plain glass,
and both being equally exposed to the sun's rays.

much increased with the elevation. At 10,000 feet in December, at 9 A.M., I saw the mercury mount to 132° with a diff. of + 94°, whilst the temperature of shaded snow hard by was 22°; at 13,100 feet, in January, at 9 A.M., it has stood at 98°, diff. + 68° 2; and at 10 A.M., at 114°, diff. + 81° 4, whilst the radiating thermometer on the snow had fallen at sunrise to 0° 7. In December, at 13,500 feet, I have seen it 110°, diff. + 84°; at 11 A.M., 11,500 feet, 122°, diff. + 82°. This is but a small selection from many instances of the extraordinary power of solar radiation in the coldest months, at great elevations.

Nocturnal and terrestrial radiation are even more difficult phenomena for the traveller to estimate than solar radiation, the danger of exposing instruments at night being always great in wild countries. I most frequently used a thermometer graduated on the glass, and placed in the focus of a parabolic reflector, and a similar one laid upon white cotton,* and found no material difference in the mean of many observations of each, though often 1° to 2° in individual ones. Avoiding radiation from surrounding objects is very difficult, especially in wooded countries. I have also tried the radiating power of grass and the earth; the temperature of the latter is generally less, and that of the former greater, than the thermometer exposed on cotton or in the reflector, but much depends on the surface of the herbage and soil.

The power of terrestrial, like that of solar radiation, increases with the elevation, but not in an equal proportion. At 7,400 feet, the mean of all my observations shows a temperature of 35° 4. During the rains, 3° to 4° is the mean maximum, but the nights being almost invariably cloudy, it is scarcely on one night out of six that there is any radiation. From October to December the amount is greater = 10° to 12°, and from January till May greater

---

* Snow radiates the most powerfully of any substance I have tried; in one instance, at 13,000 feet, in January, the thermometer on snow fell to 0·2°, which was 10·8° below the temperature at the time, the grass showing 6·7°; and on another occasion to 1·2°, when the air at the time (before sunrise) was 21·2°; the difference therefore being 20°. I have frequently made this observation, and always with a similar result; it may account for the great injury plants sustain from a thin covering of ice on their foliage, even when the temperature is but little below the freezing-point.

still, being as much as 15°. During the winter months the effect
of radiation is often felt throughout the clear days, dew forming
abundantly at 4,000 to 8,000 feet in the shaded bottoms of narrow
valleys, into which the sun does not penetrate till 10 A.M., and
from which it disappears at 3 P.M. I have seen the thermometer
in the reflector fall 12° at 10 A.M. in a shaded valley. This often
produces an anomalous effect, causing the temperature in the shade
to fall after sunrise; for the mists which condense in the bottom
of the valleys after midnight disperse after sunrise, but long before
reached by the sun, and powerful radiation ensues, lowering the
surrounding temperature: a fall of 1° to 2° after sunrise of air
in the shade is hence common in valleys in November and December.* The excessive radiation of the winter months often gives
rise to a curious phenomenon ; it causes the formation of copious dew
on the blanket of the traveller's bed, which radiates heat to the tent
roof, and this inside either an open or a closed tent. I have experienced this at various elevations, from 6,000 to 16,000 feet. Whether
the minimum temperature be as high as 50°, or but little above
zero, the effect is the same, except that hoar-frost or ice forms in
the latter case. Another remarkable effect of nocturnal radiation is
the curl of the alpine rhododendron leaves in November, which is probably due to the freezing and consequent expansion of the water in
the upper strata of cells exposed to the sky. The first curl is
generally repaired by the ensuing day's sun, but after two or three
nights the leaves become permanently curled, and remain so till they
fall in the following spring.

I have said that the nocturnal radiation in the English spring
months is the great obstacle to the cultivation of many Himalayan
plants; but it is not therefore to be inferred that there is no similar
amount of radiation in the Himalaya; for, on the contrary, in April
its amount is much greater than in England, frequently equalling
13° of difference; and I have seen 16° at 7,500 feet ; but the minimum

* Such is the explanation which I have offered of this phenomenon in the
Hort. Soc. Journal. On thinking over the matter since, I have speculated upon
the probability of this fall of temperature being due to the absorption of heat
that must become latent on the dispersion of the dense masses of white fog that
choke the valleys at sunrise.

temperature at the time is 51°, and the absolute amount of
cold therefore immaterial. The mean minimum of London is 38°,
and, when lowered 5·5° by radiation, the consequent cold is very
considerable. Mr. Daniell, in his admirable essay on the climate
of London, mentions 17° as the maximum effect of nocturnal
radiation ever observed by him. I have registered 16° in April
at Dorjiling; nearly as much at 6,000 feet in February; twice
13°, and once 14° 2 in September at 15,500 feet; and 10° in
October at 16,800 feet; nearly 13° in January at 7,000 feet;
14° 5 in February at that elevation, and, on several occasions,
14° 7 at 10,000 feet in November.

The annual rain-fall at Dorjiling averages 120 inches (or 10 feet),
but varies from 100 to 130 in different years; this is fully three
times the amount of the average English fall*, and yet not one-fourth
of what is experienced on the Khasia hills in Eastern Bengal, where
fifty feet of rain falls. The greater proportion descends between June
and September, as much as thirty inches sometimes falling in one
month. From November to February inclusive, the months are
comparatively dry; March and October are characterised by violent
storms at the equinoxes, with thunder, destructive lightning, and
hail.

The rain-gauge takes no account of the enormous deposition from
mists and fogs: these keep the atmosphere in a state of moisture,
the amount of which I have estimated at 0·88 as the saturation-point
at Dorjiling, 0·83 being that of London. In July, the dampest
month, the saturation-point is 0·97; and in December, owing to the
dryness of the air on the neighbouring plains of India, whence
dry blasts pass over Sikkim, the mean saturation-point of the month
sometimes falls as low as 0·69.

The dew-point is on the average of the year 49° 3, or 3° below the
mean temperature of the air. In the dampest month (July) the
mean dew-point is only eight-tenths of a degree below the temperature, whilst in December it sinks 10° below it. In London the

---

* The general ideas on the subject of the English rain-fall are so very vague,
that I may be pardoned for reminding my readers that in 1852, the year of extraordinary rain, the amounts varied from 28·5 inches in Essex, to 50 inches at
Cirencester, and 67·5 (average of five years) at Plympton St. Mary's, and 102·5 at
Holme, on the Dart.

dew-point is on the average 5° 6 below the temperature ; none of the English months are so wet as those of Sikkim, but none are so dry as the Sikkim December sometimes is.

On the weight of the atmosphere in Sikkim; and its effects on the human frame.

Of all the phenomena of climate, the weight of the atmosphere is the most remarkable for its elusion of direct observation, when unaided by instruments. At the level of the sea, a man of ordinary bulk and stature is pressed upon by a superincumbent weight of 30,000 pounds or 13½ tons. An inch fall or rise in the barometer shows that this load is lightened or increased, sometimes in a few hours, by nearly 1,000 pounds ; and no notice is taken of it, except by the meteorologist, or by the speculative physician, seeking the subtle causes of epidemic and endemic complaints. At Dorjiling (7,400 feet), this load is reduced to less than 22,500 pounds, with no appreciable result whatever on the frame, however suddenly it be transported to that elevation. And the observation of my own habits convinced me that I took the same amount of meat, drink, sleep, exercise and work, not only without inconvenience, but without the slightest perception of my altered circumstances. On ascending to 14,000 feet, owing to the diminished supply of oxygen, exercise brings on vertigo and headache; ascending higher still, lassitude and tension across the forehead ensue, with retching, and a sense of weight dragging down the stomach, probably due to dilatation of the air contained in that organ. Such are the all but invariable effects of high elevations; varying with most persons according to the suddenness and steepness of the ascent, the amount and duration of exertion, and the length of time previously passed at great heights. After having lived for some weeks at 15,300 feet, I have thence ascended several times to 18,500, and once above 19,000 feet, without any sensations but lassitude and quickness of pulse ; * but in these instances it required great caution to avoid painful symptoms. Residing at 15,300 feet, however, my functions were wholly undisturbed ; nor could I detect any quickness of pulse

---

* I have in a note to vol. ii. p. 160, stated that I never experienced in my own person, nor saw in others, bleeding at the ears, nose, lips, or eyelids.

or of respiration when the body was at rest, below 17,000 feet. At that elevation, after resting a party of eight men for an hour, the average of their and my pulses was above 100°, both before and after eating; in one case it was 120°, in none below 80°.

Not only is the frame of a transient visitor unaffected (when at rest) by the pressure being reduced from 30,000 to 13,000 pounds, but the Tibetan, born and constantly residing at upwards of 14,000 feet, differs in no respect that can be attributed to diminished pressure, from the native of the level of the sea. The average duration of life, and the amount of food and exercise is the same; eighty years are rarely reached by either. The Tibetan too, however inured to cold and great elevations, still suffers when he crosses passes 18,000 or 19,000 feet high, and apparently neither more nor less than I did.

Liebig remarks (in his " Animal Chemistry ") that in an equal number of respirations,* we consume a larger amount of oxygen at the level of the sea than on a mountain; and it can be shown that under ordinary circumstances at Dorjiling, 20·14 per cent. less is inhaled than on the plains of India. Yet the chest cannot expand so as to inspire more at once, nor is the respiration appreciably

---

* For the following note I am indebted to my friend, C. Muller, Esq., of Patna:—

According to Sir H. Davy, a man consumes 45,504 cubic inches of oxygen in twenty-four hours, necessitating the inspiration of 147,520 cubic inches of atmospheric air.—At pressure 23 inches, and temp. 60°, this volume of atmospheric air (dry) would weigh 35,138·75 grains.—At pressure 30 in., temp. 80°, it would weigh 43,997·63 gr.

The amount of oxygen in atmospheric air is 23·32 per cent. by weight. The oxygen, then, in 147,520 cubic inches of dry air, at pressure 23 in., temp. 60°, weighs 8,194·35 gr.; and at pressure 30 in., temp. 80°, it weighs 10,260·25 gr.

Hence the absolute quantity of oxygen in a given volume of atmospheric air, when the pressure is 23 in., and the temp. 60°, is 20·14 per cent. less than when the pressure is 30 in. and the temp. 80°.

When the air at pressure 23 in., temp. 60°, is saturated with moisture, the proportion of dry air and aqueous vapour in 100 cubic inches is as follows:—

Dry air   .   .   97·173
Vapour   .   .   2·827

At pressure 30 in., temp. 80°, the proportions are:—

Dry air   .   .   96·133
Vapour   .   .   3·867

The effect of aqueous vapour in the air on the amount of oxygen available for consumption, is very trifling; and it must not be forgotten that aqueous vapour supplies oxygen to the system as well as atmospheric air.

quickened; by either of which means nature would be enabled to make up the deficiency. It is true that it is difficult to count one's own respirations, but the average is considered in a healthy man to be eighteen in a minute; in my own case it is sixteen, an acceleration of which by three or four could not have been overlooked, in the repeated trials I made at Dorjiling, and still less the eight additional inhalations required at 15,000 feet to make up for the deficiency of oxygen in the air of that elevation.

It has long been surmised that an alpine vegetation may owe some of its peculiarities to the diminished atmospheric pressure; and that the latter being a condition which the gardener cannot supply, he can never successfully cultivate such plants in general. I know of no foundation for this hypothesis; many plants, natives of the level of the sea in other parts of the world, and some even of the hot plains of Bengal, ascend to 12,000 and even 15,000 feet on the Himalaya, unaffected by the diminished pressure. Any number of species from low countries may be cultivated, and some have been for ages, at 10,000 to 14,000 feet without change. It is the same with the lower animals; innumerable instances may with ease be adduced of pressure alone inducing no appreciable change, whilst there is absence of proof to the contrary. The phenomena that accompany diminished pressure are the real obstacles to the cultivation of alpine plants, of which cold and the excessive climate are perhaps the most formidable. Plants that grow in localities marked by sudden extremes of heat and cold, are always very variable in stature, habit, and foliage. In a state of nature we say the plants "accommodate themselves" to these changes, and so they do within certain limits; but for one that survives of all the seeds that germinate in these inhospitable localities, thousands die. In our gardens we can neither imitate the conditions of an alpine climate, nor offer others suited to the plants of such climates.

The mean height of the barometer at Mr. Hodgson's was 23·010, but varied 0·161 between July, when it was lowest, and October, when it was highest; following the monthly rise and fall of Calcutta as to period, but not as to amount (or amplitude); for the mercury at Calcutta stands in July upwards of half an inch (0·555 Prinsep) lower than it does in December.

The diurnal tide of atmosphere is as constant as to the time of its ebb and flow at Dorjiling as at Calcutta; and a number of very careful observations (made with special reference to this object) between the level of the plains of India, and 17,000 feet, would indicate that there is no very material deviation from this at any elevation in Sikkim.   These times are very nearly 9·50 A.M. and about 10 P.M. for the maxima, the 9·50 A.M. very constantly, and the 10 P.M. with more uncertainty; and 4 A.M. and 4 P.M. for the minima, the afternoon ebb being most true to its time, except during the rains.

At 9° 50 A.M. the barometer is at its highest, and falls till 4 P.M., when it stands on the average of the year 0·074 of an inch lower; during the same period the Calcutta fall is upwards of one-tenth of an inch (0·121 Prinsep).

It has been proved that at considerable elevations in Europe, the hours of periodic ebb and flow differ materially from those which prevail at the level of the sea; but this is certainly not the case in the Sikkim Himalaya.

The amplitude decreases in amount from 0·100 at the foot of the hills, to 0·074 at 7,000 feet; and the mean of 132 selected unexceptionable observations, taken at nine stations between 8,000 and 15,500 feet, at 9° 50 A.M. and 4 P.M., gives an average fall of 0·056 of an inch; a result which is confirmed by interpolation from numerous horary observations at these and many other elevations, where I could observe at the critical hours.

That the Calcutta amplitude is not exceptionally great, is shewn by the register kept at different places in the Gangetic valley and plains of India, between Saharunpore and the Bay of Bengal.   I have seen apparently trustworthy records of seven * such, and find that in all it amounts to between 0·084 and 0·120 inch, the mean of the whole being 0·101 of an inch.

The amplitude is greatest (0·088) in the spring months (March, April, and May), both at Dorjiling and Calcutta: it is least at both in June and July, (0·027 at Dorjiling), and rises again in autumn (to ·082 in September).

The horary oscillations also are as remarkably uniform at all

* Calcutta, Berampore, Benares, Nagpore, Moozufferpore, Delhi, and Saharunpore.

elevations, as the period of ebb and flow: the mercury falls slowly from 9° 50 A.M. (when it is at its highest) till noon, then rapidly till 3 P.M., and slowly again till 4 P.M.; after which there is little change until sunset; it rises rapidly between 7 and 9 P.M., and a little more till 10 P.M.; thence till 4 A.M. the fall is inconsiderable, and the great rise occurs between 7 and 9 A.M.

It is well known that these fluctuations of the barometer are due to the expansion and contraction by heat and moisture of the column of atmosphere that presses on the mercury in the cistern of the instrument: were the air dry, the effect would be a single rise and fall;* the barometer would stand highest at the hottest of the twenty-four hours, and lowest at the coldest; and such is the case in arid continental regions which are perennially dry. That such would also be the case at Calcutta and throughout the Himalaya of Sikkim, is theoretically self-evident, and proved by my horary observations taken during the rainy months of 1848. An inspection of these at the end of this section (where a column contains the pressure of dry air) shows but one maximum of pressure, which occurs at the coldest time of the twenty-four hours (early in the morning), and one minimum in the afternoon. In the table of mean temperatures of the months, also appended to this section, will also be found a column showing the pressure of dry air, whence it will be seen that there is but one maximum of the pressure of dry air, occurring at the coldest season in December, and one minimum, in July. The effect of the vapour is the same on the annual as upon the diurnal march of the pressure, producing a double maximum and minimum in the year in one case, and in the twenty-four hours in the other.

I append a meteorological register of the separate months, but at the same time must remind the reader that it does not pretend to strict accuracy. It is founded upon observations made at Dorjiling by Dr. Chapman in the year 1837, for pressure temperature and wet-bulb only; the other data and some modifications of the above are supplied from observations of my own. Those for terrestrial and

---

* This law, for which we are indebted to Professor Dove, has been clearly explained by Colonel Sabine in the appendix to his translation of Humboldt's "Cosmos," vol. i. p. 457.

nocturnal radiation are accurate as far as they go, that is to say, they are absolute temperatures taken by myself, which may, I believe, be recorded in any year, but much higher are no doubt often to be obtained. The dew-points and saturations are generally calculated from the mean of two day observations (10 A.M. and 4 P.M.) of the wet-bulb thermometer, together with the minimum, or are taken from observations of Daniell's hygrometer; and as I find the mean of the temperature of 10 A.M., 4 P.M., and the minimum, to coincide within a few tenths with the mean temperature of the whole day, I assume that the mean of the wet-bulb observations of the same hours will give a near approach to that of the twenty-four hours. The climate of Dorjiling station has been in some degree altered by extensive clearances of forest, which render it more variable, more exposed to night frosts and strong sun-heat, and to drought, the drying up of small streams being one direct consequence. My own observations were taken at Mr. Hodgson's house, elevated 7,430 feet, the position of which I have indicated at the commencement of this section, where the differences of climate due to local causes are sufficiently indicated to show that in no two spots could similar meteorological results be obtained. At Mr. Hodgson's, for instance, the uniformity of temperature and humidity is infinitely more remarkable than at Dr. Chapman's, possibly from my guarding more effectually against radiation, and from the greater forests about Mr. Hodgson's house. I have not, however, ventured to interfere with the temperature columns on this account.

DORJILING METEOROLOGICAL REGISTER.

	Pressure of Atmosphere.*	Range of Pressure.	Mean Shade.	Max. Shade.	Max. Sun.	Greatest Diff.	Mean Max. Shade.	Minim. Shade.	Minim. Rad.	Greatest Diff.	Mean Minim. Shade.	Mean Daily Range of Temp.	Sunk Therm.	Mean Dew-Point.	Mean Dryness.	Force of Vapour.	Pressure of Dry Air.	Mean Saturation.	Rain in Inches.
January	23·307	·072	40·0	56·0	119·0	72·0	47·2	29·0	16·0	12·7	32·8	14·4	46·0	34·3	5·1	·216	23·091	·84	1·72
February	·305	·061	42·1	57·0	124·0	78·0	50·0	25·5	23·0	15·3	34·2	15·8	48·0	37·2	3·9	·239	·066	·87	0·92
March	·307	·083	50·7	66·5	120·0	60·0	58·4	37·0	27·8	8·7	43·1	15·3	50·0	45·8	5·8	·323	·084	·82	1·12
April	·280	·085	55·9	68·5	125·0	66·0	63·7	38·0	33·0	16·0	48·1	15·6	58·0	49·8	6·6	·371	22·909	·80	2·52
May	·259	·088	57·6	69·0	125·0	65·0	65·3	38·0	40·0	10·0	50·0	15·3	61·0	54·4	2·7	·434	·825	·91	9·25
June	·207	·067	61·2	71·0	126·2	62·2	66·7	51·5	47·0	4·8	55·8	10·9	62·0	59·5	2·0	·515	·692	·93	26·96
July	·203	·062	61·4	69·5	130·0	62·0	65·5	56·0	52·0	3·5	57·3	8·2	62·2	60·7	0·8	·535	·668	·97	25·34
August	·230	·070	61·7	70·0	133·0	62·0	66·1	54·5	50·0	3·5	57·4	8·7	62·0	60·4	1·1	·530	·700	·96	29·45
September	·300	·082	59·9	70·0	142·0	70·0	64·7	51·5	47·5	10·0	55·2	9·5	61·0	58·5	1·4	·498	·802	·95	15·76
October	·372	·075	58·0	68·0	133·0	65·0	66·5	43·5	32·0	12·0	49·5	17·0	60·0	52·5	4·2	·407	·865	·86	8·66
November	·330	·078	50·0	63·0	123·0	68·0	56·5	38·0	30·0	12·0	43·5	13·0	55·0	46·5	3·2	·331	·999	·90	0·11
December	·365	·062	43·0	56·0	108·0	77·2	51·6	32·5	26·0	10·0	34·9	16·7	49·0	31·8	10·6	·198	23·165	·69	0·45
Mean	23·289	·074	53·5	65·4	125·7	67·3	60·2	41·3	35·4	9·9	46·8	13·4	56·2	49·4	4·0	·383	22·906	·88	Sum 122·26

* These are taken from Dr. Chapman's Table; and present a greater annual range (= 0·169) than my observations in 1848-9, taken at Mr. Hodgson's, which is higher than Dr. Chapman's; or than Mr. Muller's, which is a little lower, and very near it.

Horary Observations at Jillapahar, Dorjiling, Alt. 7,430 feet.

JULY, 1848.

No. of Observations	Hour.	Barom. corrected.	Temp. Air.	D. P.	Diff.	Tens. of Vapour.	Weight of Vapour.	Humidity.	Pressure of Dry Air.
7	1 A.M.	22·877	59·6	58·9	0·7	·504	5·65	·988	22·373
23	8	·882	62·1	60·6	1·5	·534	6·03	·950	·348
27	9	·884	62·6	61·3	1·3	·546	6·10	·960	·338
22	10	+ ·899	63·5	61·7	1·8	·554	6·12	·945	·345
20	11	·899	64·1	62·3	1·8	·565	6·27	·945	·334
26	Noon.	·884	65·0	63·1	1·9	·580	6·44	·940	·304
12	1 P.M.	·876	64·1	61·7	2·4	·566	6·13	·923	·310
11	2	·866	64·4	61·0	3·4	·541	6·00	·892	·325
25	3	·852	64·8	62·6	2·2	·571	6·32	·930	·281
23	4	·846	64·1	61·7	2·4	·554	6·13	·924	·292
13	5	— ·840	64·7	64·0	0·7	·597	6·62	·978	— ·243
10	6	·845	63·7	61·5	2·2	·549	6·12	·928	·296
6	7	·853	62·7	61·1	1·6	·542	6·03	·948	·311
6	8	·867	61·0	59·5	1·5	·515	5·74	·952	·352
22	9	·878	60·7	59·4	1·3	·512	5·72	·960	·366
6	10	·885	60·5	59·5	1·0	·514	5·75	·968	·371
6	11	+ ·887	60·2	59·2	1·0	·508	5·70	·965	·379
19	Midnt.	·887	59·8	59·1	0·7	·507	5·68	·975	+ ·382

AUGUST.

No. of Observations.	Hour.	Barom. corrected.	Temp. Air.	D. P.	Diff.	Tens. of Vapour.	Weight of Vapour.	Humidity.	Pressure of Dry Air.
15	1 A.M.	22·909	59·8	59·5	0·3	·514	5·70	·992	+ 22·395
26	8	·904	62·1	61·5	0·6	·549	6·13	·980	·355
28	9	·915	63·1	61·9	1·2	·558	6·20	·962	·357
28	10	+ ·917	64·3	62·7	1·6	·572	6·35	·950	·345
24	11	·915	64·7	63·1	1·6	·580	6·42	·948	·335
23	Noon.	·905	64·7	63·4	1·3	·586	6·50	·958	·319
21	1 P.M.	·898	65·3	63·3	2·0	·584	6·48	·940	·314
21	2	·884	65·0	63·4	1·6	·586	6·50	·950	·298
21	3	·873	64·8	63·1	1·7	·579	6·43	·943	·294
19	4	·855	63·9	62·4	1·5	·568	6·30	·952	— ·287
19	5	— ·853	63·2	61·7	1·5	·554	6·15	·952	·299
19	6	·863	62·3	60·8	1·5	·538	6·00	·952	·325
19	7	·865	61·6	60·4	1·2	·531	5·92	·962	·334
19	8	·878	61·1	60·2	0·9	·527	5·88	·970	·351
19	9	·890	60·7	60·0	0·7	·523	5·85	·976	·367
19	10	+ ·893	60·3	59·7	0·6	·518	5·78	·980	·375
19	11	·892	60·1	59·7	0·4	·517	5·79	·988	·375
19	Midnt.	·889	60·0	59·4	0·6	·513	5·73	·980	·376

SEPTEMBER.

No. of Obser- vations.	Hour.	Barom. corrected.	Temp. Air.	D. P	Diff.	Tens. of Vapour.	Weight of Vapour.	Humi- dity.	Pressure of Dry Air.
28	8 A.M.	23·000	59·2	58·1	1·1	·492	5·50	·968	·22·508
29	9	·013	60·1	58·5	1·6	·497	5·57	·945	·516
28	10	+ ·018	60·8	59·5	1·3	·514	5·77	·958	·504
24	11	·009	61·6	60·0	1·6	·523	5·83	·950	·506
23	Noon.	22·995	62·4	60·5	1·9	·533	5·93	·942	·462
23	1 P.M.	·980	62·7	60·5	2·2	·532	5·92	·942	·448
23	2	·962	62·8	60·4	2·4	·531	5 90	·925	·431
23	3	·947	62·3	60·0	2·3	·522	5·83	·924	·425
23	4	—·944	61·8	59·9	1·9	·521	5·82	·940	—·423
19	5	·944	60·3	58·6	1·7	·498	5·58	·940	·446
19	6	·948	59·4	58·4	1·0	·496	5·58	·968	·452
20	7	·958	58·7	57·4	1·3	·479	5·60	·960	·479
21	8	·975	58·2	57·0	1·2	·473	5·33	·962	·502
22	9	·986	57·8	56·6	1·2	·467	5·25	·960	·519
24	10	+ ·991	57·4	56·4	1·0	·463	5·23	·968	·528
24	11	·989	57·0	55·9	1·1	·456	5·15	·962	·533
23	Midnt.	·994	56·7	55·4	1·3	·449	5·07	·927	+ ·545

OCTOBER (22 days).

No. of Obser- vations.	Hour.	Barom. corrected.	Temp. Air.	D. P.	Diff.	Tens. of Vapour.	Weight of Vapour.	Humi- dity.	Pressure of Dry Air.
11	6–6½	23 066	54·4	52·7	1·7	·409	4·65	·943	22·657
19	7 A.M.	·072	54·3	52·3	2·0	·403	4·58	·925	+ ·669
20	8	·086	55·2	53·7	1·5	·423	4·78	·950	·663
20	9	·099	56·3	54·4	1·9	·434	4·90	·935	·665
19	10	+ ·100	57·1	55·5	1·6	·450	5·07	·942	·650
13	11	·079	57·6	55·6	2·0	·451	5·08	·935	·628
15	Noon.	·072	57·9	56·1	1·8	·459	5·15	·940	·613
13	1 P.M.	·055	58·0	56·4	1·6	·463	5·17	·950	·592
13	2	·033	57·7	56·6	1·1	·466	5 25	·962	·567
14	3	·027	57·9	56·2	1·7	·460	5·16	·940	·567
16	4	·024	57·9	56·1	1·8	·458	5·15	·940	— ·566
13	5	—·022	56·6	54·8	1·8	·439	4·98	·948	·583
6	6	·033	55·9	54·4	1·5	·433	4·90	·950	·600
7	7	·045	55·4	53·8	1·6	·424	4·80	·950	·621
3	8	·038	53·7	53·3	0·4	·417	4·75	·990	·621
7	9	·061	55·1	54·1	1·0	·429	4·83	·965	·632
14	10	+ ·072	54·6	53·0	1·6	·413	4·82	·949	·659
18	11	·067	54·5	53·0	1·5	·413	4·82	·950	·654
14	Midn.	·068	54·1	52·8	1·3	·411	4·65	·962	·657

G.

ON THE RELATIVE HUMIDITY, AND ABSOLUTE AMOUNT OF VAPOUR CONTAINED IN THE ATMOSPHERE AT DIFFERENT ELEVATIONS IN THE SIKKIM HIMALAYA.

My observations for temperature and wet-bulb being for the most part desultory, taken at different dates, and under very different conditions of exposure, &c., it is obvious that those at one station are hardly, if at all, comparative with those of another, and I have therefore selected only such as were taken at the same date and hour with others taken at the Calcutta Observatory, or as can easily be reduced; which thus afford a standard (however defective in many respects) for a comparison. I need hardly remind my reader that the vapour-charged wind of Sikkim is the southerly one, which blows over Calcutta; that in its passage northwards to Sikkim in the summer months, it traverses the heated plains at the foot of the Himalaya, and ascending that range, it discharges the greater part of its moisture (120 to 140 inches annually) over the outer Himalayan ranges, at elevations of 4000 to 8000 feet. The cooling effect of the uniform covering of forest on the Sikkim ranges is particularly favourable to this deposition, but the slope of the mountains being gradual, the ascending currents are not arrested and cooled so suddenly as in the Khasia mountains, where the discharge is consequently much greater. The heating of the atmosphere, too, over the dry plains at the foot of the outer range, increases farther its capacity for the retention of vapour, and also tends to render the rain-fall less sudden and violent than on the Khasia, where the south wind blows over the cool expanse of the Jheels. It will be seen from the following observations, that in Sikkim the relative humidity of the atmosphere remains pretty constantly very high in the summer months, and at all elevations, except in the rearward valleys; and even there a humid atmosphere prevails up to 14,000 feet, everywhere within the influence of the snowy mountains. The uniformly high temperature which prevails throughout the summer, even at elevations of 17,000 and 18,000 feet, is no doubt proximately due to the evolution of heat during the condensation of these vapours. It will be seen by the pages of my journal, that continued sunshine, and the consequent heating of the soil, is almost unknown during the summer, at any elevation on the outer or southward ranges of Dorjiling: but the sunk thermometer proves that in advancing northward into the heart of the mountains and ascending, the sun's effect is increased, the temperature of the earth becoming in summer considerably higher than that of the air.   With regard to the observations themselves, they may be depended upon as comparable with those of Calcutta, the instruments having been carefully compared, and the cases of interpolation being few. The number of observations taken at each station is recorded in a separate column; where only one is thus recorded, it is not to be regarded as a single reading, but the mean of several taken during an hour or longer period.   I have rejected all solitary observations, even when accompanied by others at Calcutta; and sundry that were, for obvious reasons, likely to mislead.   Where many observations were taken at one place, I have divided them into sets, corresponding to the hours at which alone the Calcutta temperature and wet-bulb thermometer are recorded,* in order that meteorologists may apply them to the solution of other questions relating to the distribution of heat and moisture. The Dorjiling observations, and those in the immediate neighbourhood of that station, appeared to me sufficiently numerous to render it worth while classing them in months; and keeping them in a series by themselves. The tensions of vapour are worked from the wet-bulb readings by Apjohn's formula and tables, corrected for the height of the barometer at the time. The observations, except where otherwise noted, are taken by myself.

* Sunrise; 9·50 A.M.; noon; 2·40 P.M.; 4 P.M., and sunset.

SERIES I. *Observations made at or near Dorjiling.*

JANUARY, 1849.

No. of Obs.	Place.	Elev.	Hour.	DORJILING.				CALCUTTA.			
				Tp.	D. P.	Diff.	Tens.	Tp.	D. P.	Diff.	Tens.
15	The Dale,*	6956 ft.	9·50 A.M.	42·9	32·4	10·5	·202	67·5	55·3	12·2	·446
15	Mr. Muller's.	...	Noon.	45·8	33·8	12·0	·212	72·9	55·7	17·2	·455
10	...	...	2·40 P.M.	48·3	37·4	10·9	·241	76·1	55·1	21·0	·444
8	...	...	4 P.M.	48·6	37·8	10·8	·244	75·1	54·8	20·3	·440
9	...	...	Sunset.	46·5	37·1	9·4	·238	71·8	54·9	16·9	·441
57	...	...	Mean	46·4	35·7	10·7	·227	72·7	55·2	17·5	·445

Dorjiling.—Humidity. . . . . . 0·700   Calcutta. . 0·562
    „   Vapour in cubic foot of
            atmosphere . . . 2·63 gr.   „   4·86 gr.

JANUARY, 1850.

No. of Obs.	Place.	Elev.	Hour.	DORJILING.				CALCUTTA.			
				Tp.	D.P.	Diff.	Tens.	Tp.	D. P.	Diff.	Tens.
3	Jillapahar,	7430 ft.	Sunrise	32·8	30·1	2·7	·186	51·5	48·5	3·0	·354
6	Mr.	...	9·50 A.M.	39·5	34·7	4·8	·219	66·9	55·1	11·8	·444
3	Hodgson's.	...	Noon	42·4	38·0	4·4	·246	74·1	51·7	22·4	·395
5	...	...	2·40 P.M.	41·9	37·8	4·1	·244	78·3	51·4	26·9	·391
5	...	...	4 P.M.	41·1	38·5	2·6	·250	77·4	59·5	17·9	·514
5	...	...	Sunset	38·7	35·6	3·1	·226	72·4	54·7	17·7	·438
13	...	...	Miscel.	41·9	39·9	2·0	·263	77·9	60·1	17·8	·525
4	Saddle of road at Sinchul.	7412 ft.	Do.	41·1	36·4	4·7	·233	67·7	57·2	10·5	·476
1	Pacheem.	7258 ft.	Do.	39·8	38·7	1·1	·252	71·6	50·5	21·1	·379
45	...	...	Mean	39·9	36·6	3·3	·235	70·9	54·3	16·6	·435

Dorjiling.—Humidity. . . . . . 0·890   Calcutta. . 0·580
    „   Weight of vapour . . 2·75 gr.   „   4·86 gr.

---

* Observations to which the asterisk is affixed were taken by Mr. Muller.

FEBRUARY.

No. of Obs.	Place.	Elev.	Hour.	DORJILING.				CALCUTTA.			
				Tp.	D. P.	Diff.	Tens.	Tp.	D. P.	Diff.	Tens.
6	Jillapahar,	7430 ft.	Sunrise.	36·9	34·7	2·2	·219	60·0	54·2	5·8	·431
18	1850	...	9·50	42·9	38·6	4·3	·251	72·8	58·8	14·0	·503
12	...	...	Noon.	44·8	41·3	3·5	·276	79·8	58·7	21·1	·501
12	...	...	2·40	44·8	37·4	7·4	·241	82·4	57·9	24·5	·487
17	...	...	4 P.M.	44·0	35·6	8·4	·226	81·1	58·1	23·0	·492
19	...	...	Sunset.	42·4	35·8	6·6	·228	76·3	60·7	15·6	·536
13	The Dale.*	6956	Misc.	40·8	35·1	5·7	·222	69·9	59·8	10·1	·518
97	...	...	Mean	42·4	36·9	5·4	·238	74·6	58·3	16·3	·495

Dorjiling.—Humidity. . . . . . 0·828   Calcutta. 0·590
  „   Weight of vapour. . . 2·75 gr.   „   5·40 gr.

MARCH.

No of Obs.	Place.	Elev.	Hour.	DORJILING.				CALCUTTA.			
				Tp.	D. P.	Diff.	Tens.	Tp.	D. P.	Diff.	Tens.
10	Jillapahar,	7430 ft.	9·50 A.M.	44·2	42·7	1·5	·290	81·6	64·1	17·5	·602
8	1850	...	Noon.	45·5	43·0	2·5	·293	88·2	57·0	31·2	·472
5	...	...	2·40 P.M.	46·4	44·0	2·4	·303	91·3	53·2	38·1	·416
8	...	...	4 P.M.	45·5	43·4	2·1	·297	90·1	52·0	38·1	·399
6	...	...	Sunset.	43·1	41·5	1·6	·278	82·9	63·7	19·2	·590
3	Pacheem.	7258	Misc.	44·8	44·6	0·2	·310	85·0	74·8	10·2	·848
40	...	...	Mean	44·9	43·2	1·7	·295	86·5	60·8	25·7	·555

Dorjiling.—Humidity. . . . . . 0·940   Calcutta. 0·438
  „   Weight of vapour.   3·42 gr.   „   5·72 gr.

APRIL.

No. of Obs.	Place.	Elev.	Hour.	DORJILING.				CALCUTTA.			
				Tp.	D. P.	Diff.	Tens.	Tp.	D. P.	Diff.	Tens.
3	Jillapahar,	7430 ft.	9·50 A.M.	57·0	40·2	16·8	·266	90·3	71·3	19·0	·758
3	1849.	...	Noon.	59·8	44·1	15·7	·305	97·0	64·5	32·5	·607
1	...	...	2·40 P.M.	60·2	44·4	15·8	·308	97·7	73·4	24·3	·812
7	Dr. Camp-	6932 ft.	9·50 A.M.	61·8	53·3	8·5	·417	86·7	66·3	20·4	·644
2	bell's, 1850.	...	Noon.	65·4	52·8	12·6	·411	91·3	68·8	22·5	·699
4	...	...	4 P.M.	57·5	53·7	3·8	·423	88·6	72·1	16·5	·778
3	...	...	Sunset.	56·9	51·4	5·5	·392	82·8	73·0	9·8	·800
23	...	...	Mean	59·8	48·6	11·3	·360	90·6	69·9	20·7	·728

Dorjiling.—Humidity. . . . . . 0·684   Calcutta. 0·523
   „   Weight of vapour . . . 3·98 gr.   „   7·65 gr.

MAY.

No. of Obs.	Place.	Elev.	Hour.	DORJILING.				CALCUTTA.			
				Tp.	D. P.	Diff.	Tens.	Tp.	D. P.	Diff.	Tens.
3	Smith's Hotel, 1848.	6863ft.	Miscell.	57·2	55·0	2·2	·443	88·6	78·4	10·2	·951
45	Colinton,* 1849.	7179ft.	Miscell.	60·4	57·9	2·5	·466	90·0	77·2	12·8	·917
48			Mean	58·8	56·5	2·4	·455	89·3	77·8	11·5	·934

Dorjiling.—Humidity . . 0·926   Calcutta. . 0·698
   „   Weight of Vapour 5·22 gr.   „   . 9·90 gr.

JUNE.

No. of Obs.	Place.	Elev.	Hour.	DORJILING.				CALCUTTA.			
				Tp.	D. P.	Diff.	Tens.	Tp.	D. P.	Diff.	Tens.
40	Colinton.*	7179 ft.	Miscell.	60·9	57·6	3·3	·483	85·5	78·4	7·1	·952

Dorjiling :—Humidity . 0·895   Calcutta. 0·800
   „   Weight of Vapour 5·39 gr.   „   . 10·17 gr.

JULY.

No. of Obs.	Place.	Elev	Hour.	DORJILING. Tp.	D. P.	Diff.	Tens.	CALCUTTA. Tp.	D.P.	Diff.	Tens.
18	Jillapahar,	7430 ft.	9·50 A.M.	63·2	61·4	1·8	·548	87·0	79·4	7·6	·983
25	1848.	...	Noon.	65·0	62·6	2·4	·570	89·0	80·0	9·0	1·001
24	...	...	2·40 P.M.	64·7	62·3	2·4	·565	88·1	79·4	8·7	·983
16	...	...	4	63·8	61·5	2·3	·550	87·2	79·5	7·7	·985
31	The Dale,*	6952 ft.	6 A.M.	60·2	58·7	1·5	·537	81·3	79·0	2·3	·969
31	1848.	...	2 P.M.	66·3	63·3	3·0	·621	88·0	79·6	8·4	·989
31	...	...	6	63·0	60·9	2·1	·575	84·8	79·2	5·6	·977
176	...	...	Mean	63·7	61·5	2·2	·567	86·5	79·4	7·0	·984

Dorjiling.—Humidity   . 0·929   Calcutta .   . 0·800
    „   Weight of Vapour . 6·06 gr.   „   10·45 gr

AUGUST

No. of Obs.	Place.	Elev.	Hour.	DORJILING. Tp.	D. P.	Diff.	Tens.	CALCUTTA. Tp.	D. P.	Diff.	Tens.
23	Jillapahar,	7430 ft.	9·50 A.M.	64·2	62·4	1·8	·567	85·8	79·1	6·7	·973
21	1848	...	Noon.	64·7	63·3	1·4	·584	87·2	79·2	8·0	·976
17	...	...	2·40 P.M.	64·7	62·8	1·9	·574	87·4	79·3	8·1	·979
13	...	...	4	63·9	62·5	1·4	·568	86·5	79·5	7·0	·984
31	The Dale,*	6952 ft.	6 A.M.	60·5	59·5	1·0	·551	80·8	78·8	2·0	·962
31	1848.	...	2 P.M.	65·3	63·6	1·7	·628	87·2	79·2	8·0	·976
31	...	...	6	62·8	61·8	1·0	·591	83·7	78·7	5·0	·959
167			Mean	63·7	62·3	1·5	·580	85·5	79·1	6·4	·973

Dorjiling.—Humidity   . 0·955   Calcutta . 0·818
    „·   Weight of Vapour . 6·25 gr.   „   10·35 gr.

SEPTEMBER.

No. of Obs.	Place.	Elev.	Hour.	DORJILING. Tp.	D. P.	Diff.	Tens.	CALCUTTA. Tp.	D. P.	Diff.	Tens.
28	Jillapahar,	7430 ft.	9·50 A.M.	60·8	59·3	1·5	·511	87·0	78·4	8·6	·952
23	1848.	...	Noon.	62·4	60·3	2·1	·528	88·5	78·1	10·4	·943
23	...	...	2·40 P.M.	62·4	59·6	2·8	·516	88·1	77·4	10·7	·922
21	...	...	4	62·0	59·6	2·4	·516	86·9	77·1	9·8	·914
30	The Dale,*	6952 ft.	6 A.M.	57·4	56·2	1·2	·495	80·9	78·3	2·6	·948
30	1848.	...	2 P.M.	64·9	60·8	4·1	·573	88·8	77·4	11·4	·923
30	...	...	6	60·8	59·0	1·8	·543	84·7	76·6	8·1	·899
185			Mean	61·5	59·3	2·3	·526	86·4	77·6	8·8	·929

Dorjiling.—Humidity   . 0·932   Calcutta   . 0·760
    „   Weight of Vapour . 5·72 gr.   „   . 9·88 gr.

OCTOBER.

No. of Obs.	Place.	Elev.	Hour.	Tp.	D. P.	Diff.	Tens.	Tp.	D. P.	Diff.	Tens.
		DORJILING.						CALCUTTA.			
6	Jillapahar,	7430 ft.	Noon.	55·9	55·3	0·6	·446	84·4	75·3	9·1	·863
6	1848.	...	2·40 P.M.	55·7	54·9	0·8	·440	86·0	73·3	12·7	·808
6	...	...	4 P.M.	55·6	54·9	0·7	·441	85·2	74·4	10·8	·837
4	Goong.	7436 ft.	Misc.	48·3	48·3	0·0	·352	81·2	73·7	7·5	·819
8	ditto.	7441 ft.	ditto.	51·2	50·2	1·0	·376	80·7	66·9	13·8	·657
8	The Dale.*	6952 ft.	6 A.M.	55·2	52·7	2·5	·439	76·1	74·2	1·9	·834
17	...	...	2 P.M.	61·4	56·3	5·1	·497	87·0	71·2	15·8	·756
19	...	...	6 P.M.	56·9	54·2	2·7	·463	82·8	73·9	8·9	·824
74	...	...	Mean	55·0	53·4	1·7	·432	82·9	72·9	10·1	·800

Dorjiling.—Humidity. . . . . . 0·950   Calcutta. . 0·658
,,   Weight of vapour . . . 4·74 gr.   ,,   . 8·55 gr.

NOVEMBER AND DECEMBER.

No. of Obs.	Place.	Elev.	Hour.	Tp.	D. P.	Diff.	Tens.	Tp.	D. P.	Diff.	Tens.
		DORJILING.						CALCUTTA.			
4	The Dale,*	6952 ft.	6 A.M.	45·6	41·4	4·2	·277	67·9	64·7	3·2	·610
8	Nov. & Dec.	...	2 P.M	60·0	48·3	11·7	·355	83·3	65·2	18·1	·621
6	1848.	...	6	50·6	44·7	5·9	·311	77·3	63·1	14·2	·579
9	December,	...	2	49·7	41·7	8·0	·280	79·3	59·0	20·3	·505
19	1848.	...	6	44·0	40·5	3·5	·269	75·8	62·6	13·2	·569
46	...	...	Mean	49·9	43·3	6·7	·298	76·7	62·9	13·8	·577

Dorjiling.—Humidity. . . . . . 0·798   Calcutta. . 0·640
,,   Weight of vapour . . . 3·40 gr.   ,,   . 6·27 gr.

Comparison of Dorjiling and Calcutta.

No. of Obs.	Month.	HUMIDITY.			WEIGHT OF VAPOUR IN CUBIC FOOT OF AIR.		
		Dorjiling.	Calcutta.	Diff. Dorjiling.	Dorjiling.	Calcutta.	Diff. Calcutta.
102	January . .	−·795	·571	+ ·224	− 2·68	− 4·80	+ 2·12
97	February . .	·828	·590	+ ·238	2·75	5·40	+ 2·65
40	March . . .	·940	−·438	+ ·502	3·42	5·72	+ 2·30
23	April . . .	·684	·523	+ ·161	3·98	7·65	+ 3·67
48	May . . .	·926	·698	+ ·228	5·22	9·90	+ 4·62
40	June . . .	·895	·800	+ ·095	5·39	10·17	+ 4·78
176	July . . .	·929	·800	+ ·129	6·06	10·05	+ 3·99
167	August . .	+ ·955	+ ·818	+ ·136	+ 6·25	+10·35	+ 4·10
185	September .	·932	·760	+ ·172	5·72	9·88	+ 4·16
74	October . .	·950	·658	+ ·292	4·74	8·55	+ 3·81
46	Nov. and Dec.	·798	·640	+ ·158	3·40	6·27	+ 2·87
998	Mean	0·876	0·663	+ ·212	4·51	8·07	+ 3·55

It is hence evident, from nearly 1,000 comparative observations, that the atmosphere is relatively more humid at Dorjiling than at Calcutta, throughout the year. As the southerly current, to which alone is due all the moisture of Sikkim, traverses 200 miles of land, and discharges from sixty to eighty inches of rain before arriving at Dorjiling, it follows that the whole atmospheric column is relatively drier over the Himalaya than over Calcutta; that the absolute amount of vapour, in short, is less than it would otherwise be at the elevation of Dorjiling, though the relative humidity is so great. A glance at the table at the end of this section appears to confirm this; for it is there shown that, at the base of the Himalaya, at an elevation of only 250 feet higher than Calcutta, the absolute amount of vapour is less, and of relative humidity greater, than at Calcutta.

SERIES II.—*Observations at various Stations and Elevations in the Himalaya of East Nepal and Sikkim.*

ELEVATION 735 TO 2000 FEET.

No. of Obs.	Locality.	Elev.	Month.	EAST NEPAL AND SIKKIM.				CALCUTTA.			
				Tem.	D. P.	Diff.	Tens.	Tem.	D. P.	Diff.	Tens.
3	Katong Ghat. Teesta river.	735	Dec.	60·2	55·3	4·9	·447	73·2	56·7	16·5	·468
2	Great Rungeet, at bridge .	818	April	82·8	63·5	19·3	·588	95·8	61·9	33·9	·557
1	Ditto	„	May	77·8	60·3	17·5	·528	91·7	78·3	13·4	·947
3	Tambur river, E. Nepal. .	1388	Nov.	60·6	57·0	3·6	·473	73·3	62·7	10·6	·571
1	Ditto	1457	Nov:	64·2	59·1	5·1	·507	77·3	63·4	13·9	·585
6	Bhomsong, Teesta river	1596	Dec.	58·6	52·0	6·6	·399	71·6	57·0	14·6	·474
1	Ditto	„	May	68·2	66·4	1·8	·647	82·6	77·4	5·2	·923
5	Little Rungeet . .	1672	Jan.	51·0	50·2	0·8	·377	58·5	58·0	0·5	·489
5	Pemiongchi, Great Rungeet.	1840	Dec.	54·6	53·7	0·9	·424	73·5	66·2	7·3	·642
11	Punkabaree . . .	1850	March	70·1	55·6	14·5	·472	79·2	62·6	16·6	·570
	Ditto	„	May	73·5	68·3	5·2	·687	83·7	77·9	5·8	·938
10	Guard house (Gt. Rungeet)	1864	April	73·7	63·8	9·9	·592	92·4	67·0	25·4	·660
48			Mean	66·3	58·8	7·5	·512	79·4	65·8	13·6	·652

Humidity . ·717   Calcutta . ·663
Weight of Vapour . . 5·57 gr.   . 6·88 gr.

ELEVATION 2000 TO 3000 FEET.

No. of Obs.	Locality.	Elev.	Month.	EAST NEPAL AND SIKKIM.				CALCUTTA.			
				Tem.	D. P.	Diff.	Tens.	Tem.	D. P.	Diff.	Tens.
2	Singdong . . .	2116	Dec.	60·5	53·4	7·1	·419	72·1	52·9	19·2	·411
8	Mywa Guola, E. Nepal .	2132	Nov.	66·2	57·5	8·7	·481	75·7	68·7	7·0	·697
3	Pemmi river, ditto	2256	Nov.	55·6	53·9	1·7	·426	62·9	62·3	0·6	·566
3	Tambur river, ditto .	2545	Nov.	57·3	51·6	5·7	·394	75·0	63·7	11·3	·591
2	Blingbong (Teesta) .	2684	May	72·6	64·0	8·6	·597	81·7	73·6	8·1	·817
8	Lingo ditto . .	2782	May	75·8	67·3	8·5	·666	90·7	77·7	13·0	·932
12	Serriomsa ditto .	2820	Dec.	64·1	56·8	7·3	·469	70·8	62·4	8·4	·567
8	Lingmo ditto . .	2849	May	68·6	64·6	4·0	·610	87·9	74·9	13·0	·851
3	Ditto ditto . .	2952	Dec.	56·4	53·5	2·9	·420	69·5	66·5	3·0	·647
49			Mean	64·1	58·1	6·1	·498	76·3	67·0	9·3	·675

Humidity . . ·820   Calcutta . 740
Weight of Vapour . . 5·45 gr.   7·13 gr.

ELEVATION 3000 TO 4000 FEET.

No. of Obs.	EAST NEPAL AND SIKKIM.							CALCUTTA.			
	Locality.	Elev.	Month.	Tem.	D. P.	Diff.	Tens.	Tem.	D. P.	Diff.	Tens.
5	Kulhait river . . .	3159	Jan.	49·8	47·0	2·8	·337	65·8	57·3	8·5	·477
9	Ratong river	3171	Jan.	44·2	43·0	1·2	·294	69·9	56·6	13·3	·466
3	Tambur river . . .	3201	Nov.	53·0	50·0	3·0	·373	72·9	63·2	9·7	·582
2	Chingtam	3404	Nov.	54·8	49·0	5·8	·360	74·9	73·0	1·9	·802
2	Tikbotang	3763	Dec.	56·5	53·4	3·1	·419	68·0	61·8	6·2	·555
7	Myong Valley. . .	3782	Oct.	61·4	58·4	3·0	·496	80·7	71·2	9·5	·755
7	Iwa river . . .	3783	Dec.	47·5	45·6	1·9	·321	73·3	64·7	8·6	·611
1	Ratong river	3790	Jan.	56·2	41·1	15·1	·275	75·8	53·0	22·8	·414
3	Tukcham	3849	Nov.	68·8	65·4	3·4	·625	83·7	76·8	6·9	·904
1	Pacheem village . .	3855	Jan.	54·5	46·3	8·2	·329	73·6	59·4	14·2	·513
1	Yankoong	3867	Dec.	50·0	43·6	6·4	·299	69·1	63·8	5·3	·593
2	Mikk	3912	May	66·1	63·9	2·2	·595	84·3	75·1	9·2	·856
5	Sunnook . . .	3986	Dec.	47·9	45·5	2·4	·320	69·4	61·1	8·3	·542
48			Mean	54·7	50·2	4·5	·388	74·0	64·4	9·6	·621

Humidity   ·858   Calcutta   ·732
Weight of Vapour   . 4·23 gr   ,,   6·60 gr.

ELEVATION 4000 TO 5000 FEET.

No. of Obs.	EAST NEPAL AND SIKKIM.							CALCUTTA.			
	Locality.	Elev.	Month.	Tem.	D. P.	Diff.	Tens.	Tem.	D. P.	Diff.	Tens.
3	Yangyading . . .	4111	Dec.	52·0	43·6	8·4	·300	71·1	67·2	3·9	·663
4	Gorh	4128	May	66·4	59·0	7·4	·506	85·5	74·2	11·3	·834
2	Namgah . . .	4229	Oct.	57·2	54·1	3·1	·429	80·8	73·7	7·1	·819
3	Taptiatok (Tambur) .	4283	Nov.	51·3	45·8	5·5	·323	73·3	64·8	8·5	·614
7	Myong Valley . .	4345	Oct.	59·1	57·8	1·3	·487	81·7	72·9	8·8	·797
3	Jummanoo . . .	4362	Nov.	60·4	50·0	10·4	·374	77·4	70·2	7·2	·731
6	Nampok . . .	4377	Dec.	49·6	49·1	0·5	·362	64·1	56·3	7·8	·462
7	Chakoong. . . .	4407	May	57·8	57·6	0·2	·483	83·9	76·2	7·7	·889
10	Singtam . . .	4426	May	62·4	61·7	0·7	·553	88·6	79·0	9·6	·969
5	Namten . . .	4483	Dec.	44·7	44·3	0·4	·307	64·8	58·3	6·5	·495
5	Purmiokshong . .	4521	Nov.	60·5	56·5	4·0	·466	79·2	69·5	9·7	·715
2	Rungniok . . .	4565	Jan.	54·7	44·3	10·4	·307	66·5	59·7	6·8	·517
16	Singtam . . .	4575	O.&N.	63·8	60·1	3·7	·525	82·5	76·7	5·8	·901
6	Cheadam . . .	4653	Dec.	51·4	46·6	4·8	·332	70·2	55·0	15·2	·442
4	Sablakoo . . .	4676	Dec.	50·1	44·9	5·2	·314	72·9	65·7	7·2	·632
4	Bheti	4683	Nov.	59·0	52·3	6·7	·405	78·3	66·1	12·2	·639
2	Temi	4771	May	59·8	50·1	9·7	·374	81·2	74·1	7·1	·834
4	Lingtam. . . .	4805	May	60·4	56·6	3·8	·467	80·0	73·8	6·2	·820
7	Khersiong . . .	4813	Jan.	51·0	45·2	5·8	·316	67·0	49·8	17·2	·370
6	Ditto	,,	March	53·6	45·5	8·1	·320	77·1	70·5	6·6	·738
3	Tassiding . . .	4840	Dec.	52·0	46·6	5·4	·333	79·7	60·8	18·9	·538
6	Lingcham . . .	4870	Dec.	48·5	46·1	2·4	·327	78·5	71·8	6·7	·771
11	Dikkeeling . . .	4952	Dec.	62·0	55·3	6·7	·447	80·8	62·0	18·8	·559
9	Tchonpong . . .	4978	Jan.	49·4	34·7	14·7	·219	71·0	54·7	16·3	·439
137			Mean	55·7	50·4	5·4	·387	76·5	66·8	9·7	·675

Humidity   ·837   Calcutta .   ·730
Weight of Vapour   . . 4·33 gr.   . 7·12 gr.

ELEVATION 5000 TO 6000 FEET.

No. of Obs.	Locality.		Elev.	Month.	EAST NEPAL AND SIKKIM.				CALCUTTA.			
					Tem.	D. P.	Diff.	Tens.	Tem.	D. P.	Diff.	Tens.
4	Nampok . . .	.	5075	May	65·8	60·8	5·0	·537	83·1	74·7	8·4	·845
4	Tengling. . . .	.	5257	Jan.	44·7	39·1	5·6	·257	65·4	38·1	27·3	·247
2	Choongtam, sunrise .	.	5368	May	54·9	54·7	· 0·2	·438	78·2	73·9	4·3	·826
7	,, 9·50 A.M. .	.	,,	May	71·5	58·9	12·6	·504	89·8	80·0	9·8	1·000
5	,, noon .	.	,,	May	71·0	59·4	11·6	·513	92·7	79·9	12·8	·999
3	,, 2·45 P.M. .	.	,,	May	66·4	59·4	7·0	·513	95·4	78·7	16·7	·959
4	,, 4 P.M. .	.	,,	May	63·5	59·2	4·3	·510	93·6	79·0	14·6	·971
6	,, sunset .	.	,,	May	61·4	60·5	0·9	·532	89·1	77·1	12·0	·915
8	,, 9·50 A.M. .	.	,,	Aug.	76·3	66·1	10·2	·640	85·3	78·9	6·4	·967
8	,, noon. .	.	,,	Aug.	78·8	67·8	11·0	·677	86·6	78·8	7·8	·965
7	,, 2·40 P.M. .	.	,,	Aug.	72·9	66·5	6·4	·649	86·4	78·8	7·6	·963
6	,, 4 P.M. .	.	,,	Aug.	69·5	66·8	2·7	·655	85·3	79·3	6·0	·980
8	,, sunset .	.	,,	Aug.	66·9	65·4	1·5	·627	83·6	78·5	5·1	·956
5	Sulloobong . . .	.	5277	Nov.	57·6	51·2	6·4	·390	79·4	65·8	13·6	·634
· 6	Lingdam . . .	.	5375	Dec.	44·3	43·0	1·3	·293	68·8	59·9	8·9	·521
3	Makaroumbi . .	.	5485	Nov.	52·1	48·1	4·0	·350	72·5	60·5	12·0	·532
8	Khabang . . .	.	5505	Dec.	55·1	47·3	7·8	·340	75·0	64·7	10·3	·611
6	Lingdam. . .	.	5554	Dec.	45·0	43·7	1·3	·301	71·0	56·5	14·5	·466
3	Yankutang. . .	.	5564	Dec.	43·6	41·7	1·9	·280	69·5	63·1	6·4	·579
4	Namtchi. . .	.	5608	May	67·1	61·2	5·9	·544	87·7	74·9	12·8	·850
6	Yoksun . . .	.	5619	Jan.	42·7	34·0	8·7	·214	68·2	58·1	10·1	·492
16	Ditto . . .	.	,,	Jan.	43·0	33·9	9·1	·213	66·2	51·9	14·3	·399
2	Loongtoong . .	.	5677	Nov.	45·3	42·8	2·5	·292	72·1	63·8	8·3	·595
4	Sakkiazong . .	.	5625	Nov.	54·1	50·9	3·2	·358	78·3	66·1	12·2	·639
3	Phadong 8 A.M. .	.	5946	Nov.	51·9	50·8	1·1	·383	75·0	67·5	7·5	·670
3	,, 9·50 A.M. .	.	,,	Nov.	55·9	53·0	2·9	·413	80·9	67·9	13·0	·678
3	,, noon. .	.	,,	Nov.	60·7	56·5	4·2	·465	85·6	64·8	20·8	·613
3	,, 2·40 P.M. .	.	,,	Nov.	57·4	54·7	2·7	·438	86·6	62·2	24·4	·562
2	,, 4 P.M. .	.	,,	Nov.	55·5	52·8	2·7	·410	85·5	61·9	23·6	·557
3	,, sunset .	.	,,	Nov.	53·7	52·6	1·1	·408	80·6	67·4	13·2	·667
3	Tumloong . .	.	5368	Nov.	64·2	62·6	1·6	·570	83·8	77·5	6·3	·924
22	,, 9·50 A.M. .	.	5976		54·1	50·0	4·1	·375	75·1	61·9	13·2	·557
21	,, noon. .	.	,,	Nov.	57·3	51·7	5·6	·396	79·7	60·1	19·6	·524
20	,, 2·40 P.M. .	.	,,	&	57·3	51·4	5·9	·391	81·3	58·0	23·3	·489
21	,, 4 P.M. .	.	,,	Dec.	54·7	50·5	4·2	·380	80·2	58·6	21·6	·499
21	,, sunset .	.	,,		51·8	48·5	3·3	·355	76·7	61·2	15·5	·545
260				Mean	57·7	53·3	4·5	·438	77·6	67·8	9·8	·700

Humidity   ·865   Calcutta   ·730
Weight of Vapour   4·70 gr.   ,,   7·34 gr.

ELEVATION 6000 TO 7000 FEET.

No. of Obs.	Locality	Elev.	Month.	EAST NEPAL AND SIKKIM.				CALCUTTA.			
				Tem.	D. P.	Diff.	Tens.	Tem.	D. P.	Diff.	Tens.
5	Runkpo	6008	Nov.	57·5	54·8	2·7	·440	79·5	73·4	6·1	·810
11	Leebong	6021	Feb.	47·8	43·7	4·1	·300	74·9	59·7	15·2	·517
11	Ditto	„	Jan.	47·8	43·4	4·4	·297	66·9	56·2	10·7	·460
4	Dholep	6133	May	60·5	59·9	0·6	·520	89·4	81·4	8·0	·046
2	Iwa River . . .	6159	Dec.	41·2	40·5	0·7	·269	69·6	60·2	9·4	·527
4	Dengha	6368	Aug.	66·7	64·0	2·7	·597	86·1	78·8	7·3	·962
4	Kulhait River . . .	6390	Dec.	41·9	41·9	0·0	·283	71·3	60·9	10·4	·539
3	Latong	6391	Oct.	54·0	53·2	0·8	·416	55·5	44·1	11·4	·305
1	Doobdi	6472	Jan.	46·6	36·2	10·4	·231	78·7	58·0	20·7	·490
10	Pemiongchi	6584	Jan.	40·7	35·8	4·9	·228	66·3	54·4	11·9	·434
4	Keadom	6609	Aug.	63·5	60·0	3·5	·523	79·7	77·5	2·2	·925
6	Hee-hill	6677	Dec.	40·8	34·1	6·7	·215	64·0	58·0	6·0	·489
7	Dumpook	6678	Jan.	40·2	31·8	8·4	·198	68·5	53·8	14·7	·426
4	Changachelling . . .	6828	Jan.	50·6	31·8	18·8	·198	68·3	53·6	14·8	·423
76			Mean	50·0	45·1	4·9	·337	72·8	62·1	10·6	·597

Humidity . . . 845   Calcutta . ·701

Weight of Vapour . . 3·60 gr.   „ 6·11 gr.

ELEVATION 7000 TO 8000 FEET.

No. of Obs.	Locality	Elev.	Month.	EAST NEPAL AND SIKKIM.				CALCUTTA.			
				Tem.	D. P.	Diff.	Tens.	Tem.	D. P.	Diff.	Tens.
1	Pemiongchi . . .	7083	Jan.	46·2	33·5	12·7	·210	76·8	51·8	25·0	·396
2	Goong	7216	Nov.	49·0	48·5	0·5	·355	79·7	69·1	10·6	·705
8	Kampo-Samdong .	7329	{ May / Aug. }	59·1	58·2	0·9	·493	83·6	77·4	6·2	·922
1	Hee-hill	7289	Jan.	51·3	26·4	24·9	·163	72·8	56·6	16·2	·466
1	Ratong river . .	7143	Jan.	36·5	25·3	11·2	·157	60·0	52·9	7·1	·412
4	Source of Balasun .	7436	Oct.	48·3	48·3	0·0	·352	81·2	73·7	7·5	·819
8	Goong ridge . .	7441	Do.	51·2	50·2	1·0	·376	80·7	66·9	13·8	·657
25	Dorjiling		Mean	48·8	41·5	7·3	·301	76·4	64·1	12·3	·625

From mean of above and { Humidity . ·826   Calcutta . ·668

Dorjiling   { Weight of Vapour 3·85 gr.   „ . 7·28 gr.

ELEVATION 8,000 TO 9,000 FEET.

No. of Obs.	Locality.	Elev.	Month.	EAST NEPAL AND SIKKIM.				CALCUTTA.			
				Tem.	D. P.	Diff.	Tens.	Tem.	D. P.	Diff.	Tens.
4	Sinchul	8607	Jan.	41·7	34·3	7·4	·216	66·3	56·9	9·4	·472
2	Ditto	„	April	66·8	44·6	22·2	·310	96·9	75·4	21·5	·866
1	Ascent of Tonglo .	8148	May	56·2	54·4	1·8	·434	86·8	78·9	7·9	·967
2	Tambur river . . .	8081	Nov.	38·0	33·9	4·1	·213	71·7	64·1	7·6	·599
3	Sakkiazong . . .	8353	Nov.	49·7	37·4	12·3	·241	74·0	62·4	11·6	·566
4	Chateng	8418	Oct.	43·8	43·2	0·6	·299	79·2	77·5	1·7	·926
6	Buckim	8659	Jan.	30·2	22·8	7·4	·143	68·6	49·4	19·2	·366
9	Ditto	„	Jan.	33·9	33·1	0·8	·207	69·8	52·2	17·6	·403
1	Chateng . . .	8752	May	67·2	60·7	6·5	·536	89·7	76·8	12·9	·904
11	Lachoong. 7 A.M. .	8777		53·3	51·1	2·2	·388	83·0	78·9	4·1	·967
12	„ . 9·50 A.M .	„		60·2	55·3	4·9	·447	87·1	79·9	7·2	·999
7	„ . noon .	„	Aug.	61·6	57·1	4·5	·475	90·1	79 4	10·7	·983
4	„ . 2·40 P.M.	„	&	58·1	56·4	1·7	·464	88·0	80·0	8·0	1·007
7	„ . 4 P.M. .	„	Oct.	58·6	53·8	4·8	·424	87·5	79·4	8·1	·981
10	„ . sunset .	„		55·5	54·3	1·2	·432	84·5	78·7	5·8	·959
12	„ . Miscellaneous	„		55·9	49·6	6·3	·368	85·9	75·2	10·7	·858
10	Lamteng . 6 A.M.	8884	May,	53·9	52·0	1·9	·400	59·5	56·4	3·1	·464
10	„ . 9·50 A.M. .	„	June,	62·8	56·2	6·6	·461	88·3	78·7	9·6	·959
4	„ . noon .	„	July	62·8	56·2	6·6	·461	92·0	78·0	14·0	·939
5	„ . 2·40 P.M	„	&	58·3	54·4	3·9	·435	92·2	78·4	13·8	·950
6	„ . 4 P.M. .	„	Aug.	56·2	54·7	1·5	·438	92·3	77·1	15·2	·914
8	„ . sunset .	„		53·3	52·5	0·8	·407	88·1	77·4	10·7	·922
11	Zemu Samdong . 7 A.M.	8976		55·7	55·3	0·4	·448	80·4	79·8	0·6	·997
11	„ . 9·50 A.M.	„	June	59·7	52·8	6·9	·412	86·3	79·0	7·3	·969
7	„ . noon.	„	&	63·1	57·1	6·0	·473	88·0	79·8	8·2	·994
6	„ . 2·40 P.M..	„	July	61·0	58·6	2·4	·500	89·6	78·2	11·4	·944
8	4 . sunset .	„		57·9	56·1	1·8	·459	89·3	79·0	10 3	·970
10	„ . 4 P.M	„		53·8	52·6	1·2	·407	82·7	77·3	5·4	·920
1	Goong	8999	Nov.	49·0	48·5	0·5	·355	79·7	69·1	10·6	·705
1	Tendong (top) . .	8663	May	55·5	50·0	5·5	·373	88·6	78·1	10·5	·943
193			Mean	54·5	50 0	4·5	388	83·7	73·7	9·8	·847

Humidity . . . . ·858   Calcutta . ·730

Weight of Vapour . . . 4·23 gr.   „ . 8·75 gr

ELEVATION 9,000 TO 10,000 FEET.

No. of Obs.	Localities.	Elev.	Month.	EAST NEPAL AND SIKKIM.				CALCUTTA.			
				Tem.	D. P.	Diff.	Tens.	Tem.	D. P.	Diff.	Tens.
4	Yangma Guola . .	9279	Nov.	37·8	33·1	4·7	·207	72·7	61·4	11·3	·549
8	Nanki	9320	Nov.	42·3	38·3	4·0	·249	52·2	48·3	3·9	·352
4	Singalelah . . .	9295	Dec.	36·2	35·7	0·5	·227	70·9	62·1	8·8	·560
1	Sakkiazong . . .	9322	Nov.	53·5	33·3	20·2	·209	80·0	57·3	22·7	·478
1	Zemu river . . .	9828	June	60·0	47·6	12·4	343	93·3	81·9	11·4	1·062
18			Mean	46·0	37·6	8·4	·247	73·8	62·2	11·6	·60

Humidity . . . . ·747   Calcutta . ·724

Weight of Vapour . . 2·80 gr.   „ 6·28 g

ELEVATION 10,000 TO 11,000 FEET.

No. of Obs.	Locality.	Elev.	Month.	Tem.	D. P.	Diff.	Tens.	Tem.	D. P.	Diff.	Tens.
		EAST NEPAL AND SIKKIM.						CALCUTTA.			
13	Tonglo	10,008	May	51·5	50·2	1·3	·376	88·8	80·8	8·0	1·030
3	Nanki.	10,024	Nov.	42·8	35·5	7·3	·225	79·5	65·8	13·7	·633
4	Yalloong river .	10,058	Dec.	37·7	29·6	8·1	·183	77·7	62·1	15·6	·560
2	Tonglo top . . .	10,079	May	49·9	47·9	2·0	·348	89·4	80·5	8·9	1·018
2	Yeunga	10,196	Oct.	45·9	44·7	1·2	·311	79·5	77·1	2·4	·915
4	Zemu river . . .	10,247	June	45·4	44·2	1·2	·306	84·6	75·1	9·5	·856
10	Wallanchoon . .	10,384	Nov.	37·9	30·2	7·7	·187	76·5	61·9	14·6	·558
4	Laghep.	10,423	Nov.	46·0	42·4	3·6	·287	80·9	68·0	12·9	·681
3	Ditto	,,	Nov.	37·6	37·0	0·6	·238	75·3	69·4	5·9	·712
16	Thlonok river 7 A.M. .	10,846	June	48·5	47·2	1·3	·339	79·0	75·1	3·9	·856
17	,, 9·50 A.M. .	,,	June	57·6	51·4	6·2	·392	87·4	78·8	8·6	·965
9	,, noon . .	,,	June	56·1	50·6	5·5	·382	90·0	79·3	10·7	·979
8	,, 2·40 P.M. .	,,	June	54·8	50·6	4·2	·381	88·5	79·7	8·8	·991
9	,, 4 P.M. . .	,,	June	53·4	50·6	2·8	·381	88·7	78·7	10·0	·962
15	,, sunset .	,,	June	49·8	48·9	0·9	·359	85·5	78·0	7·5	·938
4	Yangma Valley . .	10,999	Dec.	31·6	24·3	7·3	·149	74·4	61·9	12·3	·558
123			Mean	46·7	42·8	3·8	·303	82·8	73·3	9·5	·826

Humidity . . . ·878　　　　Calcutta . ·740
Weight of Vapour . . 3·35 gr.　　　　　,, . 8·70 gr.

ELEVATION 11,000 TO 12,000 FEET.

No. of Obs.	Locality.	Elev.	Month.	Tem.	D. P.	Diff.	Tens.	Tem.	D. P.	Diff.	Tens.
		EAST NEPAL AND SIKKIM.						CALCUTTA.			
3	Barfonchen . .	11,233	Nov.	36·8	31·9	4·9	·198	76·3	69·6	6·7	·719
3	Punying . . .	11,299	Aug.	50·2	49·5	0·7	·367	84·5	78·8	5·7	·963
1	Kambachen village .	11,378	Dec.	43·3	32·5	10·8	·203	80·0	61·2	18·8	·544
12	Tallum 7 A.M. .	11,482	July	50·4	47·8	2·6	·347	85·0	80·3	4·7	1·010
6	,, 9·50 A.M. .	,,	July	58·1	50·5	7·6	·380	88·1	79·7	8·4	·993
8	,, noon . .	,,	July	57·9	50·8	7·1	·384	89·7	81·3	8·4	1·043
5	,, 2·40 P.M. .	,,	July	55·7	50·2	5·5	·377	89·3	80·6	8·7	1·020
6	,, 4 P.M. . .	,,	July	54·3	50·1	4·2	·375	90·3	79·4	10·9	·981
6	,, sunset . .	,,	July	48·8	47·3	1·5	·340	86·6	80·0	6·6	1·001
2	Kambachen Valley .	11,484	Dec.	30·4	26·0	4·4	·161	69·9	59·5	10·4	·515
10	Yeumtong 7 A.M. .	11,887		44·4	43·8	0·6	·302	83·0	78·9	4·1	·967
9	,, 9·50 A.M. .	,,	Aug.	53·6	48·9	4·7	·360	87·5	78·7	8 8	·959
5	,, noon . .	,,	Sep.	54·5	48·3	6·2	·353	89·7	77·2	12·5	·917
7	,, 2·40 P.M. .	,,	&	48·8	47·4	1·4	·342	87·2	77·2	10·0	·915
4	,, 4 P.M. . .	,,	Oct.	48·4	47·1	1·3	·338	85·2	77·8	7·4	·934
10	,, sunset . .	,,		42·0	35·9	6·1	·229	60·6	58·5	2·1	·497
7	,. Miscellaneous.	,,	Oct.	43·5	37·1	6·4	·239	83·7	69·7	14·0	·720
104			Mean	48·3	43·8	4·5	·311	83·3	74·6	8·7	·865

Humidity . . . ·860　　　　Calcutta . ·760
Weight of Vapour . . 3·46 gr.　　　　　,, . 9·00 gr.

ELEVATION 12,000 TO 13,000 FEET.

No. of Obs.	Locality	Elev.	Month.	EAST NEPAL AND SIKKIM.				CALCUTTA.			
				Tem.	D.P.	Diff.	Tens.	Tem.	D.P.	Diff.	Tens.
9	Zemu river 7 A.M.	12,070	June & July	46·6	45·6	1·0	·321	80·6	77·7	2·9	·931
9	„ 9·50 A.M.	„		51·1	49·0	2·1	·362	84·5	75·1	9·4	·972
7	„ noon	„		51·1	50·2	0·9	·376	87·0	82·2	4·8	1·074
7	„ 2·40 P.M.	„		51·2	50·3	0·9	·377	86·3	80·0	6·3	1·000
7	„ 4 P.M.	„		49·7	48·9	0·8	·360	86·5	80·2	6·3	1·006
8	„ sunset	„		48·1	47·6	0·5	·344	81·4	77·5	3·9	·926
2	Yangma Valley	12,129	Nov.	34·8	22·7	12·1	·143	70·6	63·7	6·9	·592
1	Zemu river	12,422	June	49·0	46·6	2·4	·332	93·2	79·6	13·6	·989
3	Chumanako	12,590	Nov.	37·3	28·3	9·0	·174	75·1	73·8	1·3	·822
7	Tungu 7 A.M.	12,751	July	45·1	44·1	1·0	·305	80·5	78·3	2·2	·949
5	„ 9·50 A.M.	„	July	53·1	48·6	4·5	·355	87·1	79·4	7·7	·982
1	„ noon	„	July	62·3	52·7	9·6	·409	88·9	77·8	11·1	·935
1	„ 2·40 P.M.	„	July	60·0	53·8	6·2	·425	85·3	79·5	5·8	·985
6	„ sunset	„	July	46·4	45·3	1·1	·317	84·7	79·1	5·6	·974
3	„ sunrise	„	Oct.	38·2	35·0	3·2	·222	79·4	77·8	1·6	·932
4	„ 9·50 A.M.	„	Oct	46·5	42·8	3·7	·292	85·0	78·6	6·4	·957
4	„ noon	„	Oct.	46·1	42·0	4·1	·284	85·0	78·2	6·8	·944
4	„ 2·40 P.M.	„	Oct.	43·8	42·1	1·7	·285	86·4	78·8	7·6	·963
4	„ 4 P.M.	„	Oct.	42·3	40·8	1·5	·271	85·9	78·5	7·4	·956
6	„ sunset	„	Oct.	41·0	38·7	2·3	·253	83·3	78·2	5·1	·947
23	„ Miscellaneous	„	Oct.	43·2	40·8	2·4	·272	84·5	78·4	6·1	·950
13	„ Ditto	„	July	51·3	47·7	3·6	·345	85·7	79·0	6·7	·971
6	Tuquoroma	12,944	Nov.	26·0	23·4	2·6	·146	75·1	60·8	14·3	·537
140			Mean	46·3	42·9	3·4	·303	83·6	77·1	6·5	·926

Humidity . . . . . ·890　　　　Calcutta . ·815
Weight of Vapour . . . . 3·37 gr.　　　　„ . 9·75 gr.

ELEVATION 13,000 TO 14,000 FEET.

No. of Obs.	Locality	Elev.	Month.	EAST NEPAL AND SIKKIM.				CALCUTTA.			
				Tem.	D.P.	Diff.	Tens.	Tem.	D.P.	Diff.	Tens.
7	Mon Lepcha	13,090	Jan.	27·1	18·5	8·6	·122	70·0	50·8	19·2	·527
4	Ditto	13,073	Jan.	25·6	16·4	9·2	·113	71·7	49·9	21·8	·373
2	Tunkra valley	13,111	Aug.	45·0	43·5	1·5	·298	81·2	78·7	2·5	·962
21	Jongri	13,194	Jan.	22·7	10·5	12·2	·091	70·6	53·2	17·4	·417
1	Zemu river	13,281	June	46·7	46·7	0·0	·334	92·9	86·6	6·2	1·230
4	Choonjerma	13 288	Dec.	39·0	11·1	27·9	·093	69·8	61·8	8·0	·555
10	Yangma village	13,502	Nov. Dec.	33·8	18·6	15·2	·123	78·9	62·1	16·8	·561
1	Wallanchoon road	13,505	Nov.	28·0	9·5	18·5	·088	66·4	61·8	4·6	·555
3	Kambachen, below pass	13,600	Dec.	40·0	18·6	21·4	·123	72·9	62·2	10·7	·563
53			Mean	34·2	21·5	12·6	·154	74·9	63·0	11·9	·636

Humidity . . . . . ·634　　　　Calcutta . ·678
Weight of Vapour . . . . 1·61 gr.　　　　„ . . 6·28 gr.

ELEVATION 15,000 TO 16,000 FEET.

No. of Obs.	Locality.	Elev.	Month.	Tem.	D. P.	Diff.	Tens.	Tem.	D. P.	Diff.	Tens.
	EAST NEPAL AND SIKKIM.							CALCUTTA.			
1	Yangma valley	15,186	Dec.	42·2	20·7	21·5	·133	80·8	62·0	18·8	·559
1	Choonjerma pass	15,259	Dec.	34·3	10·5	23·8	·091	77·9	60·6	17·3	·534
8	Lachee-pia	15,262	Aug.	42·0	41·6	0·4	·279	85·5	79·4	6·1	·982
12	Momay, 7 A.M.	,,	Sept.	39·4	34·7	4·7	·219	80·5	78·8	1·7	966
6	,, 9·50 A.M.	,,	Sept.	50·9	41·7	9·2	·280	87·6	78·8	8·8	·963
4	,, noon	,,	Sept.	51·7	43·6	8·1	·299	89·5	79·7	9·8	·990
8	,, 2·40 P.M.	,,	Sept.	49·7	41·9	7·8	·283	90·0	78·3	11·7	·949
10	,, 4 P.M.	,,	Sept.	44·4	41·3	3·1	·276	88·7	77·6	11·1	·928
16	,, sunset	,,	Sept.	41·5	38·6	2·9	·252	84·2	78·4	5·8	·952
8	,, Miscellaneous	,,	Sept.	47·6	41·4	6·2	·277	87·4	78·6	8·8	·956
6	,, ,,	,,	Oct.	40·9	36·5	4·4	·234	83·9	69·3	14·6	·710
3	Sittong	15,372	Oct.	38·6	29·8	8·8	·184	84·0	77·5	6·5	·926
2	Palung	15,676	Oct.	44·6	39·8	4·8	·262	86·8	78·5	8·3	·954
1	Kambachen pass	15,770	Dec.	26·5	15·9	10 6	·111	78·0	58·5	19·5	·498
1	Yeumtong	15,985	Sept.	44·6	43·7	0·9	·300	88·8	80·5	8·3	1 016
87			Mean	42·6	34·8	7·8	·232	84·9	74·4	10·5	0·859

Humidity   .   .   .   .   ·763   Calcutta   .   ·719
Weight of Vapour   .   .   2·55 gr.   ,,   .   .   8·95 gr.

ELEVATION 16,000 TO 17,000 FEET.

No. of Obs.	Locality.	Elev.	Month.	Tem.	D. P.	Diff.	Tens.	Tem.	D. P.	Diff.	Tens.
	EAST NEPAL AND SIKKIM.							CALCUTTA.			
1	Kanglachem pass	16,038	Dec.	32·8	16·3	16·5	·110	80·7	61·1	19·6	·543
3	Tunkra pass	16,083	Aug.	39·8	38·7	1·1	·252	86·0	78·7	7·3	·959
1	Wallanchoon pass	16,756	Nov.	18·0	−6·0	24·0	·046	79·9	57·6	22·3	·483
5	Yeumtso	16,808	Oct.	32·4	25·1	7·3	·156	85·0	75·7	9·3	·872
6	Cholamoo lake	16,900	Oct.	31·4	20·2	11·2	·130	79·8	68·4	11·4	·690
1	Donkia mountain	16,978	Sept.	40·2	25·9	14·3	·160	87·6	78·8	8·8	·963
17			Mean	32·4	20·0	12·4	·142	83·2	70·1	13·3	·752

Humidity   .   .   .   .   ·640   Calcutta   .   ·658
Weight of Vapour   .   .   1·53 gr.   ,,   .   .   7·80 gr.

ELEVATION 17,000 TO 18,500 FEET.

No. of Obs.	Locality.	Elev.	Month.	EAST NEPAL AND SIKKIM.				CALCUTTA.			
				Tem.	D. P.	Diff.	Tens.	Tem.	D. P.	Diff.	Tens.
1	Kinchinjhow . . .	17,624	Sept.	47·5	30·9	16·6	·191	85·7	79·7	6·0	·991
1	Sebolah pass . . .	17,585	Sept.	46·5	34·6	11·9	·218	88·8	80·0	8·8	1·002
1	Donkia mountain . .	18,307	Sept.	38·8	35·3	3·5	·224	90·7	79·3	11·4	·981
3	Bhomtso	18,450	Oct.	54·0	4·4	49·6	·072	91·1	61·1	20·0	·543
2	Donkia pass . . .	18,466	Sept.	41·8	30·3	11·5	·188	84·1	78·4	5·7	·950
2	Ditto	18,466	Oct.	40·1	25·0	15·1	·155	86·5	65·5	21·0	·627
10			Mean	44·8	26·8	18·0	·175	87·8	74·0	12·2	·849

Humidity . . . . ·532		Calcutta	·648
Weight of Vapour . . . 1·90 gr.		„	. 8·78 gr

SUMMARY.

No. of Obs.	Elevations in Feet.	Sta- tions.	HUMIDITY.			WEIGHT OF VAPOUR.		
			Sikkim.	Cal- cutta.	Diff. Sikkim.	Sikkim.	Cal- cutta.	Diff. Sikkim.
48	735 to 2000	9	·717	·663	+ ·054	5·57	6·88	— 1·31
49	2000 „ 3000	9	·820	·740	·080	5·45	7·13	1·68
48	3000 „ 4000	13	·858	·732	·116	4·23	6·60	2·37
137	4000 „ 5000	23	·837	·730	·107	4·33	7·12	2·79
260	5000 „ 6000	15	·865	·730	·135	4·70	7·34	2·64
76	6000 „ 7000	13	·845	·701	·144	3·60	6·71	3·11
1023	7000 „ 8000	14	·826	·668	·158	3·85	7·28	3·43
193	8000 „ 9000	13	·858	·730	·128	4·23	8·75	4·52
18	9000 „ 10.000	5	·747	·724	·023	2·80	6·28	3·48
123	10,000 „ 11.000	10	·878	·740	·138	3·35	8·70	4·35
104	11,000 „ 12,000	6	·860	·760	·100	3·46	9·00	5·54
140	12,000 „ 13,000	6	·890	·815	·075	3·37	9·75	6·38
53	13,000 „ 14,000	9	·634	·678	— ·044	1·61	6·28	4·67
87	15,000 „ 16,000	8	·763	·719	+ ·044	2·55	8·95	6·40
17	16,000 „ 17,000	6	·640	·658	·018	1·53	7·80	6·27
10	17,000 „ 18,500	5	·532	·648	— ·116	1·90	8·78	6·88
2386		154						

Considering how desultory the observations in Sikkim are, and
how much affected by local circumstances, the above results must
be considered highly satisfactory: they prove that the relative
humidity of the atmospheric column remains pretty constant
throughout all elevations, except when these are in a Tibetan
climate; and when above 18,000 feet, elevations which I attained
in fine weather only. Up to 12,000 feet this constant humidity
is very marked; the observations made at greater elevations were

almost invariably to the north, or leeward of the great snowy peaks, and consequently in a drier climate; and there it will be seen that these proportions are occasionally inverted; and in Tibet itself a degree of relative dryness is encountered, such as is never equalled on the plains of Eastern Bengal or the Gangetic delta. Whether an isolated peak rising near Calcutta, to the elevation of 19,000 feet, would present similar results to the above, is not proven by these observations, but as the relative humidity is the same at all elevations on the outermost ranges of Sikkim, which attain 10,000 feet, and as these rise from the plains like steep islands out of the ocean, it may be presumed that the effects of elevation would be the same in both cases.

The first effect of this humid wind is to clothe Sikkim with forests, that make it moister still; and however difficult it is to separate cause from effect in such cases as those of the reciprocal action of humidity on vegetation, and vegetation on humidity, it is necessary for the observer to consider the one as the effect of the other. There is no doubt that but for the humidity of the region, the Sikkim Himalaya would not present the uniform clothing of forest that it does; and, on the other hand, that but for this vegetation, the relative humidity would not be so great.*

The great amount of relative humidity registered at 6000 to 8000

* Balloon ascents and observations on small mountainous islands, therefore, offer the best means of solving such questions: of these, the results of ballooning, under Mr. Welsh's intrepid and skilful pioneering (see Phil. Trans. for 1853), have proved most satisfactory; though, from the time for observation being short, and from the interference of belts of vapour, some anomalies have not been eliminated. Islands again are still more exposed to local influences, which may be easily eliminated in a long series of observations. I think that were two islands, as different in their physical characters as St. Helena and Ascension, selected for comparative observations, at various elevations, the laws that regulate the distribution of humidity in the upper regions might be deduced without difficulty. They are advantageous sites, from differing remarkably in their humidity. Owing partly to the indestructible nature of its component rock (a glassy basalt), the lower parts of Ascension have never yielded to the corroding effects of the moist sea air which surrounds it; which has decomposed the upper part into a deep bed of clay. Hence Ascension does not support a native tree, or even shrub, two feet high. St. Helena, on the other hand, which can hardly be considered more favourably situated for humidity, was clothed with a redundant vegetation when discovered, and trees and tree-ferns (types of humidity) still spread over its loftiest summits. Here the humidity, vegetation, and mineral and mechanical composition reciprocate their influences.

feet, arises from most of the observations having been made on the outer
range, where the atmosphere is surcharged. The majority of those at
10,000 to 12,000 feet, which also give a disproportionate amount of
humidity, were registered at the Zemu and Thlonok rivers, where the
narrowness of the valleys, the proximity of great snowy peaks, and
the rank luxuriance of the vegetation, all favour a humid atmosphere.

I would have added the relative rain-fall to the above, but this is so
very local a phenomenon, and my observations were so repeatedly
deranged by having to camp in forests, and by local obstacles of all
kinds, that I have suppressed them; their general results I have
given in Appendix F.

I here add a few observations, taken on the plains at the foot of the
Sikkim Himalaya during the spring months.

*Comparison between Temperature and Humidity of the Sikkim Terai
and Calcutta, in March and April,* 1849.

No. of Obs.	Locality.	Elev. above sea, Feet.	TEMP.		D. P.		TENSION.		SAT.	
			C	T.	C.	T.	C.	T.	C.	T.
4	Rummai . .	293	82·2	70·6	61·7	60·5	·553	·532	·517	·717
4	Belakoba ..	368	92·8	85·5	62·6	63·0	·570	·578	·382	·485
3	Rangamally .	275	84·2	75·0	68·7	62·5	·695	·568	·605	·665
3	Bhojepore	404	90·1	81·2	54·1	44·3	·429	·308	·313	·295
4	Thakyagunj .	284	84·9	77·1	61·3	60·8	·547	·537	·466	·588
3	Bhatgong	225	87·4	74·9	64·7	54·6	·611	·436	·480	·512
2	Sahibgunj .	231	80·2	68·0	66·2	53·1	·642	·414	·635	·409
8	Titalya .	362	85·5	80·0	55·4	56·1	·448	·459	·376	·459
31	Means . .	305	85·9	79·0	61·8	56·9	·562	·479	·472	·516
	May, 1850 } Kishengunj {	131	89·7	K 78·6	76·7	K 71·4	·904	K ·759	·665	K· 793

Vapour in a cubic foot—Kishengunj 8·20   Terai . 5·08
                        Calcutta . 9·52   Calcutta. 5·90
Mean difference of temperature between Terai and Calcutta, from
  31 observations in March, as above, excluding minima   . . Terai—6·9
Mean difference from 26 observations in March., including minima . „ —9·7
Mean difference of temperature at Siligoree on May 1, 1850 −10·9
        „   Kishengunj   „   −11·1

From the above, it appears that during the spring months, and
before the rains commence, the belt of sandy and grassy land along
the Himalaya, though only 3½ degrees north of Calcutta, is at least
6° or 7° colder, and always more humid relatively, though there is
absolutely less moisture suspended in the air. After the rains commence, I believe that this is in a great measure inverted, the plains

becoming excessively heated, and the temperature being higher than
at Calcutta.   This indeed follows from the well known fact that the
summer heat increases greatly in advancing north-west from the Bay
of Bengal to the trans-Sutledge regions; it is admirably expressed in
the maps of Dove's great work " On the Distribution of Heat on the
Surface of the Globe."

H.

ON THE TEMPERATURE OF THE SOIL AT VARIOUS ELEVATIONS.

These observations were taken by burying a brass tube two feet
six inches to three feet deep, in exposed soil, and sinking in it, by
a string or tied to a slip of wood, a thermometer whose bulb was
well padded with wool : this, after a few hours' rest, indicates the
temperature of the soil.   Such a tube and thermometer I usually
caused to be sunk wherever I halted, if even for one night, except
during the height of the rains, which are so heavy that they communicate to the earth a temperature considerably above that of the air.

The results proved that the temperature of the soil at Dorjiling
varies with that of the month, from 46° to 62·2°, but is hardly
affected by the diurnal variation, except in extreme cases.   In summer,
throughout the rains, May to October, the temperature is that of the
month, which is imparted by the rain to the depth of eleven feet
during heavy continued falls (of six to twelve inches a day), on which
occasions I have seen the buried thermometer indicating a temperature above the mean of the month.   Again, in the winter months,
December and January, it stands 5° above the monthly mean ; in
November and February 4° to 5°; in March a few degrees below the
mean temperature of the month, and in October above it; April
and May being sunny, it stands above their mean ; June to
September a little below the mean temperature of each respectively.

The temperature of the soil is affected by :—1.   The exposure of
the surface ; 2. The nature of the soil ; 3. Its permeability by rain,
and the presence of underground springs ; 4. The sun's declination ;
5. The elevation above the sea, and consequently the heating power
of the sun's rays : and, 6, The amount of cloud and sunshine.

The appended observations, though taken at sixty-seven places, are far from being sufficient to supply data for the exact estimation of the effects of the sun on the soil at any elevation or locality; they, however, indicate with tolerable certainty the main features of this phenomenon, and these are in entire conformity with more ample series obtained elsewhere. The result, which at first sight appears the most anomalous, is, that the mean temperature of the soil, at two or three feet depth, is almost throughout the year in India above that of the surrounding atmosphere. This has been also ascertained to be the case in England by several observers, and the carefully-conducted observations of Mr. Robert Thompson at the Horticultural Society's Gardens at Chiswick, show that the temperature of the soil at that place is, on the mean of six years, at the depth of one foot, 1° above that of the air, and at two feet 1½°. During the winter months the soil is considerably (1° to 3°) warmer than the air, and during summer the soil is a fraction of a degree cooler than the air.

In India, the sun's declination being greater, these effects are much exaggerated, the soil on the plains being in winter sometimes 9° hotter than the air; and at considerable elevations in the Himalaya very much more than that; in summer also, the temperature of the soil seldom falls below that of the air, except where copious rain-falls communicate a low temperature, or where forests interfere with the sun's rays.

At considerable elevations these effects are so greatly increased, that it is extremely probable that at certain localities the mean temperature of the soil may be even 10° warmer than that of the air; thus, at Jongri, elevation 13,194 feet, the soil in January was 34·5°, or 19·2° above the mean temperature of the month, immediately before the ground became covered with snow for the remainder of the winter; during the three succeeding months, therefore, the temperature of the soil probably does not fall below that of the snow, whilst the mean temperature of the air in January may be estimated at about 20°, February 22°, March 30°, and April 35°. If, again, we assume the temperature of the soil of Jongri to be that of other Sikkim localities between 10,000 and 14,000 feet, we may assume the soil to be warmer by 10° in July (see Tungu observations), by 8° or 9° in September (see Yeumtong); by 10° in October (see Tungu); and by 7° to 10° in November (see Wallanchoon and Nanki) These temperatures,

however, vary extremely according to exposure and amount of
sunshine; and I should expect that the greatest differences would
be found in the sunny climate of Tibet, where the sun's heat is most
powerful. Were nocturnal or terrestrial radiation as constant and
powerful as solar, the effects of the latter would be neutralised; but
such is not the case at any elevation in Sikkim.

This accumulated heat in the upper strata of soil must have a very
powerful effect upon vegetation, preventing the delicate rootlets of
shrubs from becoming frozen, and preserving vitality in the more
fleshy roots, such as those of the large rhubarbs and small orchids,
whose spongy cellular tissues would no doubt be ruptured by severe
frosts. To the burrowing rodents, the hares, marmots, and rats,
which abound at 15,000 to 17,000 feet in Tibet, this phenomenon is
even more conspicuously important; for were the soil in winter to
acquire the mean temperature of the air, it would take very long to
heat after the melting of the snow, and indeed the latter phenomenon
would be greatly retarded. The rapid development of vegetation
after the disappearance of the snow, is no doubt also proximately
due to the heat of the soil, quite as much as to the increased strength
of the sun's direct rays in lofty regions.

I have given in the column following that containing the temperature of the sunk thermometer, first the extreme temperatures of
the air recorded during the time the instrument was sunk; and in
the next following, the mean temperature of the air during the
same period, so far as I could ascertain it from my own observations.

SERIES I.—*Soane Valley.*

Locality.	Date.	Elevation.	Depth.	Temp. of sunk Therm.	Extreme Temperature of Air observed.	Approximate Mean Temp. of Air deduced.	Diff. between Air and sunk Therm.
		ft.	ft. in.				
Muddunpore .	Feb. 11 to 12	440	3 4	71·5	62·0 to 77·5	67·0	+4·5
Nourunga . .	Feb. 12 „ 13	340	3 8	71·7	57·0 „ 71·0	67·3	3·4
Baroon . .	Feb. 13 „ 14	345	2 4	68·5	53·5 „ 76·0	67·6	1·9
Tilotho . . .	Feb. 15 „ 16	395	4 6	76·5	58·5 „ 80·0	67·8	8·7
Akbarpore . .	Feb. 17 „ 19	400	{2 ther. 4 6 / „ 5 6}	76·0	56·9 „ 79·5	68·0	8·0

SERIES II.—*Himalaya of East Nepal and Sikkim.*

Locality.	Date.	Elevation.	Depth.	Temper. of sunk Therm.	Extreme Temperature of Air observed.	Approximate Mean Temp. of Air deduced.	Diff. between Air and sunk Therm.
		ft.	ft. in.				
Base of Tonglo	May 19	3,000	2 0	78·	67·5 to 67·0		
Simsibong	„ 20	7,000	2 0	61·7	59·0 „ 59·5		
Tonglo saddle	„ 21 to 22	10,008	2 6	50·7*	47·5 „ 57·5	52·5	− 1·8
„ summit	„ 23	10,079	2 6	49·7	47·5 „ 53·2	„	„
Simonbong	„ 24	5,000	2 6	69·7	51·2 „ 55·5	„	„
Nanki	Nov. 4 to 5	9,300	3 0	51·5	33·0 „ 50·5	41·2	+ 9·7
Sakkiazong	„ 9 „ 10	8,353	3 0	53·2	37·8 „ 55·0	46·1	+ 7·1
Mywa guola	„ 17 „ 18	2,132	3 0	73·0	41·0 „ 85·0	63·4	+ 9·6
Banks of Tambur	„ 18 „ 19	2,545	3 0	71·0	48·0 „ 65·0	55·6	+15·4
„ higher up river.	„ 19 „ 20	3,201	3 0	64·5	44·3 „ 60·0	51·6	+12·9
Wallanchoon	„ 23 „ 25	10,386	2 0	43·5 to 45·0	25·0 „ 49·7	37·4	+ 7·6
Yangma village	Nov. 30 Dec. 3	13,502	2 0	37·3 „ 38·0	20·0 „ 46·0	33·0	+ 4·7
„ river.	Dec. 2 to 3	10,999	2 7	41·4 „ 42·0	23·0 „ 40·0	27·9	+ 3·6
Bhomsong	„ 24 „ 25	1,596	2 7	64·5 „ 65·0	42·8 „ 71·3	57·1	+ 6·6
Tchonpong	Jan. 4	4,978	2 7	55·0	33·0 „ 54·8	43·9	+11·1
Jongri	„ 10 to 11	13,194	2 7	34·5	3·7 „ 34·0	15·3	+19·2
Buckeem	„ 12	8,665	2 7	43·2	40·0 „ 29·8	32·4	+10·8
Choongtam	May 19 to 25	5,268	2 7	62·5 to 62·7	48·0 „ 78·3	63·2	− 0·6
Junction of Thlonok and Zemu	June 13 „ 16	10,846	2 7	51·2	38·2 „ 57·2	49·8	+ 1·4
Tungu	July 26 „ 30	12,751	2 5	59·0 to 56·5	38·0 „ 62·3	50·0	+ 7·7
„	Oct. 10 „ 15	12,751	2 7	50·8 „ 52·5	34·5 „ 53·3	41·1	+10·7
Lamteng	Aug. 1 „ 3	8,884	2 7	62·2 „ 62·5	47·5 „ 78·2	57·0	+ 5·3
Choongtam	„ 13 „ 15	5,268	2 7	72·1	54·8 „ 82·0	72·0	+ 0·1
Lachoong	„ 17 „ 19	8,712	2 7	66·3 „ 66·0	43·5 „ 68·7	57·0	+ 9·2
Yeumtong	Sept. 2 „ 8	11,919	2 7	55·5 „ 56·1	39·5 „ 59·5	47·2	+ 8·6
Momay	„ 10 „ 14	15,362	2 7	52·5 „ 51·5	31·0 „ 62·5	41·6	+10·4
Yeumtso	Oct. 16 „ 18	16,808	2 7	43·5 „ 43·0	4·0 „ 52·0	30·6	+12·6
Lachoong	„ 24 „ 25	8,712	2 7	60·2	39·0 „ 62·6	52·0	+ 8·2
Great Rungeet	Feb. 11 „ 13	818	2 7	65·0	56·0 „ 71·0	63·5	+ 1·5
Leebong	„ 14 „ 15	6,000	2 7	50·8 „ 52·0	41·5 „ 56·0	46·0	+ 5·4
Kursiong	Apr. 16	4,813	2 7	64·5	63·0 „ 60·0	63·0	+ 1·5
Leebong	„ 22	6,000	2 7	61·8 „ 62·0	54·0 „ 67·8	60·0	+ 1·9
Punkabaree	May 1	1,850	2 7	80·0	68·2 „ 78·0	76·0	+ 4·0
	Aug. 15 to 16	7,430	5 0	62·0 „ 62·8	58·0 „ 66·0	61·5	+ 0·9
	„ 15 „ 16	„	7 7	61·5 „ 62·3			+ 0·4
	„ 20 „ 22	„	5 0	61·6 „ 61·7	58·7 „ 67·8	61·7	− 0·1
	„ 20 „ 22	„	7 7	60·7	„ „	„	− 1·0
	Sept. 9	„	5 0	60·2	56·2 „ 65·0	60·0	+ 0·2
Jillapahar (Mr. Hodgson's)	„ 9	„	7 7	60·5			+ 0·5
	Oct. 6	„	7 7	60·0	52·0 „ 61·0	58·5	+ 1·5
	„ 20	„	7 7	58·5	49·7 „ 55·2	56·5	+ 2·0
	Feb. 18 to 28	„	2 7	46·0 „ 46·7	36·0 „ 52·8	43·0	+ 6·4
	March 1 „ 13	„	2 7	46·3 „ 48·3	34·5 „ 53·3	46·0	+ 1·3
	April 18 „ 20	„	2 7	55·3 „ 56·0	46·0 „ 61·3	54·0	+ 1·7
	„ 30	„	2 7	57·4	„ „	55·0	+ 2·4
Superinten. house	„ 21 to 30	6,932	2 7	58·8 „ 60·2	48·5 „ 65·8	58·0	+ 1·5

* Sheltered by trees, ground spongy and wet.

SERIES III.—*Plains of Bengal.*

Locality.	Date.	Elevation.	Depth.	Temp. of sunk Therm.	Extreme Temp. of Air observed.	Approximate Mean Temp. of Air deduced.	Diff. between Air and sunk Therm.
		Feet.	Ft. In.				
Kishengunj	May 3 to 4	131	2 7	§82·8 to 83·0	70·0 to 85·7	82·0	+0·8
Dulalgunj	„ 7	130	„	§81·3 „	74·3 „ 90·3	82·0	-0·7
Banks of Mahanuddy river	„ 8	100	„	+79·3 „	75·0 „ 91·5	83·0	-3·7
„ „	„ 9	100	„	+87·5 „	77·8 „ 92·5	83·0	-4·5
„ „	„ 10	100	„	+88·0 „	78·5 „ 91·5	82·3	-5·7
Maldah	„ 11	100	„	+88·8 „	75·3 „ 91·3	82·3	-6·5
Mahanuddy river	„ 14	100	„	+87·8 „	71·0 „ 91·7	82·3	-4·5
Ganges	„ 15	100	„	+88·0 „	73·0 „ 87·8	82·3	-5·7
Bauleah	„ 16 to 18	130	„	87·5 „ 89·8	78·0 „ 106·5	80·5	+7·3
Dacca	„ 28 „ 30	72	„	84·0 „ 84·3	75·3 „ 95·5	83·3	+0·9

SERIES IV.—*Khasia Mountains.*

Locality.	Date.	Elevation.	Depth.	Temp. of sunk Therm.	Extreme Temp. of Air observed.	Approximate Mean Temp. of Air deduced.	Diff. between Air and sunk Therm.
		Feet.	Ft. In.				
Churra	June 23 to 25	4,226	2 7	*71·8 to 72·3	64·8 to 72·2	69·9	+2·2
„	Oct.29 Nov.16			68·3 „ 64·0	70·7 „ 49·3	61·7	+4·5
Kala-panee	June 28 to 29	5,302	„	69·2	64·2 „ 71·2	67·2	+2·0
„	Aug. 5 „ 7			70·0 „ 70·4	72·2 „ 61·8	64·9	+5·2
„	Sept. 13 „ 14			*70·2	65·5 „ 69·8	66·0	+4·2
„	Oct. 27 „ 28			*66·3	64·0 „ 56·0	60·0	+6·3
Moflong	June 30 July 4	6,062	„	65·0 „ 67·3	61·0 „ 68·3	64·0	+2·2
„	July 30 Aug.4			67·3	64·0 „ 75·8	68·5	-1·2
„	Oct. 25 „ 27			63·2	63·7 „ 55·7	64·1	-0·9
Syong	July 29 to 30	5,725	„	69·2 „ 69·3	60·0 „ 78·5	69·2	+0·1
„	Oct. 11 „ 12			67·0	65·7 „ 55·5	62·8	+4·2
Myrung	July 9 „ 10	5,647	„	66·2 „ 66·3	60·0 „ 73·8	67·5	-1·2
„	„ 26 „ 29			68·3	78·0 „ 64·2	71·1	-2·8
„	Oct. 12 „ 17			66·0 „ 64·8	70·0 „ 55·5	63·0	+2·4
„	„ 21 „ 25			64·8 „ 64·0	66·0 „ 53·0	60·5	+3·9
Nunklow	July 11 „ 26	4,688	„	70·5 „ 71·3	65·5 „ 81·5	71·5	-0·5
„	Oct. 17 „ 21			68·8 „ 68·3	75·7 „ 58·0	66·1	+2·5
Pomrang	Sept. 15 „ 23	5,143	„	70·3 „ 68·5	73·0 „ 57·0	65·5	+3·9
„	Oct. 6 „ 10			68·3	73·7 „ 58·2	65·0	+3·3

* Hole full of rain-water.   † Soil, a moist sand.   § Dry sand.

SERIES V.—*Jheels, Gangetic Delta, and Chittagong.*

Locality.	Date.	Elevation.	Depth.	Temp. of sunk Therm.	Extreme Temp. of Air observed.	Approximate Mean Temp. of Air deduced.	Diff. between Air and sunk Therm.
		Feet.	Ft. In.				
Silchar . . .	Nov. 27 to 30	116	2 7	77·7 to 75·8	55·0 to 81·7	69·1	+7·7
Silhet	Dec. 3 „ 7	133		73·5 „ 73·7	63·0 „ 74·5	69·5	+3·1
Noacolly . . .	„ 18 „ 19	20		73·3	58·5 „ 76·5	69·5	+3·8
Chittagong . . .	„ 23 „ 31	191		72·5 „ 73·0	53·2 „ 75·0	63·8	+9·0
„ . . .	Jan. 14 „ 16			73·3 „ 73·7	61·3 „ 78·7	65·5	+8·3
„ flagstaff hill .	Dec. 28 „ 30	151		72·0 „ 71·8	55·2 „ 74·2	65·3	+6·6
Hat-hazaree . .	Jan. 4 „ 5	20		71·3	50·5 „ 62·0	65·0	+6·3
Sidhee	„ 5 „ 6	20		71·0	52·7 „ 70·2	65·0	+6·0
Hattiah . . .	„ 6 „ 9	20		*67·7	50·2 „ 77·5	64·5	+3·2
Seetakoond . . .	„ 9 „ 14	20		73·3 „ 73·7	55·2 „ 79·5	70·2	+3·3
Calcutta** . .	Jan. 16 Feb. 5	18		76·0 „ 77·0	§56·5 „ 82·0	69·3	+7·2

* Shaded by trees.   ** Observations at the Mint, &c., by Mr. Muller.
§ Observations for temperature of air, taken at the Observatory.

I.

ON THE DECREMENT OF TEMPERATURE IN ASCENDING THE SIKKIM HIMALAYA MOUNTAINS AND KHASIA MOUNTAINS.

I HAVE selected as many of my observations for temperature of the air as appeared to be trustworthy, and which, also, were taken contemporaneously with others at Calcutta, and I have compared them with the Calcutta observations, in order to find the ratio of decrement of heat to an increase of elevation. The results of several sets of observations are grouped together, but show so great an amount of discrepancy, that it is evident that a long series of months and the selection of several stations are necessary in a mountain country to arrive at any accurate results. Even at the stations where the most numerous and the most trustworthy observations were recorded, the results of different months differ extremely; and with regard to tne other stations, where few observations were taken, each one is affected differently from another at the same level with it, by the presence or proximity of forest, by exposure to the east or west, to ascending or descending currents in the valleys, and to cloud or sunshine. Other and still more important modifying influences are to be traced to the monthly variations in the amount of humidity in the air and the strength of its currents, to radiation, and to the evolution of heat which accompanies condensation raising the temperature of elevated regions during the rainy season. The proximity of large masses of snow has not the influence I should have expected in lowering the temperature of the surrounding atmosphere, partly no doubt because of the more rapid condensation of vapours which it effects, and partly because of the free circulation of the currents around it. The difference between the temperatures of adjacent grassy and naked or rocky spots, on the other hand, is very great indeed, the former soon becoming powerfully heated in lofty regions where the sun's rays pass through a rarefied atmosphere, and the rocks especially radiating much of the heat thus accumulated, for long after sunset. In various parts of my journals I have alluded to other disturbing causes, which being all more or less familiar to meteorologists, I need not recapitulate here. Their combined effects raise all the summer temperatures above what they should theoretically be.

In taking Calcutta as a standard of comparison, I have been guided by two circumstances; first, the necessity of selecting a spot where observations were regularly and accurately made; and secondly, the being able to satisfy myself by a comparison of my instruments that the results should be so far strictly comparable.

I have allowed 1° Fahr. for every degree in latitude intervening between Sikkim and Calcutta, as the probable ratio of diminution of temperature. So far as my observations made in east Bengal and in various parts of the Gangetic delta afford a means of solving this question, this is a near approximation to the truth. The spring observations however which I have made at the foot of the Sikkim Himalaya would indicate a much more rapid decrement; the mean temperature of Titalya and other parts of the plains south of the forests, between March and May being certainly 6°—9° lower than Calcutta: this period however is marked by north-west and north-east winds, and by a strong haze which prevents the sun's rays from impinging on the soil with any effect. During the southerly

winds, the same region is probably hotter than Calcutta, there being
but scanty vegetation, and the rain-fall being moderate.

In the following observations solitary readings are always rejected.

I.—*Summer or Rainy Season observations at Dorjiling.*

Observations taken during the rainy season of 1848, at Mr. Hodgson's (Jillapahar,
Dorjiling) alt. 7,430 feet, exposure free to the north east and west, the slopes
all round covered with heavy timber; much mist hence hangs over the
station. The mean temperatures of the month at Jillapahar are deduced
from horary observations, and those of Calcutta from the mean of the daily
maximum and minimum.

Month.	No. of Obs. at Jillapahar.	Temp.	Temp. Calcutta.	Equiv of 1° Fahr.
July　.　.　.　.	284	61·7	86·6	364 feet
August　.　.　.　.	378	61·7	85·7	346　,,
September　.　.　.	407	58·9	84·7	348　,,
October　.　.　.　.	255	55·3	83·3	316　,,
	1,324	...	Mean	344 feet

II.—*Winter or dry season observations at Dorjiling.*

1. Observations taken at Mr. J. Muller's, and chiefly by himself, at
   "the Dale;" elev. 6,956 feet; a sheltered spot, with no forest
   near, and a free west exposure. 103 observations. Months:
   November, December, January, and February　.　.　.　. 1° = 313 ft.
2. Observations at Dr. Campbell's (Superintendent's) house in April;
   elev. 6,950 feet; similar exposure to the last. 13 observations
   in April　.　.　.　.　.　.　.　.　.　.　.　. 1° = 308 ft.
3. Observations by Mr. Muller at Colinton; elev. 7,179 feet; free
   exposure to north-west; much forest about the station, and a
   high ridge to east and south. 38 observations in winter months 1° = 290 ft.
4. Miscellaneous (11) observations at Leebong; elev. 6000 feet; in
   February; free exposure all round　.　.　.　.　.　. 1° = 266 ft.
5. Miscellaneous observations at "Smith's Hotel," Dorjiling, on a
   cleared ridge; exposed all round; elev 6,863 feet. April and
   May　.　.　.　.　.　.　.　.　.　.　.　. 1° = 252 ft.

Mean of winter observations　.　.　. 1° = 286
Mean of summer　　,,　　.　.　. 1° = 344

Mean　　　310

III.—*Miscellaneous observations taken at different places in Dorjiling, elevations 6,900 to 7,400 feet, with the differences of temperature between Calcutta and Dorjiling.*

Month.	Number of Observ.	Difference of Temperature.	Equivalent.
January . . .	27	30·4	1°=287 ft.
February . . .	84	32·8	1°=265
March	37	41·9	1°=196
April	7	36·0	1°=236
March and April . .	29	37·3	1°=224
July	83	23·6	1°=389
August . . .	74	22·4	1°=415
September . . .	95	25·7	1°=350
October . . .	18	29·5	1°=297
	Sum 454	Mean 31·1	Mean 1°=295 ft.

These, it will be seen, give a result which approximates to that of the sets I and II. Being deduced from observations at different exposures, the effects of these may be supposed to be eliminated. It is to be observed that the probable results of the addition of November and December's observations, would be balanced by those of May and June, which are hot moist months.

IV.—*Miscellaneous cold weather observations made at various elevations between 1000 and 17,000 feet, during my journey into east Nepal and Sikkim, in November to January 1848 and 1849 The equivalent to 1° Fahr. was deduced from the mean of all the observations at each station, and these being arranged in sets corresponding to their elevations, gave the following results.*

Elevation.	Number of Stations.	Number of Observations.	Equivalent.
1,000 to 4,000 ft.	27	111	1°=215 ft.
4,000 to 8,000	52	197	1°=315
8,000 to 12,000	20	84	1°=327
12,000 to 17,000	14	54	1°=377
	Sum 113	Sum 446	Mean 1°=308 ft.

The total number of comparative observations taken during that journey, amounted to 563, and the mean equivalent was $1° = 303$ feet, but I rejected many of the observations that were obviously unworthy of confidence.

V.—*Miscellaneous observations (chiefly during the rainy season) taken during my journey into Sikkim and the frontier of Tibet, between May 2nd and December 25th, 1848. The observations were reduced as in the previous instance. The rains on this occasion were unusually protracted, and cannot be said to have ceased till mid-winter, which partly accounts for the very high temperatures.*

Elevations.	No. of Stations.	No. of Observations.	Equivalent.
1,000 to 4,000 ft.	10	45	$1°=422$ ft.
4,000 to 8,000	21	283	$1°=336$
8,000 to 12,000	18	343	$1°=355$
12,000 to 18,000	29	219	$1°=417$
	Sum 78	Sum 890	$1°=383$ ft.

The great elevation of the temperature in the lowest elevations is accounted for by the heating of the valleys wherein these observations were taken, and especially of the rocks on their floors. The increase with the elevation, of the three succeeding sets, arises from the fact that the loftier regions are far within the mountain region, and are less forest clad and more sunny than the outer Himalaya.

A considerable number of observations were taken during this journey at night, when none are recorded at Calcutta, but which are comparable with contemporaneous observations taken by Mr. Muller at Dorjiling. These being all taken during the three most rainy months, when the temperature varies but very little during the whole twenty-four hours, I expected satisfactory results, but they proved very irregular and anomalous.

The means were—

At 21 stations of greater elevation than Dorjiling . . $1°=348$ feet.
At 17 „ lower in elevation „ „ . . $1°=447$ „

VI.—*Sixty-four contemporaneous observations at Jillapahar, 7,430 feet, and the bed of the Great Rungeet river, 818 feet ; taken in January and February, give.* . . . $1° = 322$ feet.

VII.—*Observations taken by burying a thermometer two and a half to three feet deep, in a brass tube, at Dorjiling and at various elevations near that station.*

Month.	Upper Stations.	Lower Stations.	
February and March	Jillapahar, 7,430 feet.	Leebong, 6000 feet .	1°=269 ft.
February . . .	Do., do., „	{ Guard-house, Great Rungeet, 1,864 feet }	1°=298 „
April . . .	Leebong, 6000 „	Guard-house, do., „ .	1°=297 „
April	Jillapahar, 7,430 „	Khersiong, 4,813 „ .	1°=297 „
March and April .	Khersiong 4,813 „	Punkabaree, 1850 „ .	1°=223 „
March, April, May .	Jillapahar, 7,430 „	Do., do., „ .	1°=253 „
		Mean,	1°=273 ft.

The above results would seem to indicate that up to an elevation of 7,500 feet, the temperature diminishes rather more than 1° Fahr. for every 300 feet of ascent or thereabouts; that this decrement is much less in the summer than in the winter months; and I may add that it is less by day than by night. There is much discrepancy between the results obtained at greater or less elevations than 7000 feet; but a careful study of these, which I have arranged in every possible way, leads me to the conclusion that the proportion may be roughly indicated thus:—

$1° = 300$ feet, for elevations from   1000 to   8000 feet.
$1° = 320$ „   „   „   8000 „ 10,000 „
$1° = 350$ „   „   „   10,000 „ 14,000 „
$1° = 400$ „   „   „   14,000 „ 18,000 „

VIII.—*Khasia mountain observations.*

Date.	Calcutta Observations.	Number of Observations.	Churra Observations.	Number of Observations.		Altitude above the Sea.
Churra Poonji, June 13—26	86·3	63	70·1	67	1 =300 feet.	4,069 ft.
„ Aug. 7 to Sept. 4.	84·6	196	69·2	214	1 =331 „	4,225 „
„ Oct. 29 to Nov. 16	80·7	85	63·1	133	1 =282 „	4,225 „
		354		414	Mean, 304 feet	

Date.	Calcutta Observations.	Number of Observations.	Khasia Observations.	Number of Observations.		Altitude above the Sea.
Kala-panee, June, Aug., Sept.	85°5	35	67°4	35	1°=345 feet.	5,302 ft.
Moflong, June, July, Aug. Oct.	85·9	73	68·8	74	1°=373 „	6,062 „
Syong	85·1	4	65·0	6	1°=332 „	5,734 „
Myrung, August . . .	89·1	42	69·7	41	1°=343 „	5,632 „
„ October . . .	82·9	21	63·2	58	1°=336 „	5,632 „
Nunklow	86·4	139	70·9	139	1°=372 „	4,688 „
Mooshye, September 23 . .	78·5	9	66·3	12	1°=499 „	4,863 „
Pomrang, „ . . .	82·7	51	65·8	51	1°=369 „	5,143 „
Amwee „ . . .	79·9	15	67·1	11	1°=396 „	4,105 „
Joowy „ . . .	79·5	11	69·0	7	1°=567 „	4,387 „
		400		434	1°=385 feet.	

The equivalent thus deduced is far greater than that brought out
by the Sikkim observations. It indicates a considerably higher
temperature of the atmosphere, and is probably attributable to the
evolution of heat during extraordinary rain-fall, and to the formation
of the surface, which is a very undulating table-land, and everywhere
traversed by broad deep valleys, with very steep, often precipitous
flanks; these get heated by the powerful sun, and from them,
powerful currents ascend. The scanty covering of herbage too over
a great amount of the surface, and the consequent radiation of heat
from the earth, must have a sensible influence on the mean temperature of the summer months.

J.

ON THE MEASUREMENT OF ALTITUDES BY THE BOILING-POINT THERMOMETER.

THE use of the boiling-point thermometer for the determination of elevations in mountainous countries appearing to me to be much underrated, I have collected the observations which I was enabled to take, and compared their results with barometrical ones.

I had always three boiling-point thermometers in use, and for several months five ; the instruments were constructed by Newman, Dollond, Troughton, and Simms, and Jones, and though all in one sense good instruments, differed much from one another, and from the truth. Mr. Welsh has had the kindness to compare the three best instruments with the standards at the Kew Observatory at various temperatures between 180° and the boiling-point ; from which comparison it appears, that an error of 1½° may be found at some parts of the scale of instruments most confidently vouched for by admirable makers. Dollond's thermometer, which Dr. Thomson had used throughout his extensive west Tibetan journeys, deviated but little from the truth at all ordinary temperatures. All were so far good, that the errors, which were almost entirely attributable to care-lessness in the adjustments, were constant, or increased at a constant ratio throughout all parts of the scale ; so that the results of the different instruments have, after correction, proved strictly comparable.

The kettle used was a copper one, supplied by Newman, with free escape for the steam; it answered perfectly for all but very high elevations indeed, where, from the water boiling at very low tempe-ratures, the metal of the kettle, and consequently of the thermo-meter, often got heated above the temperature of the boiling water.

I found that no confidence could be placed in observations taken at great elevations, by plunging the thermometer in open vessels of boiling water, however large or deep, the abstraction of heat from the surface being so rapid, that the water, though boiling below, and hence bubbling above, is not uniformly of the same temperature throughout.

In the Himalaya I invariably used distilled, or snow or rain-water;

but often as I have tried common river-water for comparison, I never
found that it made any difference in the temperature of the boiling-
point. Even the mineral-spring water at Yeumtong, and the
detritus-charged glacial streams, gave no difference, and I am hence
satisfied that no objection can be urged against river waters of
ordinary purity.

On several occasions I found anomalous rises and falls in the
column of mercury, for which I could not account, except theoretically, by assuming breaks in the column, which I failed to detect
on lifting the instrument out of the water; at other times, I
observed that the column remained for several minutes stationary,
below the true temperature of the boiling water, and then suddenly
rose to it. These are no doubt instrumental defects, which I only
mention as being sources of error against which the observer must
be on the watch: they can only be guarded against by the use of two
instruments.

With regard to the formula employed for deducing the altitude
from a boiling-point observation, the same corrections are to a great
extent necessary as with barometric observations: if no account is
taken of the probable state of atmospheric pressure at the level of
the sea at or near the place of observation, for the hour of the day
and month of the year, or for the latitude, it is obvious that errors
of 600 to 1000 feet may be accumulated. I have elsewhere stated
that the pressure at Calcutta varies nearly one inch (1000 feet),
between July and January; that the daily tide amounts to one-tenth
of an inch (= 100 feet); that the multiplier for temperature is too
great in the hot season and too small in the cold; and I have experimentally proved that more accuracy is to be obtained in measuring
heights in Sikkim, by assuming the observed Calcutta pressure and
temperature to accord with that of the level of the sea in the latitude
of Sikkim, than by employing a theoretical pressure and temperature
for the lower station.

In the following observations, the tables I used were those
printed by Lieutenant-Colonel Boileau for the East India Company's
Magnetic Observatory at Simla, which are based upon Regnault's
Table of the 'Elastic Force of Vapour.' The mean height of
the barometrical column is assumed (from Bessel's formula) to

be 29·924 at temp. 32°, in lat. 45°, which, differing only ·002 from
the barometric height corresponding to 212° Fahrenheit, as determined experimentally by Regnault, gives 29·921 as the pressure
corresponding to 212° at the level of the sea.

The approximate height in feet corresponding to each degree of
the boiling-point, is derived from Oltmann's tables.   The multipliers
for the mean temperature of the strata of atmosphere passed through,
are computed for every degree Fahrenheit, by the formula for
expansion usually employed, and given in Baily's Astronomical
Tables and Biot's Astronomie Physique.

For practical purposes it may be assumed that the traveller, in
countries where boiling-point observations are most desired, has
never the advantage of a contemporaneous boiling-point observation
at a lower station.   The approximate difference in height is hence,
in most cases, deduced from the assumption, that the boiling-point
temperature at the level of the sea, at the place of observation, is
212°, and that the corresponding temperature of the air at the
level of the sea is hotter by one degree for every 330 feet of difference
in elevation. As, however, the temperature of boiling water at the level
of the sea varies at Calcutta between July and January almost from
210°·7 to 212°·6, I always took the Calcutta barometer observation at
the day and hour of my boiling-point observation, and corrected my
approximate height by as many feet as correspond to the difference
between the observed height of the barometer at Calcutta and
29·921 ; this correction was almost invariably (always normally)
subtractive in the summer, often amounting to upwards of 400 feet :
it was additive in winter, and towards the equinoxes it was very
trifling.

For practical purposes I found it sufficient to assume the Calcutta
temperature of the air at the day and hour of observation to be that
of the level of the sea at the place of observation, and to take out the
multiplier, from the mean of this and of the temperature at the upper
station.   As, however, 330 feet is a near approach to what I have
shown (Appendix I.) to be the mean equivalent of 1° for all
elevations between 6000 and 18,000 feet ; and as the majority of
my observations were taken between these elevations, it results that
the mean of all the multipliers employed in Sikkim for forty-four

observations amounts to 65°·1 Fahrenheit, using the Calcutta and upper station observations, and 65°·3 on the assumption of a fall of 1° for every 330 feet.   To show, however, how great an error may accrue in individual cases from using the formula of 1° to 330, I may mention that on one occasion, being at an elevation of 12,000 feet, with a temperature of the air of 70°, the error amounted to upwards of 220 feet ; and as the same temperature may be recorded at much greater elevations, it follows that in such cases the formula should not be employed without modification.

A multitude of smaller errors, arising from anomalies in the distribution of temperature, will be apparent on consulting my observations on the temperature at various elevations in Sikkim; practically these are unavoidable.   I have also calculated all my observations according to Professor J. Forbes's formula of 1° difference of temperature of boiling-water, being the equivalent of 550 feet at all elevations.   (See Ed. Phil. Trans., vol xv. p. 405.)   The formula is certainly not applicable to the Sikkim Himalaya; on the contrary, my observations show that the formula employed for Boileau's tables gives at all ordinary elevations so very close an approach to accuracy on the mean of many observations, that no material improvement in its construction is to be anticipated.

At elevations below 4000 feet, elevations calculated from the boiling-point are not to be depended on ; and Dr. Thomson remarked the same in north-west India : above 17,000 feet also the observations are hazardous, except good shelter and a very steady fire is obtainable, owing to the heating of the metal above that of the water.   At all other elevations a mean error of 100 feet is on the average what is to be expected in ordinary cases.   For the elevation of great mountain masses, and continuously elevated areas, I conceive that the results are as good as barometrical ones; for the general purposes of botanical geography, the boiling-point thermometer supersedes the barometer in point of practical utility, for under every advantage, the transport of a glass tube full of mercury, nearly three feet long, and cased in metal, is a great drawback to the unrestrained motion of the traveller.

In the Khasia mountains I found, from the mean of twelve stations and twenty-three observations, the multiplier as derived from the

mean of the temperature at the upper station and at Calcutta, to be
75°·2, and as deduced from the formula to be 73°·1. Here, however,
the equivalent in feet for 1° temp. is in summer very high, being
1° = 385 feet. (See Appendix I.) The mean of all the elevations
worked by the boiling-point is upwards of 140 feet below those
worked by the barometer.

The following observations are selected as having at the time been
considered trustworthy, owing to the care with which they were
taken, their repetition in several cases, and the presumed accuracy
of the barometrical or trigonometrical elevation with which they are
compared.   A small correction for the humidity of the air might
have been introduced with advantage, but as in most barometrical
observations, the calculations proceed on the assumption that the
column of air is in a mean state of saturation; as the climate of the
upper station was always very moist, and as most of the observations
were taken during the rains, this correction would be always additive,
and would never exceed sixty feet.

It must be borne in mind that the comparative results given
below afford by no means a fair idea of the accuracy to be obtained
by the .boiling-point.   Some of the differences in elevation are
probably due to the barometer.   In other cases I may have read
off the scale wrong, for however simple it seems to read off an instrument, those practically acquainted with their use know well how
some errors almost become chronic, how with a certain familiar
instrument the chance of error is very great at one particular
part of the scale, and how confusing it is to read off through steam
alternately from several instruments whose scales are of different
dimensions, are differently divided, and differently lettered; such
causes of error are constitutional in individual observers.   Again,
these observations are selected without any reference to other considerations but what I have stated above; the worst have been put in
with the best.   Had I been dependent on the boiling-point for
determining my elevations, I should have observed it oftener, or at
stated periods whenever in camp, worked the greater elevations from
the intermediate ones, as well as from Calcutta, and resorted to every
system of interpolation.   Even the following observations would be
amended considerably were I to have deduced the elevation by

observations of the boiling-point at my camp, and added the height of my camp, either from the boiling-point observations there, or by barometer, but I thought it better to select the most independent method of observation, and to make the level of the sea at Calcutta the only datum for a lower station.

Series I.—*Sikkim Observations.*

Place.	Month.	Elev. by Barom. or Trigonom.	Tem. B. P.	Air.	Elevation. by B. P.	Error.
Great Rungeet river . .	Feb.	B. 818 ft.	210·7	56·3	904 ft.	+ 86 ft.
Bhomsong	Dec.	1,544	210·2	58·0	1,321	— 223
Guard House, Gt. Rungeet	April	1,864	208·1	72·7	2,049	+ 185
Choongtam . . .	Aug.	5,268	202·6	65·0	5,175	— 93
Dengha	Aug.	6,368	200·6	68·0	6,246	— 122
Mr. Muller's (Dorjiling) .	Feb.	Tr. 6,925	199·4	41·3	7,122	+ 197
Dr. Campbell's (do.) . .	April	6,932	200·1	59·5	6,745	— 187
Mr. Hodgson's (do.) . .	Feb.	B. 7,429	199·4	47·6	7,318	— 111
Sinchul	Jan.	Tr. 8,607	197·0	41·7	8,529	— 78
Lachoong	Aug.	B. 8,712	196·4	54·6	8,777	+ 65
Lamteng	Aug.	8,884	196·3	77·0	8,937	+ 53
Zemu Samdong . . .	July	8,976	196·1	58·6	8,916	— 60
Mainom	Dec.	Tr. 10,702	193·4	38·0	10,516	— 186
Junction of Zemu & Thlonok	July	B. 10,846	193·6	52·0	10,872	+ 26
Tallum	July	11,482	191·8	54·6	11,451	— 31
Yeumtong	Sept.	11,919	191·3	52·2	11,887	— 32
Zemu river	June	12,070	190·4	48·5	12,139	+ 69
Tungu	{ July & Oct. }	12,751	189·7	43·4	12,696	— 55
Jongri	Jan.	13,194	188·8	26·0	13,151	— 43
Zemu river	June	13,281	188·5	47·0	13,360	+ 79
Lachee-pia . . .	Aug.	15,262	186·0	42·8	14,912	— 350
Momay	Sept.	15,362	186·1	48·6	14,960	— 402
Palung	Oct.	15,620	185·4	45·8	15,437	— 183
Kongra Lama . . .	July	15,694	184·1	41·5	16,041	+ 347
Snow-bed above Yeumtong	Sept.	15,985	184·6	44·5	15,816	— 169
Tunkra pass . . .	Aug.	16,083	164·1	39·0	16,137	+ 54
Yeumtso	Oct.	16,808	183·1	15·0	16,279	— 529
Donkia	Sept.	16,978	182·4	41·0	17,049	+ 71
Mountain above Momay .	Sept.	17,394	181·9	47·8	17,470	+ 76
Sebolah pass . . .	Sept.	17,585	181·9	46·5	17,517	— 68
Kinchinjhow . . .	Sept.	17,624	181·0	47·5	18,026	+ 402
Donkia Mountain . .	Sept.	18,510	180·6	37·1	18,143	— 367
Ditto . . .	Sept.	18,307	179·9	38·8	18,597	+ 290
Bhomtso	Oct.	18,450	181·2	52·0	18,305	— 145
Donkia pass . . .	Sept.	18,466	181·2	45·5	17,866	— 600
	Mean					— 58

SERIES II.—*Khasia Mountains.*

Place.	Month.	Elev. Bar.	B. P.	Tm.Air.	Elev. by B. P.	Diff.
Churra . .	June	4,069 ft.	204·4	70·3	4,036 ft.	— 33 ft.
Amwee . . .	September	4,105	205·1	67·7	4,041	— 64
Nurtiung . .	October	4,178	205·0	70·0	4,071	—107
Nunklow . .	July	4,688	203·9	69·8	4,333	—355
Kala-panee . .	Jun., Jul., Sep., Oct.	5,302	202·2	65·8	5,202	—100
Myrung . . .	July	5,647	201·9	69·4	5,559	— 88
Syong . .	July	5,725	201·8	70·8	5,632	— 93
Moflong . . .	Jul., Aug., Oct., Nov.	6,062	201·4	64·8	5,973	— 89
Chillong . .	November	6,662	201·2	62·3	6,308	—354
	Mean	5,160 ft.			5·016 ft.	—143

K.

ACTINOMETER OBSERVATIONS.

THE few actinometer observations which I was enabled to record, were made with two of these instruments constructed by Barrow, and had the bulbs of their thermometers plunged into the fluid of the chamber. They were taken with the greatest care, in conformity with all the rules laid down in the "Admiralty Guide," and may, I think, be depended upon. In the Sikkim Himalaya, a cloudless day, and one admitting of more than a few hours' consecutive observations, never occurs—a day fit for any observation at all is very rare indeed. I may mention here that a small stock of ammonia-sulphate of copper in crystals should be supplied with this instrument, also a wire and brush for cleaning, and a bottle with liquid ammonia: all of which might be packed in the box.

Actine 6·568. Time always mean.

Jillapahur, Dorjiling, Elev. 7430 *feet, Lat.* 27° 3' *N., Long.* 88° 13' *E.*

A.—APRIL 19TH, 1850.　*Watch slow* 1' 15" *mean time.*

Hour.	Act.	Tem. Act.	Act. Reduced.	Barom.	Air.	D. P.	Diff.	Sat.	Black Bulb.	
A.M. 8·0 to 8·13	11·1	65·5	9·9900	22·960	53·5	33·8	19·7	·505	88·0	Day unexcep-
8·15 „ 8·28	15·0	69·5	12·2645						111·5	tionable, wind
9·0 „ 9·13	17·7	71·5	14·5140	22·948	56·0	37·2	18·8	·513	110·0	S.W., after 10
10·0 „ 10·13	19·1	72·5	15·4710	22·947	57·0	39·7	17·3	·550	121·0	A.M. squally.
11·0 „ 11·13	19·0	75·0	14·9150	22·946	58·5	38·2	20·3	·500	125·0	
P.M. 0·0 „ 0·13	18·8	75·0	12·7600	22·944	60·3	44·8	15·5	·592	120·0	
1·0 „ 1·13	17·2	73·3	13·8976	22·939	59·4	40·7	18·7	·546	122·0	Dense haze over
2·0 „ 2·13	17·4	74·0	13·8330	22·914	60·3	44·1	16·2	·577	108·0	snowy Mts.

B.—APRIL 20TH.

Hour.	Act.	Tem. Act.	Act. Reduced.	Barom.	Air.	D. P.	Diff.	Sat.	Black Bulb.	
A.M. 8·0 to 8·13	11·8	64·0	10·9150	22·969	54·2	43·4	10·8	·691	74·0	Dense haze, S.E.
9·0 „ 9·13	17·8	73·3	14·2750	22·974	56·2	44·1	12·1	·662	92·0	wind, cloud-
10·0 „ 10·13	18·8	65·0	14·7580	22·985	57·0	42·5	14·5	·609	92·0	less sky.

Superintendent's House, Dorjiling.　Elev. 6932 *feet.*

C —APRIL 21ST.　*Watch slow* 1' *mean time.*

Hour.	Act.	Tem. Act.	Act. Reduced.	Barom.	Air.	D. P.	Diff.	Sat.	Black Bulb.	
A.M. 8·35 to 8·48	17·3	65·0	15·7084		56·4	47·6	8·8	·741	97·0	Day very fine,
9 7 „ 9·20	20·9	72·7	16·8872	23·447	63·8	49·9	13·9	·628	100·0	snowy Mts. in
10·0 „ 10·13	23·9	77·3	18·3791		60·8	49·2	11·6	·677	109·0	dull red haze,
11·0 „ 11·13	24·4	81·0	17·8864						107·5	wind S.E. faint.

Rampore Bauleah (Ganges).　Elev. 130 *feet, Lat.* 24° 24' *N., Long.* 88° 40' *E.*

MAY 17TH, 1850,　*Watch slow* 15" *mean time.*

Hour.	Act.	Tem. Act.	Act. Reduced.	Barom.	Air.	D. P.	Diff.	Sat.	Black Bulb.	
A.M. 7·51 to 8·13	13·0	88·0	8·8790	29·698	87·5	80·1	7·4	·793	·91·0	S.E. wind, very
9·3 „ 9·16	19·5	96·0	12·5190		92·0	81·2	10·8	·715	83·8	hazy to west,
9·20 „ 9·33	21·2	107·0	12·7836	29·615	92·3	80·2	12·1	·687	132·0	sky pale blue.
11·15 „ 11·28	21·1	105·0	12·8490		98·5	74·8	23·7	·478	98·5	Wind west,
11·32 „ 11·45	16·5	108·7	9·8770	29·620	98·3	74·3	24·0	·475	142·0	rising.
P.M. 1·20 „ 1·33	21·6	108·5	12·9348		104·5	76·7	27·8	·425	144·0	
1·40 „ 1·53	21·4	113·7	12·4976		105·8	72·2	33·6	·355	134·0	

Churra, Khasia Mountains. Elev. 4225 feet, Lat. 25° 15' N., Long. 91° 47' E.

A.—NOVEMBER 4TH, 1850. *Watch slow 7' mean time.*

Hour.	.	Act.	Tem. Act.	Act. Reduced.	Barom.	Air.	D. P.	Diff.	Sat.	Black Bulb	
A.M. 6·20 to 6·30		5·0	63·7	4·6400	25·781	57·8	53·1	4·7	·850	75·0	Sky faint blue,
6·32 „ 6·42		7·4	65·4	6·6896		59·0	54·8	4·2	·870	83·0	cloudless, wind
7 55 „ 8·5		20·0	77·5	15·2400		63·5	56·9	6·6	·806	108·0	S.W., clouding.
8·8 „ 8·18		21·0	82·0	15·2040		64·4	57·3	7·1	·790	106·5	
8·20 „ 8·30		24·2	85·8	16·8432		64·8	59·5	5·3	·837	113·5	

B.—NOVEMBER 5TH. *Watch slow 7' mean time.*

Hour.	Act.	Tem. Act.	Act. Reduced.	Air.	D. P.	Diff.	Sat.	Black Bulb.	
A.M. 6·39 to 6·49	11·2	70·2	9 3408	59·4	57·6	1·8	·940		Wind S.W.,
6·51 „ 7·1	13·4	72·8	10·8138	60·5	57·8	2·7	·918		clouds rise and
7·56 „ 8·6	18·4	73·2	15·0161	61·7	57 7	4·0	·875		disperse. Sky
8·8 „ 8·21	20·4	77·7	15·4836	63·3	58·7	4·6	·860		pale.
9·26 „ 9·36	23·8	79·5	17·8072						
9·37 „ 9·47	25·1	84·0	17·7959						
10·57 „ 11·7	29·0	89·5	19·5460	66·7	60·8	5·9	8·28	126·0	

C.—NOVEMBER 6TH. *Watch slow 7' mean time.*

Hour.	Act.	Tem. Act.	Act. Reduced.	Barom.	Air.	D. P.	Diff.	Sat.	Black Bulb.	
A.M. 6·5 to 6·18	2·6	62·0	2·4986	25·781	56·5	54·5	2·0	·935		Sunrise, 6, pale
6 22 „ 6·35	6·5	63·5	6·0710		57·0	55·1	1·9	·935		yellow red,
6·38 „ 6·51	9·6	66·7	8·5152		61·0	57 4	3·6	·888		cloudless.
8·27 „ 8·37	21·7	78·8	16·2750		64·2	59·3	4·9	·855	100·0	Cirrhus below.
8·39 „ 8·52	23·0	81·7	19·4750		64·5	59·4	5·1	·847	105·0	

D.—NOVEMBER 14TH.

Hour.	Act.	Tem. Act.	Act. Reduced.	Bar.	Air.	D. P.	Diff.	Sat.	Black Bulb.	
A.M. 6·12 to 6·22	2·9	60·6	3·5988	25·783	51·5	49·4	2·1	·930		Thick cumulus
6·24 „ 6·37	6·1	66·0	5·4472		52·7	50·3	2·4	·925		low on plains.
7·13 „ 7·23	12·4	70·8	10·2672		56·5	52·3	4·2	·900	98·0	Sunrise yel-
7·24 „ 7·34	14·7	76·0	11·4025		57·8	53·1	4·7	·855	104·0	low red.
8·34 „ 8·44	19·9	82·8	14·2653		59·8	50·8	9·0	·742	117·0	Sunrise yel-
8·47 „ 9·0	21·7	88·8	14·7343		60·5	51·6	8·9	·730	121·0	Cloudless.
9·53 „ 10·3	23·5	86·6	16·2620	25·832	67·2	61·6	5·6	·832	127·0	
10·4 „ 10·17	25·3	89·5	17·0775		67·0	58·8	8·2	778	133·0	
11·24 „ 11·31	33·3	111·5	20·7014	25·819	64·6	59·0	5·6	·832	130·0	Clouds rise.

E.—NOVEMBER 15TH.

Hour.	Act.	Tem. Act.	Act. Reduced.	Bar.	Air.	D. P.	Diff.	Sat.	
A.M. 9·53 to 10·6	25·8	78·0	17·5306	25·854	63·0	55·3	8·7	·772	Sky cloudless. Wind
10·50 ,, 11·3	26·1	80·5	19·1835		64·0	52·8	11·2	·690	N.E.
11·31 ,, 11·44	28·5	84·0	20·2065		65·3	51·9	13·4	·638	
P.M. 0·33 ,, 0·46	30·9	91·5	20·4267	25·844	65·8	51·2	14·6	·620	
1·7 ,, 1·21	29·1	90·5	20·4388		67·0	49·6	17·4	·560	
2·47 ,, 3·0	21·1	75·0	16·5635	25·808	67·2	56·6	10·6	·708	
3·48 ,, 4·0	16·7	73·0	13·4435		62·0	50·8	11·2	·690	
4·3 ,, 4·16	16·2	75·0	12·7170	25·803	61·5	50·5	11·0	·692	

Silchar (Cachar), Elev. 116 feet, Lat. 24° 30′ N., Long. 93° E. (approximate).

NOV. 26TH, 1850.　*Watch slow 13′ 39″ mean time.*

Hour.	Act.	Tem. Act.	Act. Reduced.	Bar.	Air.	D. P.	Diff.	Sat.	
A.M. 9·11 to 9·24	19·4	69·0	16·4706		66·3	63·5	2·8	·860	Dense fog till 7·30
9·34 ,, 9·41	22·7	81·0	16·5937						A.M. Wind north.
9·50 ,, 9·57	25·3	87·5	17·3558	29·999	68·7	61·5	7·2	·788	Clear.
10·7 ,, 10·14	26·5	91·5	17·5695		70·3	62·7	7·6	·780	
11·3 ,, 11·16	26·3	89·0	17·5251		73·2	60·3	12·9	·657	Wind N. E. Light
P.M. 0·0 ,, 0·13	26·4	90·0	17·8144	29·967	74·5	61·7	12·8	·658	cirrhus low.
0·58 ,, 1·11	27·6	94·0	17·9676		76·8	60·3	16·5	·586	
2·51 ,, 3·4	23·0	93·0	15·0880	29·892	78·5	62·1	16·4	·588	Streaks of cirrhus
3·55 ,, 4·8	17·6	91·5	11·6688		79·5	57·0	22·5	·480	aloft.
4·9 ,, 4·22	15·5	93·5	11·0215	29·881	79·4	62·1	17·3	·570	
4·23 ,, 4·36	12·0	93·7	7·8360		78·5	62·1	16·4	·588	Sun sets in hazy cirrhus.

Chittagong, Elev. 200 feet, Lat. 22° 20′ N., Long. 91° 55 E.

A.—DECEMBER 31ST, 1850.　*Watch slow 3′ 45″ mean time.*

Hour.	Act.	Tem. Act.	Act. Reduced.	Bar.	Air.	D. P.	Diff.	Sat.	Black Bulb.	
A.M. 7·39 to 7·52	10·0	70·0	8·3700		57·0	55·7	1·3	·960		Cloudless. Moun-
8·40 ,, 8·53	21·3	91·5	14·1219	29·874	59·5	57·2	2·3	·920	127·0	tains clear. Wind
9·4 ,, 9·8	23·2	89·5	15·6136		63·3	59·7	3·6	·890		E.N.E. Cool.
9·52 ,, 9·56	24·3	87·3	16·7341	29·923	64·5	61·3	3·2	·900	142·0	Wind N.W.
10·2 ,, 10·6	25·1	90·5	16·7668		65·7	60·4	5·3	·840	148·0	
11·16 ,, 11·29	24·3	84·5	17·1558		68·5	58·6	9·9	·722	150·0	
11·52 ,, 11·56	26·6	92·6	17·5028	29·892	69·5	59·2	10·3	·710		Wind S.W.
P.M. 1·38 ,, 1·41	24·7	84·0	17·5123		71·7	61·8	9·9	·720		
1·47 ,, 1·51	25·4	90·7	16·8418							
3·10 ,, 3·17	21·1	86·0	14·4645	29·831	71·0	60·5	10·5	710		Clouds about in
3·18 ,, 3·25	19·3	89·3	13·0468							patches.

B.—JANUARY 1, 1851. *Watch slow 3′ 45″ mean time.*

Hour.	Act.	Tem. Act.	Act. Reduced.	Barom.	Air.	D. P.	Diff.	Sat.	Black bulb.	
A.M. 7·34 to 7·41	10·0	69·4	8·4200	29·948	55·4	54·0	1·4	·953		Mist rises and
8·38 „ 8·45	16·0	70·0	13·3920		58·9	57·7	1·2	·970	104·5	drifts west-
9·44 „ 9·51	19·5	74·7	15·3660	29·891	63·2	61·7	1·5	·960	115·0	ward till
10·46 „ 10·53	21·0	78·2	15·8550		66·7	62·4	4·3	·870	120·0	7·30 A.M.
11·50 „ 11·57	21·5	81·2	15·6950		69·8	58·3	11·5	·688	117·0	Wind N.W.,
P.M. 0·6 „ 0·13	24·1	88·0	16·4603	29·850	70·3	56·0	14·3	·625	122·5	clouds rise.
0·58 „ 1·2	23·9	87·2	16·4432		71·0	56·7	14·3	·625		
1·45 „ 1·52	21·4	84·5	15·0870		71·3	57·5	13·8	·633	117·0	
3·15 „ 3 22	18·1	82·5	13·0320	29·798	71·3	57·1	14·2	·625		
4·27 „ 4·34	10·2	82·0	7·3746		70·0	59·5	10·5	·708		
4·36 „ 4·43	9·8	84·0	6·9482							
4·45 „ 4·52	8·5	85·0	5·9670							
4·56 „ 5·9	5·6	85·0	3·9312		67·5	62·7	4·8	·855		Sunset cloud-
5·12 „ 5·18	3·8	84·0	2·6942	29·778	68·7	62·2	6·5	·810		less.

C.—JANUARY 2, 1851. *Watch slow 3′ mean time.*

Hour.	Act.	Tem. Act.	Act. Reduced.	Barom.	Air.	D. P.	Diff.	Sat.	Black bulb.	
A.M. 10·2 to 10·9	19·2	71·0	15·8592		64·5	60·6	3·9	·878	116·0	Low, dense fog
10·20 „ 10·24	22·6	79·0	16·9048	29·861	65·6	61·4	4·2	·872		at sunrise,
P.M. 0·3 „ 0·10	24·7	89·2	16·6972	29·858	69·0	59·3	9·7	·728	119·0	clear at 9 A.M.
0·22 „ 0·25	25·9	95·5	18·6796		70·7	57·5	3·2	·650		Hills hazy &
2·4 „ 2·8	23·3	91·5	15·4479		71·2	61·0	10·2	·718	112·0	horizon grey.
2·10 „ 2·14	23·8	93·0	15·6128							

L.

TABLE OF ELEVATIONS.

In the following tables I have given the elevations of 300 places, chiefly computed from barometric data. For the computations such observations alone were selected as were comparable with contemporaneous ones taken at the Calcutta Observatory, or as could, by interpolation, be reduced to these, with considerable accuracy: the Calcutta temperatures have been assumed as those of the level of the sea, and eighteen feet have been added for the height of the Calcutta Observatory above the sea. I have introduced two standards of comparison where attainable; namely, 1. A few trigonometrical data, chiefly of positions around Dorjiling, measured by Lieutenant-Colonel Waugh, the Surveyor-General, also a few measured by Mr. Muller and myself, in which we can put full confidence: and, 2. A number of elevations in Sikkim and East Nepal, computed by simultaneous barometer observations, taken by Mr. Muller at Dorjiling. As the Dorjiling barometer was in bad repair, I do not place so much confidence in these comparisons as in those with Calcutta. The coincidence, however, between the mean of all the elevations computed by each method is very remarkable; the difference amounting to only thirty feet in ninety-three elevations; the excess being in favour of those worked by Dorjiling. As the Dorjiling observations were generally taken at night, or early in the morning, when the temperature is below the mean of the day, this excess in the resulting elevations would appear to prove, that the temperature correction derived from assuming the Calcutta observations to correspond with eighteen feet above the level of the sea at Sikkim, has not practically given rise to much error.

I have not added the boiling-point observations, which afford a further means of testing the accuracy of the barometric computations; and which will be found in section J of this Appendix.

The elevation of Jillapahar is given as computed by observations taken in different months, and at different hours of the day; from which there will be seen, that owing to the low temperature of sunrise in the one case, and of January and October in the others, the result for these times is always lowest.

Most of the computations have been made by means of Oltmann's tables, as drawn up by Lieutenant-Colonel Boileau, and printed at the Magnetic Observatory, Simla; very many were worked also by Bessell's tables in Taylor's "Scientific Memoirs," which, however, I found to give rather too high a result on the averages; and I have therefore rejected most of them, except in cases of great elevation and of remarkable humidity or dryness, when the mean saturation point is an element that should not be disregarded in the computation. To these the letter B is prefixed. By far the majority of these elevations are not capable of verification within a few feet; many of them being of villages, which occupy several hundred feet of a hill slope: in such cases the introduction of the refinement of the humidity correction was not worth the while.

SERIES I.—*Elevations on the Grand Trunk-road. February*, 1848.

No. of Obs.	Name of Locality.	Elevation. Feet.
1	Burdwan	93
2	Gyra	630
3	Fitcoree	860
2	Tofe Choney	912
4	Maddaobund	1230
1	Paras-nath saddle	B.4231
2	„ east peak	4215
1	„ flagstaff	4428
1	„ lower limit of *Clematis* and *Berberis*	3162
1	Doomree	996
1	Highest point on grand trunk-road	1446
4	Belcuppee	1219
1	Hill 236th mile-stone	1361
3	Burree	1169
1	Hill 243rd mile-stone	1339
3	Chorparun	1322
3	Dunwah	625
1	Bahra	479
1	284th mile-stone	474
2	Sheergotty	460
4	Muddunpore	402
1	312th mile-stone	365
3	Naurungabad	337
4	Baroon (on Soane)	344
4	Dearee ., ..	332

SERIES II.—*Elevations in the Soane Valley.*   *March*, 1848.

No. of Obs.	Name of Locality.	Elevation. Feet.
3	Tilotho	395
6	Akbarpore	403
2	Rotas palace	1489
4	Tura	453
3	Soane-pore	462
6	Kosdera	445
4	Panchadurma	492
1	Bed of Soane above Panchadurma . . .	482
3	Pepura	587
1	Bed of Soane river	400
9	Chahnchee	499
4	Hirrah	531
4	Kotah	541
4	Kunch	561
7	Sulkun	684

SERIES III.—*Elevations on the Kymore Hills.*   *March*, 1848.

No. of Obs.	Name of Locality.	Elevation. Feet.
2	Roump	1090
9	Shahgunj	1102
1	Amoee	818
1	Goorawul	905
9	Mirzapore (on the Ganges)	362

SERIES IV.—*Elevations near Dorjiling.* 1848 *to* 1850.

Number of Obs.	Name of Locality.	Elevation. Feet.
	Jillapahar (Mr. Hodgson's house)	
.9	,, ,, sunrise . . .	7301
110	,, ,, 9·50 P.M. . .	7443
104	,, ,, noon . . .	7457
99	,, ,, 2·40 P.M. . .	7477
93	,, ,, 4 P.M. . . .	7447
37	,, ,, sunset . . .	7447
Sum 452	Mean	**7429**
	Ditto by Monthly observations.	
27	January	7400
84	February	7445
37	March	7517
7	April	7582
83	July	7412
74	August	7421
95	September	7454
18	October	7351
Sum 434	Mean	**7448**
103	The Dale (Mr. Muller's)	B. 6957
	,, by trigonometry . . .	6952
16	Superintendent's house	B. 6932
	,, ,, by trigonometry. .	6932
38	Colinton (Mr. Muller's)	B. 7179
25	Leebong	B. 5993
	,, by trigonometry . . .	6021*
2	Summit of Jillapahar	B. 7896
2	Smith's hotel	6872
7	Monastery hill below the Dale . . .	B. 214·1
	The Dale by barometer	6952
		7166
	Monastery hill by trigonometry . .	7165·3
1	Ging (measured from Dale) . . .	B. 5156
12	Guard-house at Great Rungeet . .	B. 1864
2	Bed of Great Rungeet at cane-bridge .	818
5	Guard-house at Little Rungeet . .	1672
8	Sinchul top	8655
	,, by trigonometry . .	8607
4	Saddle of road over shoulder of Sinchul	7412
4	Senadah (Pacheem) bungalow . .	7258
1	Pacheem village	3855
13	Kursiong bungalow	B. 4813
13	Punkabaree	1815
2	Rungniok village	B. 4565
2	Tonglo, summit	B.10·078
	,, ,, by trigonometry . . .	10·079·4
13	,, Saddle below summit . .	B.10·008
1	,, Rocks on ascent of . .	B. 8148
4	Source of Balasun	7436
	,, ,, by Dorjiling . .	7451
8	Goong ridge	7441

* To summit of chimney, which may be assumed to be 30 feet above where the barometer was hung.   H H 2

Series V.—*Elevations in East Nepal, October to December,* 1848.

No. of Obs.	Name of Locality.	By Calcutta Barom.	By Dorjiling Barom.
		Feet.	Feet.
1	Source of Myong river	4,798	
7	Myong valley, camp in	4,345	4,345
7	Myong valley	3,801	3,763
5	Purmiokzong	4,507	4,535
2	Shoulder of Nanki	7,216	
1	„ „ Shepherds' huts on do.	8,999	
3	Summit of Nanki	9,994	10,045
8	„ „ Camp on Nanki	9,315	9,324
3	Jummanoo	4,320	4,404
5	Sulloobong	5,244	5,311
4	Bheti village	4,683	
4	Sakkiazong village	5,804	5,847
3	Camp on ridge of mountain	8,315	8,391
1	Peak on Sakkiazong	9,356	9,289
3	Makarumbi	5,444	5,525
3	Pemmi river	2,149	2,262
3	Tambur river at junction with Pemmi	1,289	1,487
1	Camp on Tambur, Nov. 13	1,418	1,496·
3	„ „ Nov. 14	1,600	
2	Chintam village	3,404	
8	Mywa Guola	2.079	2,185
3	Tambur river, Nov. 18	2,515	2,574
3	„ „ Nov. 19	3,113	3,289
3	Taptiatok village	4.207	4,359
2	Loontoong village	5,615	5,738
2	Tambur river, Nov. 23	8,066	8,096
10	Wallanchoon village	10,384	10,389
6	Tuquoroma	12,889	12,999
1	Wallanchoon pass	B.16,764	16,748
1	Foot of pass-road	13,501	13,518
4	Yangma Guola	9,236	9,322
2	Base of great moraine	12,098	12,199
2	Top of moraine above ditto	B. 679	
9	Yangma village camp	B.13,516	13,488
1	Lake bed in valley	15,186	
1	Upper ditto (Pabuk)	B.16,038	
4	Yangma valley camp, Dec. 2	10,997	11,001
1	Kambachen pass	B.15,770	
3	Camp below ditto	11,643	11,611
1	Kambachen village	11,378	
2	Camp in valley	11,454	11,514
1	Choonjerma pass	B.15,259	
4	Camp below ditto	13,289	13,287
1	Yalloong river-terrace	10,449	
4	Camp side of valley	10,080	10,035
3	Yankutang village	5,530	5,598
1	Saddle on road south of Khabili	5,746	
8	Khabang village	5,495	5,515
1	Spur of Sidingbah, crossed Nov. 10	6.057	5,980
3	Yangyading village	4,082	4,145
4	Sablakoo	4.635	4,718
7	Iwa river, Dec. 12	3,747	3,818
2	„ „ Dec. 13	6,134	6,184
4	Singalelah, camp on	9,263	9,328
1	Islumbo pass	10,388	

SERIES VI.—*Elevations in Sikkim, December, 1848, and January, 1849.*

No. of Obs.	Name of locality.	By Calcutta Barometer.	By Dorjiling Barometer.
		Feet.	Feet.
4	Kulhait valley, camp in	6,406	6.374
6	Lingcham village	4,892	4,848
5	Bed of Great Rungeet, December 20	1.805	1,874
6	Lingdam village, December 21.	5.552	5,556
6	Nampok village	4,354	4,501
7	Bhomsong	1,556	1,533
8	Mainom top	Tr. 10,702	B. 10,613
1	Neon-gong Goompa	5,225	
1	Pass from Teesta to Rungeet	6,824	
6	Lingdam village	5,349	5,401
1	Great Rungeet below Tassiding	2,030	
	Tassiding temples	4,840	
5	Sunnook, camp on	3,955	4,018
1	Bed of Ratong	2,481	
1	Pemiongchi temple	7,083	
10	Camp at Pemiongchi village	6,551	6,616
9	Tchonpong village	4,952	5,003
1	Bed of Rungbi river	3,165	
9	Camp on Ratong river	3,100	3,242
1	Doobdi Goompa	6,493	6,451
22	Yoksun	5,600	5,635
7	Dumpook	6,646	6,710
15	Buckim	8,625	8,693
7	Mon Lepcha top	13,090	13,045
21	Jongri	B. 13,170	13.184
1	Ratong below Mon Lepcha	7,069	7,217
1	„ below Yoksun	3,729	3,851
1	Catsuperri lake	6,068	6,009
1	„ temple	6,493	6,476
4	Tengling village	5,295	5,219
5	Rungbee river bed	3,230	3,350
5	Changachelling temple	6,805	6,850
5	Kulhait river	3,075	3,243
1	Saddle of Hee hill	7,289	
6	Camp on Hee hill.	6,609	6,744

SERIES VII.—*Elevations in the Sikkim Terai and Plains of India, Gangetic Delta and Jheels.*

No. of Obs.	Name of Locality.	Elevation. Feet.
3	Siligoree Bungalow	302
12	Titalya ., ...	326
3	Sahibgunj (west of Titalya)	231
4	Bhatgong . . . . : .	225
4	Thakya-gunj	284
4	Bhojepore	404
5	Rummai	293
5	Rangamally	262
5	Belakoba	368
1	Mela-meli	337
6	Kishengunj	131
43	Mahanuddy river between Kishengunj and Maldah . . .	153
24	„ „ Maldah and Rampore Bauleah .	98
12	Rampore (Mr. Bell's)	130
13	Dacca (Mr. Atherton's)	72
54	Jheels, Dacca to Pundua	*−·003
33	Megna river (June 1st-6th)	+·008
13	Soormah (June 9th)	+·048
4	Pundua (June 10th and 11th) . . . ., . . .	+·018
3	„ (Sept. 7th)	—·016
5	„ (Nov. 16th and 17th) :	—0·66

SERIES VIII.—*Elevations in Sikkim, May to December, 1849.*

No. of Obs.	Name of Locality.	By Calcutta Barometer.	By Dorjiling Barometer.
		Feet.	Feet.
2	Mik, on Tendong	3,912	
4	Namtchi, camp on spur	5,608	
1	Tendong summit	B. 8,671	Tr. 8,663
2	Temi, Teesta valley	4,771	
4	Nampok „	B. 5,138	5,033
8	Lingmo „	„ 2,861	2,838
4	Lingtam spur, Teesta valley	„ 4,743	4,867
4	Gorh „	„ 4,061	4,195
2	Bling-bong „	„ 2,657	2,711
8	Lingo village „ . . .	„ 2,724	2,839
10	Singtam, May 14 to 16	„ 4,435	4,477
16	„ (higher on hill) Oct. 30 to Nov. 2 . .	„ 4,575	

* The observations marked thus * are the differences in inches between the readings of my barometer at the station, and that at the Calcutta observatory, which is 18 feet above the sea-level.

SERIES VIII.—(*Continued.*)

No. of Obs.	Name of Locality.	By Calcutta Barometer.	By Dorjiling Barometer.
		Feet.	Feet.
5	Niong.		3,954
2	Namgah.	4,229	
7	Chakoong	4,371	4,443
27	Choongtam, May	5,245	5,284
37	„ August	5,247	5,297
4	Dholep, Lachen	6,120	6,145
4	Dengha „	6,337	6,399
3	Latong „	6,471	6,310
8	Kampo Samdong	7,315	7,344
1	Chateng.	8,819	8,695
1	„ lower on spur	8,493	8,343
33	Lamteng village	8,900	8,867
53	Zemu Samdong	9,026	8,926
1	Snow bed across Zemu river	9,828	
4	Camp on banks of Zemu	10,223	10,271
74	Junction of Thlonok and Zemu	10,864	10,828
47	Camp on banks of Zemu river	12,064	12,074
1	Zemu river, June 13	12,422	
1	„ higher up, June 13	13,281	
2	Yeunga (Lachen valley)	10,196	
43	Tallum Samdong	11,540	11,424
20	Tungu, July	12,779	12,723
30	„ October.	12,799	12,747
1	Palung plains.	15,697	
3	Sitong	15,372	
2	Kongra Lama pass.	15,745	15,642
5	Yeumtso (in Tibet)	16,808	
2	Bhomtso do.	18,590	
6	Cholamoo lakes do.	16,900	
2	Donkia pass, October	18.589	
2	„ September	18,387	
56	Momay Samdong	15,362	15,069
			Measured rom Momay.
1	Donkia, September 13.	16,876	17,079
1	Kinchinjhow, September 14	17,495	17.656
1	Sebolah pass	17,604	17.567
1	South shoulder of Donkia, September 20	18,257	18,357
1	Mountain north of Momay, September 17.		B. 17,394
1	West shoulder of Donkia mountain, Sept. 26.		„ 18,510
	The following were measured trigonometrically.		
	Forked Donkia mountain		Ir. 20,870
	Kinchinjhow mountain		„ 22,750
	Tomo-chamo, east top of Kinchinjhow		„ 21,000
	Thlonok mount, Peak on		„ 20,000
	Chango-khang mountain		„ 20,600
	Tukcham mountain, from Dorjiling		„ 19,472
	Chomiomo mountain.		„ 22,700

SERIES VIII.—(*Continued.*)

No. of Obs.	Name of Locality.	By Calcutta Barometer.	Measured by Trigonometry, &c.
		Feet.	Feet.
	The following were measured trigonometrically.		
	Summit? of Donkia (from Donkia pass and Bhomtso)		Tr. 22,650
	Tunkra Mountain, from Dorjiling . . .		„ 18,250
			By Dorjiling Barometer.
48	Yeumtong	11,933	11,839
7	„ October	11,951	
			By Yeumtong Barometer.
2	Snow bed above Yeumtong	B. 15,971	16,000
3	Punying	„ 11,299	
			By Der. Bar.
51	Lachoong village, August	„ 8,712	8,474
12	„ „ October	„ 8,705	
8	Lacheepia	„ 15,293	15,231
2	Tunkra pass	„ 16,083	
3	Rock on ascent to ditto	„ 13,078	13,144
4	Keadom	„ 6,609	
3	Tukcham village	„ 3,849	
5	Rinkpo village	„ 6,008	
7	Laghep	„ 10,423	
1	Phieungoong	„ 12,422	
3	Barfonchen	„ 11,233	
1	Chola pass	„ 14,925	
3	Chumanako	„ 12,590	
17	Phadong	„ 5,946	
3	Tumloong, Nov. 3rd and 4th . . .	„ 5,368	
105	Higher on hill, Nov. 16th to Dec. 9th. .	„ 5,976	
1	Yankoong	„ 3,867	
2	Tikbotang	„ 3,763	
3	Camp, Dec. 11th	„ 2,952	
12	Serriomsa	„ 2,820	
11	Dikkeeling	„ 4,952	
2	Singdong	„ 2,116	
3	Katong ghat, Teesta	„ 735	
5	Namten	„ 4,483	
6	Cheadam	„ 4,653	

SERIES IX.—*Khasia Mountains, June to November*, 1850.

No. of Obs.	Name of Locality.	Elevation. Feet.
36	Churra (Mr. Inglis's)	4,069
167	„ bungalow opposite church, August	4,193
102	„ „ Oct., Nov.	4,258
25	Kala-panee bungalow	5,302
63	Moflong „	6,062
1	Chillong hill	6,662
9	Syong bungalow	5,725
1	Hill south of ditto	6,050
32	Myrung bungalow, July	5,647
6	„ „ Sept.	5,709
9	Chela	80
63	Nunklow	4,688
6	Nonkreem	5,601
10	Mooshye	4,863
35	Pomrang	5,143
12	Amwee	4,105
9	Joowye	4,387
3	Nurtiung	4,178

SERIES X.—*Soormah, Silhet, Megna, Chittagong, &c.*

No. of Obs.	Name of Locality.	Elevation.
27	Silhet (Mr. Stainforth's)	133 ft.
38	Soormah river, between Silhet and Megna	46 „
36	Silchar	116 „
24	Megna river	+ ·020*
12	Noacolly (Dr. Baker's)	—·039
10	„ on voyage to Chittagong ·	·000†
72	Chittagong (Mr. Sconce's)	191 ft.
8	„ flagstaff-hill at south head of harbour . .	151 „
2	Seetakoond hill	1,136 „
16	„ bungalow	—·069*
3	Hat-Hazaree	—·039
12	Hattiah	—·049
4	Sidhee	—·039
17	Chittagong to Megna	— ·014†·
10	Eastern Sunderbunds	+ ·002

* Difference between barometer at station and Calcutta barometer.

† The observations were taken only when the boat was high and dry, and above the mean level of the waters.

INDEX.

THE END.

LONDON:
BRADBURY AND EVANS, PRINTERS, WHITEFRIARS.

Printed in the United States
By Bookmasters